Technological Advancements for Deep Sea Ecosystem Conservation and Exploration

U. Vignesh
Vellore Institute of Technology, Chennai, India

Parvathi R.
Vellore Institute of Technology, Chennai, India

IGI Global
Scientific Publishing
Publishing Tomorrow's Research Today

Vice President of Editorial Melissa Wagner
Managing Editor of Acquisitions Mikaela Felty
Managing Editor of Book Development Jocelynn Hessler
Production Manager Mike Brehm
Cover Design Phillip Shickler

Published in the United States of America by
IGI Global Scientific Publishing
701 East Chocolate Avenue
Hershey, PA, 17033, USA
Tel: 717-533-8845
Fax: 717-533-8661
Website: https://www.igi-global.com E-mail: cust@igi-global.com

Library of Congress Cataloging-in-Publication Data

Names: Umapathi, Vignesh, 1989- editor. | Parvathi, R., 1972- editor.
Title: Technological advancements for deep sea ecosystem conservation and exploration / edited by U. Vignesh, Parvathi R.
Description: Hershey, PA : IGI Global Scientific Publishing, [2025] | Includes bibliographical references and index. | Summary: "This book delves into the intricacies and significance of deep-sea ecosystems, employing interdisciplinary approaches to unveil their mysteries"-- Provided by publisher.
Identifiers: LCCN 2024053016 (print) | LCCN 2024053017 (ebook) | ISBN 9798369366707 (h/c) | ISBN 9798369366714 (s/c) | ISBN 9798369366721 (ebook)
Subjects: LCSH: Ocean engineering. | Marine resources conservation. | Underwater exploration. | Marine ecology.
Classification: LCC TC1645 .T43 2025 (print) | LCC TC1645 (ebook) | DDC 620/.4162--dc23/eng/20250212
LC record available at https://lccn.loc.gov/2024053016
LC ebook record available at https://lccn.loc.gov/2024053017

British Cataloguing in Publication Data
A Cataloguing in Publication record for this book is available from the British Library.

Table of Contents

Detailed Table of Contents

The deep sea, a vast and enigmatic realm beneath the ocean's surface, holds some of the planet's greatest mysteries. These harsh conditions create a unique and diverse ecosystem, teeming with life forms adapted to survive in ways unimaginable to terrestrial beings. The journey into the deep sea begins at the continental slope, which descends steeply from the continental shelf into the abyssal plain. This transition zone marks the boundary between the relatively shallow coastal waters and the deep ocean basins. As one descends deeper, light from the surface diminishes rapidly, giving way to the twilight zone at depths of 200 to 1,000 meters. Despite the absence of sunlight, the deep sea is a dynamic and vibrant ecosystem. It hosts a myriad of species, many of which are specially adapted to the extreme conditions. Bioluminescence, the production and emission of light by living organisms, is a common adaptation, serving various purposes such as predation, camouflage, and communication. Creatures like the anglerfish, with its luminescent lure, and the vampire squid, which uses bioluminescent displays to evade predators, exemplify the ingenious adaptations that have evolved in the deep sea. The deep sea floor is dotted with hydrothermal vents and cold seeps, which are hotspots of biological activity. Hydrothermal vents, first discovered in the late 1970s, are formed by volcanic activity on the ocean floor, where seawater percolates through the Earth's crust, heats up, and re-emerges laden with minerals. These vents support unique ecosystems, relying not on sunlight for energy but on chemosynthesis, a process by which bacteria convert chemical energy from vent minerals into organic matter. The organisms inhabiting cold seeps, such as mussels, clams, and various types of worms, have adapted to exploit the chemical energy available in these otherwise inhospitable environments. Understanding the mechanisms and efficiency of the biological pump is crucial for predicting the deep sea's role in mitigating climate change.

This chapter discusses the design and management of deep-sea biodiversity with a focus on sustainable conservation principles. Deep-sea biodiversity, as a valuable yet poorly understood natural resource, requires ecosystem exploration, species identification and the development of evidence-based conservation strategies. This research identifies key drivers such as climate change, pollution and human activities that influence management and conservation policies. Cutting-edge technologies, including GIS, remote sensors and big data analytics, are integrated to improve monitoring and conservation planning. The chapter also outlines challenges to conservation implementation, such as legal constraints, resistance to change, and technological limitations. Successful and failed case studies provide valuable insights for future strategies. The chapter concludes that holistic approaches, technological innovation and stakeholder engagement are necessary to maintain the balance and sustainability of deep-sea ecosystems.

Environmental conditions of ocean are crucial for predicting harmful events, which gives an impact on marine ecosystems and human well-being. Various threats, such as ocean acidification, harmful algal blooms, and coral bleaching, need early monitoring of variables, including temperature, acidity, pollution levels, and biodiversity, to mitigate the adverse effects. This proactive approach enables conservation efforts that safeguard marine life, fisheries, and coastal communities while promoting sustainable ocean stewardship. Deep learning techniques play a pivotal role in this endeavour by processing underwater acoustic recordings. It gives accurate results to identify marine species, detect changes in their behavior, and predict environmental conditions based on sound patterns. The proposed work suggests that, combination of Long Short Term Memory(LSTM) and Graphical Neural Network(GNN) which is used to predict the harmful changes over the ocean. LSTM and GNN architectures establish an indirect relationship, shedding light on emerging trends and potential threats.

Deep sea mining is a potential industry in discovering enormous mineral resources, such as polymetallic nodules, hydrothermal vent deposits, and cobalt-rich crusts. These minerals are essential for high-tech companies and renewable energy applications. However, resource exploitation presents substantial obstacles and threats to deep marine ecosystem, are least studied. The geographical areas chosen for deep-sea mining are frequently rich in biodiversity, that could be adversely damaged by mining operations. Deep sea mining has important legal and societal ramifications, with disputes centred on regulatory frameworks, ownership rights, and the possible socioeconomic benefits. Current legislation is evolving to address these issues, but comprehensive worldwide guidelines are urgently required to ensure environmentally and socially acceptable mining practices. Future directions in deep-sea mining will likely focus on increasing operational sustainability, improving environmental monitoring, and creating technologies that reduce ecological footprints.

One enormous, unexplored source of renewable energy is the ocean. Sound energy produced by sound waves, especially those produced by marine life, has the potential to be used effectively to create electricity. Materials known as piezoelectric materials generate electricity when they experience mechanical stress, such as vibrations from sound waves. It is therefore feasible to generate power using the movement of whales by putting piezoelectric materials in places where whales congregate, such as along migration routes. Whale movements in the water induce vibrations in piezoelectric materials, which then transform mechanical stress into electrical energy for use in a variety of applications. The piezoelectric materials are selected in such a way which are sensitive to low frequency vibrations caused by whales and also environment friendly and protecting marine life. This approach will serve as an efficient way to collect energy from natural environment without harming marine life.

Concrete is widely acknowledged as the paramount composite construction material due to its versatility and durability. Its properties and characteristics exhibit variability contingent upon environmental conditions. Moreover, altering the proportions of concrete ingredients, significantly influences its strength and performance attributes. Admixtures, compounds to concrete during mixing to modify its properties, play a pivotal role in augmenting concrete strength and enhancing properties. Notably, construction activities conducted in deep-sea environments entail distinct challenges compared to those on conventional land surfaces, necessitating specialized techniques and considerations. Through the integration of sensor data and machine learning algorithms, construction stakeholders can gain valuable insights into the performance of concrete structures in marine settings, thereby ensuring the durability and resilience of infrastructure projects amidst diverse environmental pressures.

This study explores a novel application of tidal energy using two opposing repelling neodymium magnets to operate tiny underwater turbines, harnessing ocean currents for sustainable energy. By leveraging magnetic repulsion, the design aims for simplicity, efficiency, and environmental sustainability. Neodymium magnets are either embedded in or mounted on turbine blades, increasing magnetic interaction and generating a strong repulsive force that accelerates rotation. The turbines convert

promising avenues for unravelling the mysteries of the deep and safeguarding its fragile ecosystems. This abstract presents an overview of the manifold applications of ML in deep-sea ecosystem analysis, focusing on key areas such as species identification, habitat characterization, and event detection. Primarily, ML algorithms, including Convolutional Neural Networks (CNNs), Support Vector Machines (SVMs), and Random Forests, are deployed for automated species identification, leveraging vast repositories of underwater imagery. By discerning intricate visual patterns, these algorithms facilitate rapid biodiversity assessment and aid in monitoring shifts in species distributions. Moreover, ML techniques, such as Decision Trees, K-Nearest Neighbour (KNN), and deep learning architectures, are instrumental in habitat classification endeavours. Through the analysis of multifaceted sensor data, these algorithms enable the delineation of distinct seafloor habitats, empowering conservationists with invaluable insights into ecosystem structure and function. Furthermore, ML-based anomaly detection algorithms, Hidden Markov Models (HMMs), and reinforcement learning (RL) strategies contribute to event detection efforts in the deep sea. By scrutinizing temporal data streams from underwater sensors, these algorithms detect aberrations indicative of natural phenomena or anthropogenic disturbances, thus facilitating timely interventions and mitigative actions. In conclusion, the integration of ML techniques into deep-sea ecosystem analysis represents a paradigm shift in our approach to understanding and preserving these enigmatic realms. By automating labor-intensive tasks, enhancing data processing capabilities, and fostering real-time monitoring, ML empowers scientists and conservationists to unlock the secrets of the deep and embark on a journey towards sustainable stewardship of these invaluable ecosystems.

D. Chitradevi, SRM Institute of Science and Technology, India
K. Jayanthi, SRM Institute of Science and Technology, India
S. Prabha, Mayo Clinic, Florida, USA
M. Muthukamatchi, Hindustan Institute of Technology and Science, India

The marine benthic zone is host to a vast array of species that are vital to both ecosystem function and biodiversity conservation. Understanding these underwater communities is essential for ecological research, conservation efforts, and sustainable resource management. However, traditional surveying techniques face significant challenges in the underwater environment. This research extends strategies for underwater biological monitoring by offering a reliable and useful method for detecting benthos in marine benthic zones. Ultimately, by promoting conservation initiatives and sustainable management techniques in marine settings, the proposed

Preface

This book delves into the intricacies and significance of deep-sea ecosystems, employing interdisciplinary approaches to unveil their mysteries.The deep sea, a vast and largely unexplored frontier, harbors intricate ecosystems that play a critical role in maintaining the health of our planet. This book embarks on an interdisciplinary journey to illuminate the significance and complexities of these abyssal realms. By integrating insights from biology, ecology, oceanography, and technology, we aim to comprehensively investigate the dynamics and global importance of deep-sea ecosystems.

The deep sea is divided into distinct zones, each with unique features and inhabitants. From the sunlit epipelagic zone to the darkest hadal zone, these regions support an astonishing array of life. The mesopelagic zone, often referred to as the "twilight zone," is characterized by limited sunlight and an abundance of bioluminescent organisms. In contrast, the bathypelagic zone is marked by near-total darkness and the presence of deep-sea fish with enlarged eyes.

Hydrothermal vents, underwater springs that emit hot, mineral-rich fluids, support vibrant communities of organisms. Giant tube worms, vent crabs, and other species thrive in these unique environments, often exhibiting symbiotic relationships with microorganisms. Beyond vents, deep-sea trenches, coral reefs, and seamounts harbor diverse assemblages of fish, invertebrates, and microorganisms. These ecosystems are not only fascinating but also provide valuable insights into the evolution of life on Earth.

Remotely Operated Vehicles (ROVs) and Autonomous Underwater Vehicles (AUVs) have revolutionized deep-sea exploration. These technologies enable scientists to collect data, conduct experiments, and observe phenomena in real-time. Advanced sensors and sampling equipment allow researchers to study water chemistry, currents, and marine life in unprecedented detail. This technological advancement has significantly expanded our understanding of deep-sea ecosystems and their interconnectedness with the global ocean.

However, deep-sea mining and climate change pose significant threats to these fragile ecosystems. Mining operations can disrupt habitat-forming organisms and release toxic chemicals, while climate change alters ocean chemistry and circulation patterns. To mitigate these impacts, sustainable conservation strategies are imperative. Establishing Marine Protected Areas (MPAs) can safeguard critical habitats and prevent mining and fishing activities. Ecosystem-Based Management integrates ecological principles into decision-making processes, ensuring consideration of deep-sea ecosystems' interconnectedness. International cooperation among nations, organizations, and industries is essential for effective conservation.

Deep-sea environments play a pivotal role in global biogeochemical cycles. Deep-sea sediments and organisms absorb and store carbon dioxide, contributing to carbon sequestration. Deep-sea processes influence nutrient availability for phytoplankton and marine food webs. Furthermore, deep-sea circulation patterns regulate global climate and weather patterns, underscoring the intrinsic value of these ecosystems.

Through interdisciplinary research and advocacy, we strive to promote awareness and facilitate informed decision-making. By recognizing the global significance of deep-sea ecosystems, we can ensure their preservation and responsible management. Our findings underscore the urgent need for sustainable conservation strategies to safeguard these vulnerable ecosystems. As we continue to explore and understand the complexities of the deep sea, we must prioritize its protection for the benefit of our planet and future generations.

The preservation of deep-sea ecosystems requires a collective effort, involving scientists, policymakers, industry leaders, and the general public. By working together, we can address the challenges facing these ecosystems and ensure their integrity for generations to come. This book serves as a call to action, highlighting the importance of deep-sea ecosystems and the need for responsible stewardship of our planet's greatest natural resource – the ocean.

Ultimately, the exploration and conservation of deep-sea ecosystems represent a critical frontier in our quest to understand and protect the natural world. As we venture into the unknown, we must balance scientific inquiry with environmental responsibility, ensuring that our pursuit of knowledge does not come at the expense of the planet's well-being. By embracing this challenge, we can unlock the secrets of the deep sea and safeguard its wonders for posterity.

We invite you to explore the chapters ahead and engage with the rich drapery of research and applications presented. Exploring'Technological Advancements for Deep Sea Ecosystem Conservation and Exploration' through all chapters provide insight into complexities and significance of deep-sea ecosystems.

Chapter 1, "A comparative study on technologies deployed for marine ecosystem monitoring and restoration," focuses on novel technologies and methods deployed to restore deep-sea ecosystems and habitats. A comparative study is performed on

Human Underwater Vehicles (HOV), Autonomous Underwater Vehicles (AUVs), Remotely Operated Vehicles (ROVs), Crawlers, Cabled observatories, Soft robotic bio-inspired machines and other exploration vehicles, which vary in their spatial and temporal monitoring capabilities. It also includes a study of computer algorithms such as graph neural networks, visual simultaneous localization and mapping (vSLAM) and the advanced sensors used to obtain better data from our ocean's ecosystems, along with the techniques used to analyse the gathered data.

In Chapter 2, "Anthropogenic Waste Classification in Deep Sea Environments," imparts how protecting the ecosystems is essential for sustainable use and to the ecological integrity of deep-sea ecosystems which is seriously threatened by anthropogenic waste.The efforts to control and lessen this problem are discussed in this chapter which suggests creating a systematic framework for classification using advanced deep learning techniques which integrated with a user-friendly web application for effective classification.

Chapter 3, "Casestudy on Integrating Advanced Technologies for the Conservation and Exploration of Deep Sea Ecosystems", deals with the world's deep-sea ecosystems face unprecedented threats, necessitating innovative solutions. This chapter presents a compelling case study on leveraging cutting-edge technologies – remote sensing, AUVs, ROVs, and genetic sequencing – to revolutionize deep-sea conservation and exploration. Expert insights reveal the transformative power of technology in protecting these vital ecosystems.

Chapter 4, "Climate Change and the Deep Sea", anlyzes: The deep sea, Earth's largest ecosystem, faces unprecedented threats from climate change. This chapter delves into the uncharted world beneath the ocean's surface, exploring the impacts of rising temperatures, ocean acidification, and altered circulation patterns on the deep sea's unique biodiversity and ecosystem processes, and the far-reaching consequences for our planet.

Chapter 5, "Designing and Managing Deep Sea Biodiversity", discusses the design and management of deep-sea biodiversity with a focus on sustainable conservation principles. As deep-sea biodiversity faces escalating threats, effective conservation strategies are crucial. This chapter presents a comprehensive framework for designing and managing deep-sea ecosystems, integrating cutting-edge technologies and sustainable principles. Expert insights highlight key drivers, challenges, and successful case studies, informing evidence-based policies for preserving these vital, yet vulnerable, ecosystems.

In Chapter 6, "Early Prediction of Harmful Environmental Changes Based on Bioacoustics Analysis", a proactive approach to monitoring and mitigating these hazards using deep learning techniques were discussed. Leveraging underwater acoustic recordings, Long Short Term Memory (LSTM) and Graphical Neural

Network (GNN) architectures accurately predict harmful changes, enabling timely conservation efforts and sustainable ocean stewardship.

In Chapter 7, "Exploring the Depths-Comprehensive Insights into Deep Sea Mining and its Implications for Ecosystems, Technologies, and Sustainability", delves into the complexities of deep-sea mining, exploring its potential, pitfalls, and future directions.Deep-sea mining, a nascent industry, promises to unlock vast mineral resources, including polymetallic nodules, hydrothermal vent deposits, and cobalt-rich crusts, crucial for high-tech and renewable energy applications. This chapter provides knowledge on the lack of comprehensive regulations heightens risks to deep marine ecosystems, sparking debates on ownership rights, socioeconomic benefits, and ecological sustainability and ways to overcome them.

Chapter 8, "Generating Electricity from Sound Vibrations Underwater", explores an innovative approach: harnessing electricity from whale movements using piezo-electric materials. By leveraging vibrations from whale migrations, this eco-friendly technology generates power without harming marine life, offering a sustainable solution for our energy needs.

Chapter 9, "Generative Artificial Intelligence Enabled Climate Change and Deep-Sea Food Webs", examines the part of generative artificial intelligence (AI) in understanding and relieving the impacts of climate change on deep-sea food webs. This chapter analyzes the potential applications of generative AI, such as machine learning calculations and simulation strategies, in modeling the complex intuitive among species and environmental factors in deep-sea ecosystems. Through a combination of hypothetical examination and experimental case thinks about, this book illustrates how generative AI can improve our understanding of the drivers of alter in deep-sea nourishment networks and illuminate preservation and administration methodologies.

Chapter 10, "Predicting Concrete Strength of Structures under Deep Sea using Machine Learning Techniques", sheds light on about the Concrete's versatility and durability make it a cornerstone of construction. However, its properties vary with environmental conditions, and precise ingredient proportions are crucial. This chapter explores the role of admixtures in enhancing concrete strength and performance, particularly in deep-sea environments, where innovative sensing and machine learning techniques optimize infrastructure resilience.

Chapter 11, "Sea Spin Magnets",explores a novel application of tidal energy using two opposing repelling neodymium magnets to operate tiny underwater turbines, harnessing ocean currents for sustainable energy.Leveraging magnetic repulsion, this innovative design prioritizes simplicity, efficiency, and environmental sustainability, offering a promising breakthrough in renewable marine energy generation technology.

In Chapter 12, "The Future of Deep Sea Technologies: Opportunities and Challenges",the authors, provides an in-depth examination of the significant opportunities and challenges associated with deep-sea technologies, emphasizing their transformative potential across scientific, economic, and ecological domains. As technological advancements continue to progress, they are reshaping our capabilities to explore and understand the deep sea, offering new insights and opportunities that were once beyond reach.

Chapter 13, "Unveiling the Depths - Exploring Deep-Sea Ecosystems through Advanced Data Analysis and Machine Learning", provides knowledge about how the deep sea, Earth's most mysterious ecosystem, is now accessible through cutting-edge machine learning (ML) technologies. This chapter explores ML's transformative potential in deep-sea ecosystem analysis, leveraging Convolutional Neural Networks, Support Vector Machines, and Random Forests for automated species identification, habitat characterization, and event detection, revolutionizing our understanding and conservation of these fragile environments.

In Chapter 14, "Utilizing Single Shot Multi Box Detector for Detecting Underwater Benthos in the Marine Benthic Zone", the authors introduces about the marine benthic zone harbors immense biodiversity, crucial for ecosystem function and conservation. However, traditional survey methods face significant challenges. This chapter presents a breakthrough: a Single Shot Detector (SSD)-based system for efficient and reliable benthos detection, revolutionizing underwater biological monitoring and informing sustainable management practices.

Chapter 15, "Utilizing Temperature Difference for Sustainable Maritime Transportation", deals with the Sustainable Maritime Transportation. As maritime transportation seeks cleaner propulsion methods, renewable energy solutions become paramount. This chapter presents an innovative approach: leveraging ocean thermal energy conversion (OTEC) to generate electricity. By exploiting temperature differences between surface and underwater waters, a novel heat exchanger system enables sustainable energy production, reducing greenhouse gas emissions and paving the way for eco-friendly maritime operations.

These chapters collectively form a landmark volume, showcasing the latest technological breakthroughs and strategic insights transforming deep sea ecosystem conservation and exploration.

As we culminate our journey through "Technological Advancements for Deep Sea Ecosystem Conservation", we gain a profound appreciation for the intricate dynamics and vital global significance of deep-sea ecosystems.

Chapter 1
A Comparative Study on Technologies Deployed for Marine Ecosystem Monitoring and Restoration

U. Vignesh
Vellore Institute of Technology, Chennai, India

R. Parvathi
Vellore Institute of Technology, Chennai, India

Pranav Swarup Kumar
Vellore Institute of Technology, Chennai, India

Abdulkareem Sh. Mahdi Al-Obaidi
https://orcid.org/0000-0003-2575-0441
Taylor's University, Malaysia

ABSTRACT

Deep-sea ecosystems play a significant role in climate change mitigation. To preserve these delicate ecosystems, innovative technologies are being deployed to safeguard deep-sea environments. Marine ecosystem restoration presents unique challenges due to the remoteness and complexity of these delicate habitats. This research review focuses on novel technologies and methods deployed to restore deep-sea ecosystems and habitats. A comparative study is performed on Human Underwater Vehicles (HOV), Autonomous Underwater Vehicles (AUVs), Remotely Operated Vehicles

DOI: 10.4018/979-8-3693-6670-7.ch001

(ROVs), Crawlers, Cabled observatories, Soft robotic bio-inspired machines and other exploration vehicles, which vary in their spatial and temporal monitoring capabilities. The review also includes a study of computer algorithms such as graph neural networks, visual simultaneous localization and mapping (vSLAM) and the advanced sensors used to obtain better data from our ocean's ecosystems, along with the techniques used to analyse the gathered data.

I. INTRODUCTION

More than 95 per cent of our planet's living space exists in the oceans. However, deep-sea ecosystems are some of the least explored ecosystems on Earth, with less than 0.0001% physically sampled or observed (Ramirez Londra et al., 2010). Seas which extend below 200m from surface level, are classified as 'deep' seas. Below this depth, solar energy usually does not support primary productivity through photosynthesis (Ramirez Londra, 2010). Prior beliefs were that the oceans were a vast barren landscape with little to no biodiversity. However, with recent research, we now know that various organisms thrive in these environments and form complex interconnected ecosystems in the dark.

In recent decades, the interest in exploiting newly unlocked resources from the seas has risen, leading to adverse anthropological impacts on these ecosystems such as littering, oil and gas extraction and deep-sea mining. Combined with other large-scale stressors such as climate change, de-oxygenation and accumulating pollutants, the decline in biodiversity of the oceans is being accelerated (Ramirez Londra, 2010). These losses threaten the well-being of humankind as a whole since we heavily depend on the services of marine ecosystems such as climate regulation, for food and medicinal products, coastal protection, recreation, and other livelihoods (Ward et al., 2022).

Apart from the inherent value in safeguarding these ecosystems, the organisms that exist in these hostile, alien conditions have adapted mechanisms and biochemical pathways which may catalyse the progress of biological research as a whole. A great example is the discovery of the Thermus aquaticus bacteria whose extremophile behaviour led to the discovery of an enzyme - DNA polymerase "Taq" and the TaqI restriction enzyme. This enzyme became the key to the invention of the Polymerase Chain Reaction (PCR) (Brock, 1997). A famous technique used to amplify gene expressions and thus identify diseased patients (popularised during the Covid-19 pandemic).

Due to the remoteness of these ecosystems, the exploration and hence conservation of these vast ecosystems remains a challenging feat and requires the use of advanced technologies. Surveying tools must withstand high pressures and low temperatures and resist corrosion, while efficiently collecting data and samples.

Since the 1960s, three types of underwater exploration vehicles have been in use. Other vehicles such as dredges, bottom landers, crawlers and cabled observatories are used for marine exploration too. However, common classification contains Human Operated Vehicles (HOVs), Remote Operated Vehicles (ROVs) and Autonomous Operated Vehicles (AUVs), each of which complements the other's functionality in deep-sea exploration (Woods Hole Oceanographic Institution, n.d.a). This paper gives a brief about these vehicles with a comparative approach, highlighting the strengths and limitations of each type of vessel. A review has been done on newly emerging technologies which seek to mimic biological creatures with soft robotic bio-inspired for varying undersea applications. This paper also explores technological advancements in navigation and newly proposed systems with a global underwater positioning system. Novel sensing technologies for seafloor mapping, microimaging with Raman spectroscopy and ocean floor surveys are discussed along with innovative ways machine learning and artificial intelligence are used in processing research data.

II. SUBMERSIBLES

2.1 HUMAN OPERATED VEHICLES (HOV)

Human Operated Vehicles or HOVs contain a compact pressurised sphere fit for accommodating human pilots within the vessel itself. HOVs contain exterior lights, sensors and cameras for monitoring. They also contain manipulator arms to obtain samples and perform in situ tests. Thrusters allow manoeuvring over rugged submerged terrain and allow the vehicle to hover above the seafloor. These HOVs are often used along with Remotely operated vehicles ROVs. These vehicles are comparatively harder to create and deploy, owing to the human component in their design.

Though remotely operated vehicles can carry out similar tasks, having a scientist or an expert present within the target site itself, can streamline the mission, allowing for real-time, faster and better decisions to be taken on the spot. However, an HOV does not operate individually and is deployed from a surface ship with the use of a launch and recovery system (LARS). When the vessel is underwater, the data and voice communication with the deployment ship are essential for monitoring the health of the HOV systems (Ramadass et al., 2020).

2.2 Remotely Operated Vehicles (ROV)

Remotely Operated Vehicles or ROVs are vehicles tethered to a surface base or ship that enable researchers to collect samples, monitor the research site and pilot the vessel remotely from the surface. Since they contain no human pilot inside, they can stay submerged for extended periods and eliminate the risk of sending pilots to dangerous environments (Feng et al., 2022).

Raw data from sensors and images from cameras can be sent to the surface with minimal processing through the tether, reducing the need for heavy processing power onboard the vehicle. Instructional data from the surface is sent to the vehicle to operate the manipulator arms, thrusters and sensors to carry out precise experiments underwater. With a high endurance, temporal changes in the study can be measured over comparatively longer time scales (Lessard-Pilon et al., 2010).

ROV imaging allows the investigation of density and population structure at smaller scales but with higher efficiency and in a non-destructive manner (Chimienti et al., 2018).

The types of ROVs differ vastly based on the specific need. Work class ROVs interact directly with their surroundings and contain heavier machinery. However, purely observation/inspection or portable ROVs are built on a smaller frame, are less bulkier in size and mostly do not require a Launch and Recovery System (LARS) (Reach Robotics, n.d.).

2.3 Autonomous Underwater Vehicles (AUVs)

Autonomous Underwater Vehicles or AUVs are non-piloted vehicles, programmed to carry out monitoring tasks independently (Fig 1). They are not tethered to a surface base and function with their internal power source. They may be designed to drift, drive, or glide through the ocean without manual control by operators (Woods Hole Oceanographic Institution, n.d.b.). AUVs contain an onboard navigation system and a computer responsible for making real-time decisions based on external conditions or the gathered data. An advantage the AUVs hold over ROVs is the absence of a tether to a surface base.

Though the ROVs can draw more power and transmit large quantities of data faster through a wired connection to the surface, they are limited in mobility and range of exploration (Wynn et al., 2014). Here, AUVs can be used for mission profiles involving surveying and monitoring over large spatial or temporal scales. However, they commonly do not contain tools to interact directly with their surroundings such as a manipulator arm or a sample basket and HOVs or ROVs better suited.

Figure 1. The REMUS 600 AUV

(Image Source: https://oceanexplorer.noaa.gov/; B. Eakins, CU Boulder and NOAA NCEI)

Hybrid Remotely Operated Vehicles (HROVs) are a hybrid type of vehicle which combine the advantages of ROVs and AUVs. They can be free-swimming AUVs and can also be converted aboard ship into an ROV, attached to a surface ship via a lightweight, fibre-optic tether (Woods Hole Oceanographic Institution, n.d.b.).

In 2018, a cruise in the Izu-Ogasawara area deployed multiple low-cost AUVs which cooperatively discovered the hydrothermal chimney from topography and optical images (Ramdass et al., 2020). Such flexibility and customizability increase the efficiency and chance of success of subsea missions in unexplored areas.

Table 1. Comparative analysis of Vehicle Types

Vehicle Type	**Spatial Coverage**	**Temporal Coverage**
HOVs	Poor	Poor
ROVs	Poor	Average
AUVs	Good	Good

2.4 Soft Bio-Inspired Robots

Certain creatures have evolved to thrive in seemingly hostile conditions to deployed machinery and systems. After centuries of evolution and natural selection, these creatures have developed highly specialised systems fit for their habitats. A form of robotic fish movement that is similar to the carangiform style of swimming seen in nature was found to be reasonably accurate and useful for modelling thrust generation (Mason and Burdick, 2000).

Traditional modes of movement used in vessels on the surface and at shallow depths are bulky, rigid and prone to break down in extreme atmospheres. An article highlighted a group of researchers who created a robot inspired by the body features of a snailfish. A dielectric elastomer (DEA) material was used to create the 'musculature' of the robot (Fig. 2). The robot was shown to have actuated its wings freely in the Mariana Trench at a depth of over 10 kilometres. In such conditions, soft, lightweight devices prove to have an edge over conventional exploration tools (Li et al., 2021).

Soft robots, as opposed to conventional rigid vessels, hold the key to many advantages and emerging technologies. The concept of morphological intelligence, observed in creatures like the octopus, can simplify control, and improve algorithms and efficiency by cutting down computing costs, which may be vital in hostile conditions (Li et al., 2023). Sample collection and physical intervention can be performed in a highly controlled and dexterous way, ensuring a non-destructive approach to bio-exploration and restoration.

Figure 2. Snailfish mimicking bioinspired robot. (Image by Baosheng Wu, in the Article, 'Self-powered soft robot in the Mariana Trench' Nature. 591. 66-71)

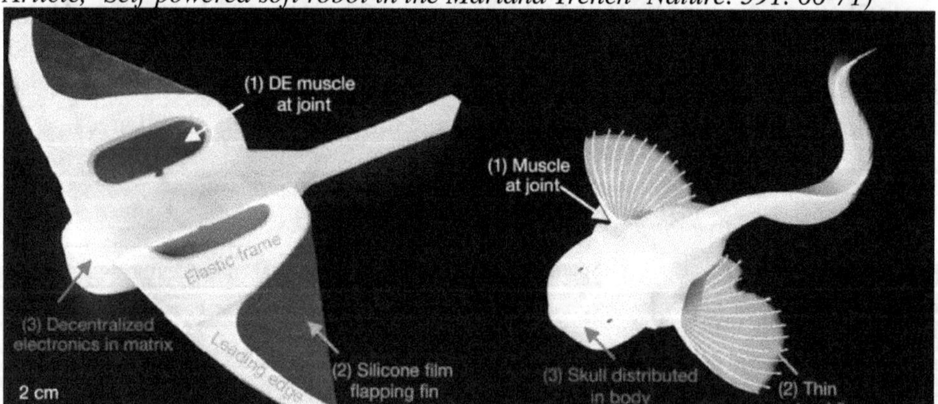

New technologies can be integrated to carry out specialised mission profiles. Energy from hydrothermal vents may serve as a potential source of power in deep-sea robots wrapped in thermal energy harvesting skins. Sensor skins may replace bulky sensing units and provide high-level autonomous adaptations (Li et al., 2023).

Findings on self-healing materials can prove to be indispensable for soft actuators and robots that operate in dynamic and real-world environments (Pena-Francesch et al., 2020). Interdisciplinary research with experts in marine biology, mechanics, materials research, oceanography and other fields can lead to breakthroughs in deep-sea exploration.

III. NAVIGATION TECHNOLOGIES USED IN EXPLORATION VESSELS

AUVs and other Unmanned Underwater Vehicles (UUVs) require strong navigation technology considering their independent wireless operation. Conventional global navigation satellite systems (GNSS) cannot be used as water blocks electromagnetic signals from navigation satellites. Setting up localisation equipment several kilometres subsea is not practical either (Song et al., 2022). Acoustic navigation is a popular method used to achieve absolute positioning through beacons, Based on the deployment of beacons, it can be classified as the long-baseline system (LBL), short baseline system (SBL), and ultra-short baseline system (USBL) Out of these, the LBL is the most accurate with centimetre accuracy (Sun et al., 2021).

However, the transponders need to be placed on the seafloor, reducing the ease of operation of this mode of navigation. In USBL, the transducer array is attached to the bottom of the mothership and thus its operation is more straightforward (Zhang et al., 2018). USBL is a popular positioning tool for manned submersibles and HOVs.

These navigation methods need to be tailored to the specific mission type. For short dives to frequently visited exploration sites, LBL transducers can be permanently installed on the site's seafloor. New innovations such as a battery-free navigation system called the Underwater Backscatter Localization (UBL) reflect modulated signals from its environment rather than emitting its own, at net-zero energy consumption (Massachusetts Institute of Technology, 2020).

3.1 Global Multipurpose Ocean Acoustic Network

This technology was recently proposed (Sun et al., 2021), drawing from a direct analogy to the GPS. Low-frequency acoustic sources, when placed judiciously, can provide an understanding of ecological processes in the ocean with large-scale measurements of a high temporal resolution. An underwater geo-positioning (UW-GPS) service enables the operation of floats, gliders, and AUVs (Miksis-Olds et al., 2021) which can all link to the UW-GPS as a common anchor for navigation and communications. This technology aims to kindle a multi-disciplinary discovery by improving the operation of exploration and restoration vessels (Miksis-Olds et al., 2021).

IV. SENSING TECHNOLOGIES

Advancements in sensing technologies help make underwater research vessels more efficient and resourceful in their missions. Underwater environments vary from surface conditions due to their limited accessibility, insufficient illumination, scattering and diffusion for cameras (Massot-Campos and Oliver-Codina, 2015). Moreover, cameras need to be at a very close range for lighting the terrain and getting pictures but underwater acoustic imaging can gather data from a range of more than 100 metres (Centre for Maritime Research and Experimentation, 2019).

The tradeoff, however, is the loss of quality with acoustic imaging. A novel technique proposed is a fusion between optical data and acoustic imaging. In Williams and Mahon (2004), the researchers used a pencil beam sonar to aid navigation. The quick data retrievals from the pencil beam sonar were used to instantiate Simultaneous Localization and Mapping (SLAM) features that could be tracked in the optical camera. In another example, side scan sonar data was fused with the optical system's data. This optical data was used to build a textured three-dimensional scene. The sonar data could produce a depth/bathymetric profile by shading in the 3D scene (Moroni et al., 2012).

4.1 Underwater Microimaging

Microscale processes play a major role in understanding the deep sea, especially in learning how ecosystems exist. These processes must be analysed within their environments, thus calling for in-situ analysis. Raman spectroscopy is a common tool used for microscopic identification. This module, when attached to a submersible ROV or HOV, can assist in the recognition of microscale targets (Liu et al., 2022).

A group of researchers came up with a deep-sea microscopic imaging system, MICROcean, to perform microimaging in the deep sea. The system integrated an optical zoom lens and illumination to provide light to the imaging area. It also uses an underwater Raman spectroscope to recognise and analyse samples.

However, portable spectrometry equipment has a large performance gap as opposed to lab-grade equipment. A paper highlighted efforts to create a compact and flexible Raman spectroscope (Liu et al., 2021). Keeping in mind applications on lightweight underwater platforms and vessels. It was successfully deployed near an underwater hydrothermal vent and cold seep to detect targets.

4.2 Seafloor Mapping Techniques

The ocean covers 70% of all livable space on Earth, but as of June 2024, only 26.1% of the global seafloor had been mapped with modern high-resolution technology (National Oceanic and Atmospheric Administration, n.d.).

Sound navigation and ranging, or SONAR is the most popular method used for bathymetry today. Single-beam echo sounding (SBES) was the first to be applied for seafloor mapping. A transducer emits a cone of sound, which travels underwater and reflects after making contact with the seafloor. The returned echo is sensed by the ship's transducer. Based on the duration the pulse of sound travels underwater, the depth is calculated (Woods Hole Oceanographic Institution, n.d.c.).

Multibeam echo sounding (MBES) utilises multiple beams at once both directly under the vessel and to the sides of it (Sun et al., 2021). This way, the system measures not only depth but also backscatter. Backscatter gives scientists information about the geological composition of the seafloor. A comparative study of SBES and MBES highlighted that the MBES system offers a higher resolution and coverage. This considerable advantage over SBES systems was shown to exist due to backscatter characteristics from oblique angles present only with multiple beams of sonar along the hull (Parnum et al., 2009). Another configuration in the form of side beam sonars has two transducers on either side of the ship, providing coverage of up to 240° below the vessel. Since the two sensors do not interfere with each other, this method optimises the range of exploration (Sun et al., 2021).

The above-discussed surveying methods of echo sounding have a high cost of operation and maintenance when deployed in shallow or remote areas such as hidden reefs (Sagawa et al., 2019). An alternative space-borne data acquisition technique being researched today is satellite-derived bathymetry or SDB.

A group of researchers successfully implemented Machine Learning models to perform bathymetry estimations from synthetic satellite images (Al Najar et al., 2023). Another paper showed SDB used in the Hramina Bay to gain a positional accuracy of 10 and 30 metres. The depth gradients and coastline were well surveyed by using the SDB method (Duplančić et al., 2019).

V. APPLICATIONS OF ARTIFICIAL INTELLIGENCE IN THE DEEP-SEA

Machine learning (ML) has been pivotal in analysing image, and video data. Sensors, cameras and other data-gathering tools sent on deep-sea missions are made to withstand adverse environments.

Thus, the data gathered by these tools has a lot of scope for enhancement using machine learning algorithms. Moreover, ROVs and AUVs can significantly benefit from enhanced image recognition techniques. The accuracy can surpass that of human eyes, especially when tuned to conditions where images are not sharp and the contrast is low.

Machine Learning and its subset, deep learning, has been highly popular for more than a decade. Convolutional neural networks or CNNs are popularly used for finding patterns in images to recognize objects and categorise them. Thus they also suit exploratory scenarios. This is due to CNNs requiring minimal human intervention, as they identify relevant features on their own from the data set.

Figure 3. CNN for image classification (Source: Leoni et al. (2019).

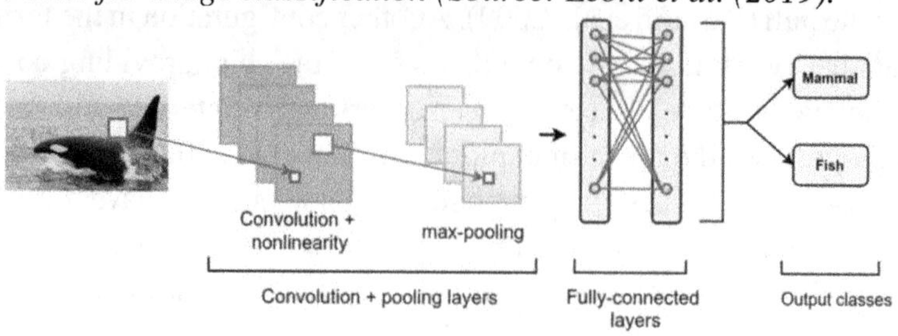

In one study, region-based CNNs were used to identify sea creatures from ROV video data. They make use of isolating a proposed region from the original image. Then, features are extracted from that region on which the trained model identifies the species (Han et al., 2020).

In another case, artificial intelligence (AI) has been used to transform a cabled observatory's camera into a biological sensor for observing an ecosystem. This AI pipeline was used for the automatic detection, classification, and counting of sablefish in the northeast Pacific region (Bonofiglio et al., 2022).

A paper highlighted the development of an application program interface (API) to facilitate a web-based interface (Chen et al., 2024). This interface hosts an AI model which recognises marine specimens.

Another paper showed the use of Machine learning to identify ship wake patterns. When applied to the ship wake detection problem, it showed an average accuracy of ~93.40% given a very limited and unbalanced dataset (Luo et al., 2022).

VI. CONCLUSION

Underwater ecosystems and habitats contain immense scope to be understood and researched. Due to their remoteness and harsh conditions, the deep sea flora and fauna remain relatively unexplored. At the heart of exploration is technology, which is being advanced in numerous aspects of deep-sea exploration such as submersibles, systems for mapping, sensing, positioning, image processing etc. Data on new ecosystems, species and their interactions are being uncovered in increasing magnitudes. With rising anthropogenic ill effects on marine life, the need for conservation is a paramount concern.

This paper reviews the fundamental submersible vessels i.e. the HOVs, ROVs and AUVs. Novel methods of mimicking live creatures to improve efficiency through design have also been explored. This is followed by a short study on common approaches to sub-sea navigation and certain proposed ideas related to it.

Key sensing technologies have also been reviewed. These tools are used in the field with exploratory vessels and robots for a diverse range of applications such as optical imagery, oceanic surveys and in-situ analysis. Finally, a few instances of innovative applications of deep learning and artificial intelligence have been reviewed.

Exploratory data complements and guides the conservation of delicate balances in deep-sea ecosystems. By adopting an ecosystem-oriented active restoration approach, future damage to underwater biology can be mitigated while understanding how bioprocesses occur. To facilitate this, technological study focussed on restoring deep sea ecosystems must be furthered to navigate these harsh environments and impact them positively.

VII. REFERENCES

Al Najar, M., Thoumyre, G., Bergsma, E. W. J., & Schenke, H. W. (2023). Satellite-derived bathymetry using deep learning. *Machine Learning*, 112(4), 1107–1130. DOI: 10.1007/s10994-021-05977-w

Bonofiglio, F., De Leo, F., Yee, C., Chatzievangelou, D., Aguzzi, J., & Marini, S. (2022). Machine learning applied to big data from marine cabled observatories: A case study of sablefish monitoring in the NE Pacific. *Frontiers in Marine Science*, 9, 842946. Advance online publication. DOI: 10.3389/fmars.2022.842946

Brock, T. D. (1997). The value of basic research: Discovery of Thermus aquaticus and other extreme thermophiles. *Genetics*, 146(4), 1207–1210. DOI: 10.1093/genetics/146.4.1207 PMID: 9258667

. Centre for Maritime Research and Experimentation. (2019). CMRE-PR-2019-078.

Chen, V. Y., Lu, D.-J., & Han, Y.-S. (2024). Hybrid intelligence for marine biodiversity: Integrating citizen science with AI for enhanced intertidal conservation efforts at Cape Santiago, Taiwan. *Sustainability (Basel)*, 16(1), 454. DOI: 10.3390/su16010454

Chimienti, G., Angeletti, L., Rizzo, L., Tursi, A., & Mastrototaro, F. (2018). ROV vs trawling approaches in the study of benthic communities: The case of Pennatula rubra (Cnidaria: Pennatulacea). *Journal of the Marine Biological Association of the United Kingdom*, 98(8), 1859–1869. DOI: 10.1017/S0025315418000851

Duplančić Leder, T., Leder, N., & Peroš, J. (2019). Satellite-derived bathymetry survey method—Example of Hramina Bay. *Transactions on Maritime Science*, 8(1), 99–108. DOI: 10.7225/toms.v08.n01.010

Feng, J.-C., Liang, J., Cai, Y., Zhang, S., Xue, J., & Yang, Z. (2022). Deep-sea organism research oriented by deep-sea technologies development. *Science Bulletin*, 67(17), 1802–1816. Advance online publication. DOI: 10.1016/j.scib.2022.07.016 PMID: 36546066

Han, F., Yao, J., Zhu, H., & Wang, C. (2020). Marine organism detection and classification from underwater vision based on the deep CNN method. *Mathematical Problems in Engineering*, 3937580, 1–11. Advance online publication. DOI: 10.1155/2020/3937580

Leoni, G., Endo, P., Monteiro, K., Rocha, E., Silva, I., & Lynn, T. (2019). Accelerometer-based human fall detection using convolutional neural networks. *Sensors (Basel)*, 19(7), 1644. DOI: 10.3390/s19071644 PMID: 30959877

Lessard-Pilon, S., Porter, M. D., Cordes, E. E., MacDonald, I., & Fisher, C. R. (2010). Community composition and temporal change at deep Gulf of Mexico cold seeps. *Deep-sea Research. Part II, Topical Studies in Oceanography*, 57(21–23), 1891–1903. DOI: 10.1016/j.dsr2.2010.05.012

Li, G., Chen, X., Zhou, F., Liang, Y., Xiao, Y., Cao, X., Zhang, Z., Zhang, M., Wu, B., Yin, S., Xu, Y., Fan, H., Chen, Z., Song, W., Yang, W., Pan, B., Hou, J., Zou, W., He, S., & Yang, W. (2021). Self-powered soft robot in the Mariana Trench. *Nature*, 591(7848), 66–71. DOI: 10.1038/s41586-020-03153-z PMID: 33658693

Li, G., Wong, T. W., Shih, B., Guo, C., Wang, L., Liu, J., Wang, T., Liu, X., Yan, J., Wu, B., Yu, F., Chen, Y., Liang, Y., Xue, Y., Wang, C., He, S., Wen, L., Tolley, M. T., Zhang, A.-M., & Li, T. (2023). Bioinspired soft robots for deep-sea exploration. *Nature Communications*, 14(1), 7097. DOI: 10.1038/s41467-023-42882-3 PMID: 37925504

Liu, Q., Guo, J., Lu, Y., Wei, Z., Liu, S., Wu, L., Ye, W., Zheng, R., & Zhang, X. (2022). Underwater Raman microscopy—A novel in situ tool for deep-sea microscale target studies. *Frontiers in Marine Science*, 9, 1018042. Advance online publication. DOI: 10.3389/fmars.2022.1018042

Liu, Q., Guo, J., Ye, W., Cheng, K., Qi, F., Zheng, R., Sun, Z., & Zhang, X. (2021). Development of an easy-to-operate underwater Raman system for deep-sea cold seep and hydrothermal vent in situ detection. *Sensors (Basel)*, 21(15), 5090. DOI: 10.3390/s21155090 PMID: 34372326

Luo, Y., Zhang, C., Liu, J., Xing, H., Zhou, F., Wang, D., Long, X., Wang, S., Wang, W., & Shi, F. (2022). Identifying ship-wakes in a shallow estuary using machine learning. *Ocean Engineering*, 246, 110456. DOI: 10.1016/j.oceaneng.2021.110456

Mason, R., & Burdick, J. W. (2000). Experiments in carangiform robotic fish locomotion. In Proceedings of the 2000 ICRA. Millennium Conference. IEEE International Conference on Robotics and Automation (Vol. 1, pp. 428–435). IEEE. DOI: 10.1109/ROBOT.2000.844093

Massachusetts Institute of Technology. (2020, November 2). A GPS-like system to track underwater locations. MIT News. https://news.mit.edu/2020/underwater-gps-navigation-1102

Massot-Campos, M., & Oliver-Codina, G. (2015). Optical sensors and methods for underwater 3D reconstruction. *Sensors (Basel)*, 15(12), 31525–31557. DOI: 10.3390/s151229864 PMID: 26694389

Miksis-Olds, J., Rehm, E., Howe, B., Worcester, P., Haralabus, G., & Sagen, H. (2021). Envisioning a global multi-purpose ocean acoustic network. *Marine Technology Society Journal*, 55(3), 78–79. DOI: 10.4031/MTSJ.55.3.27

Moroni, D., Pascali, M., Reggiannini, M., & Salvetti, O. (2012). Underwater scene understanding by optical and acoustic data integration. *Proceedings of Meetings on Acoustics Acoustical Society of America*, 17, 070085. Advance online publication. DOI: 10.1121/1.4792225

National Oceanic and Atmospheric Administration. (n.d.). How much of the ocean have we explored? NOAA Ocean Exploration. https://oceanexplorer.noaa.gov/facts/explored.html

. Parnum, I., Siwabessy, P. J., Gavrilov, A., & Parsons, M. (2009). A comparison of single-beam and multibeam sonar systems in seafloor habitat mapping.

Pena-Francesch, A., Jung, H., Demirel, M. C., & Sitti, M. (2020). Biosynthetic self-healing materials for soft machines. *Nature Materials*, 19(11), 1230–1235. DOI: 10.1038/s41563-020-0736-2 PMID: 32719508

Ramadass, G. A., Vedachalam, N., Sethuraman, R., Narayanan, S., Subramanian, A., Raju, R., Chowdhury, T., Pranesh, S. B., & Atmanand, M. (2020). Challenges in developing deep-water human occupied vehicles. *Current Science*, 118(11), 1687–1693. DOI: 10.18520/cs/v118/i11/1687-1693

Ramirez-Llodra, E. (2020). Chapter 2: Deep-sea ecosystems: Biodiversity and anthropogenic impacts. In The Law of the Seabed. Leiden, The Netherlands: Brill | Nijhoff. DOI: 10.1163/9789004391567_004

Ramirez Llodra, E., Brandt, A., Danovaro, R., De Mol, B., Escobar, E., German, C. R., Levin, L. A., Martinez-Arbizu, P., Menot, L., Buhl-Mortensen, P., Narayanaswamy, B. E., Smith, C. R., Tittensor, D. P., Tyler, P. A., Vanreusel, A., & Vecchione, M. (2010). Deep, diverse and definitely different: Unique attributes of the world's largest ecosystem. *Biogeosciences*, 7(9), 2851–2899. DOI: 10.5194/bg-7-2851-2010

Reach Robotics. (n.d.). Reach Robotics. Retrieved from https://reachrobotics.com

Sagawa, T., Yamashita, Y., Okumura, T., & Yamanokuchi, T. (2019). Satellite-derived bathymetry using machine learning and multi-temporal satellite images. *Remote Sensing (Basel)*, 11(10), 1155. DOI: 10.3390/rs11101155

Sato, T., ... (2019). Exploring hydrothermal deposits with multiple autonomous underwater vehicles. In 2019 IEEE Underwater Technology (UT) (pp. 1–5). IEEE. DOI: 10.1109/UT.2019.8734460

Song, Y., Nakath, D., She, M., & Köser, K. (2022). Optical imaging and image restoration techniques for deep ocean mapping: A comprehensive survey. PFG –. *Journal of Photogrammetry, Remote Sensing and Geoinformation Science*, 90(3), 243–267. DOI: 10.1007/s41064-022-00206-y

Sun, K., Cui, W., & Chen, C. (2021). Review of underwater sensing technologies and applications. *Sensors (Basel)*, 21(15), 7849. DOI: 10.3390/s21237849 PMID: 34883851

Ward, D., Melbourne-Thomas, J., Pecl, G. T., Evans, K., Green, M., McCormack, P. C., Novaglio, C., Trebilco, R., Bax, N., Brasier, M. J., Cavan, E. L., Edgar, G., Hunt, H. L., Jansen, J., Jones, R., Lea, M. A., Makomere, R., Mull, C., Semmens, J. M., & Layton, C. (2022). Safeguarding marine life: Conservation of biodiversity and ecosystems. *Reviews in Fish Biology and Fisheries*, 32(1), 65–100. DOI: 10.1007/s11160-022-09700-3 PMID: 35280238

Williams, S., & Mahon, I. (2004). Simultaneous localisation and mapping on the Great Barrier Reef. *Proceedings of the 2004 IEEE International Conference on Robotics and Automation*, 1771–1776. DOI: 10.1109/ROBOT.2004.1308080

Woods Hole Oceanographic Institution. (n.d.a). Three types of vehicles. Dive and Discover. Retrieved from https://divediscover.whoi.edu/underwater-vehicles/three-types-of-vehicles/

Woods Hole Oceanographic Institution. (n.d.b). Know Your Ocean: Ocean Tech - Acoustics & Sonar. Retrieved from https://www.whoi.edu/

Woods Hole Oceanographic Institution. (n.d.c). Sonar: Single-beam and multibeam. WHOI Know Your Ocean. https://www.whoi.edu/know-your-ocean/ocean-topics/ocean-tech/acoustics/sonar-single-beam/

Wynn, R. B., Huvenne, V. A. I., Le Bas, T. P., Murton, B. J., Connelly, D. P., Bett, B. J., Ruhl, H. A., Morris, K. J., Peakall, J., Parsons, D. R., Sumner, E. J., Darby, S. E., Dorrell, R. M., & Hunt, J. E. (2014). Autonomous underwater vehicles (AUVs): Their past, present and future contributions to the advancement of marine geoscience. *Marine Geology*, 352, 451–468. DOI: 10.1016/j.margeo.2014.03.012

Zhang, T., Liu, B., & Liu, Y. (2018). Positioning systems for Jiaolong deep-sea manned submersible: Sea trial and application. *IEEE Access : Practical Innovations, Open Solutions*, 1, 71644–71650. Advance online publication. DOI: 10.1109/ACCESS.2018.2881390

Chapter 2
Anthropogenic Waste Classification in Deep Sea Environments

Lakshmi Harika Palivela
Vellore Institute of Technology, Chennai, India

V. R. Balasaraswathi
Vellore Institute of Technology, Chennai, India

Karthika Veeramani
Vellore Institute of Technology, Chennai, India

Swati Kumari
Vellore Institute of Technology, Chennai, India

Yasir Hamid
Abu Dhabi Polytechnic, UAE

ABSTRACT

Deep-sea ecosystems play a significant role in regulating the Earth's climate by capturing carbon dioxide through photosynthesis and calcium carbonate deposition. They also play a crucial role in global nutrient cycling, recycling organic matter from surface waters. Economically, deep-sea ecosystems provide valuable services like fisheries resources, pharmaceutical compounds, and minerals and metals. Protecting these ecosystems is essential for sustainable use and maintaining the livelihoods of communities that depend on them. The ecological integrity of deep-sea ecosystems is seriously threatened by anthropogenic waste, but efforts to control and lessen this problem are hampered by the absence of thorough classification frameworks. In

DOI: 10.4018/979-8-3693-6670-7.ch002

order to improve our knowledge of the origins, distribution, and effects of anthropogenic waste in deep marine habitats, this chapter suggests creating a systematic framework for classification of anthropogenic waste using advanced deep learning techniques. Integrated with a user-friendly web application, our proposed work allows users to effortlessly upload images of waste materials and receive real-time classification results with 99% accuracy.

1. INTRODUCTION

Deep Sea, Earth's last frontier, is made up of enormous regions of ocean that are not visible from the surface of the sun. With its intense pressures, extremely cold temperatures, and deep depths, this enigmatic world has long captured the interest of explorers and scientists alike. The deep sea, which was once thought to be isolated and unaffected by human activity, is currently facing a subtle but ubiquitous threat: marine anthropogenic waste. As more and more data show how much human activity is affecting these delicate ecosystems, waste management in the deep sea has become a critical global issue. In contrast to the popular perception of the deep sea as a pristine wilderness, the deep sea is now acknowledged to be a holding place for plastic debris, fishing gear that has been abandoned, and other types of human waste. The deep sea's natural beauty is being ruined by an influx of anthropogenic waste that also puts the lives of its residents in danger and jeopardizes the ecosystem's stability.

An ecosystem is a group of living things, including plants, animals, and microorganisms, that interact with each other and their non-living surroundings. Biotic components, including producers (autotrophs), consumers (heterotrophs), and decomposers (detritivores), are essential elements of an ecosystem. Producers make their own food through chemosynthesis or photosynthesis, while consumers rely on other organisms for sustenance. The existence of anthropogenic waste in oceanic environments has several harmful effects. Ecosystems are the intricate connections between living organisms and physical environments, sustaining life through energy flow, nutrient cycle, and ecological succession. Understanding ecosystems is crucial for biodiversity preservation and understanding natural environments (Orcutt et al., 2020). The term "ecosystem" is used 160 times in Alibaba's 2014 IPO prospectus and by technology.

1.1 Types of Waste:

The ecosystem waste can be mostly classified into natural and anthropogenic wastes.

1.1.2 Natural Causes of Waste in Ecosystems

Decomposition is the regular interruption of organic matter, such as lifeless plants and animals, creates organic waste. Decomposers similar to bacteria, fungi, and detritivores break down this matter, reusing nutrients hindmost into the ecosystem. Animal Excretion is the animals wastes produced through processes like excretion and urination. This organic waste is finally broken down and blend back into the ecosystem (Wear et al., 2021). Plant Litter are the dropping leaves, branches, flowers, and fruits contribute to plant litter, which decays over time and turn out to be part of the soil.

1.1.3Anthropogenic Waste

Anthropogenic waste is the human - made waste that are landed up in the deep sea or the ocean. This waste arises from a variety of sources and has abundant influences on aquatic ecosystems (Häder, Donat-P., et al., 2020). The anthropogenic waste can be roughly classified into Plastics, Metals, Glass, Textiles, Chemicals, Electronic Waste(E-Waste), Radioactive waste, Construction and Demolition Debris, Organic Waste, Construction and Demolition Debris, Miscellaneous.

Plastics: Plastics can be classified into microplastics, macroplastics and fishing gear. Microplastics are the small plastic particles which are less than 5 mm in diameter, including fragments from larger plastic debris and microbeads from personal care products. Fishing Gear is referred as the missing, or unwanted fishing nets, lines, and traps

Metals: Beverage, foodstuff cans, and metal containers. Industrial Waste are the scrapings and bits and pieces from industrial activities. Shipwreck Debris are the metal parts from the sunken ships and from the other marine vehicles.

Glass: Bottles and Jars from the beverage and food containers. Industrial Glass are trash and waste from industrial processes.

Textiles: Clothing and Fabrics are from the discarded garments and cloth materials. Ropes and nets are used in aquatic and fishing industries.

Chemicals: Persistent Organic Pollutants (POPs) are the long-lasting chemicals that can be collected in marine organisms. Heavy Metals like Mercury, lead, cadmium, and other poisonous metals. Oil and Petroleum Products are from spills and leaks from marine containers and offshore drilling.

Electronic Waste (E-waste): Thrown away electronic devices and their components, together with batteries and circuit boards.

Radioactive Waste: Materials from disposed nuclear materials and underwater test sites.

Construction and Demolition Debris: Waste from construction and demolition activities. Metal, plastic, and other types of pipes and fittings.

Organic Waste: Waste from human and animal sources. Agricultural runoff includes fertilizers and pesticides that spread the deep sea.

Miscellaneous: Personal Items like lost or discarded items such as toys, tools, and household goods. Military Debris like Fragments from military operations, including weapons and vehicles.

1.2 Environmental Impact and Mitigation

Entanglement, assimilation, and habitation disruption for oceanic organisms. Bioaccumulation and biomagnification of deadly substances in aquatic food webs. Variations in deposit composition and seafloor environments.

1.2.1 Mitigation Strategies

- Educating waste disposal and reusing systems to avoid waste from reaching the ocean.
- Imposing rules and international agreements to decrease marine pollution.
- Establishing deep-sea cleanup operations and helping research to detect and eradicate waste.
- Educating the community and industries about the influences of waste on marine environments and promising sustainable practices (Ukaogo, Prince O., et al., 2020)

The ecological integrity of deep-sea ecosystems is seriously threatened by anthropogenic waste, but efforts to control and lessen this problem are hampered by the absence of thorough classification frameworks. In order to improve our knowledge of the origins, distribution, and effects of anthropogenic waste in deep marine habitats, this chapter suggests creating a systematic framework for classification. Based on its origin, composition, and environmental fate, anthropogenic trash can be classified in the deep sea using the Deep Learning techniques. It includes many kinds of trash, each with unique characteristics and ways of building up, such as plastics, metals, chemicals, and organic materials.

The suggested classification scheme highlights how crucial it is to consider both direct and indirect human activity inputs, such as air deposition and land-based pollution, as well as activities like shipping, fishing, and offshore drilling. With the use of these techniques, policy interventions and targeted interventions can be made to lessen the negative effects of anthropogenic waste in the deep sea by identifying its key origins and channels. It highlights the necessity for customized management techniques by examining the ecological and socioeconomic effects of various anthropogenic waste types in deep-sea settings. The effects of anthropogenic waste are numerous and intricate, ranging from the entanglement and ingestion of marine creatures to the modification of deep-sea environments and the poisoning of seafood supplies. In order to manage anthropogenic waste in deep marine settings, this proposed work intends to provide guidance for future study, policy formulation, and conservation initiatives through the use of a systematic classification strategy.

The swift pace of industrialization and urban expansion has triggered a concerning surge in solid waste production, presenting pressing environmental and health dilemmas. Traditional waste management practices, like incineration, exacerbate air pollution and hazardous material dispersion, necessitating immediate action toward waste recycling and separation. Currently, manual labor primarily drives waste segregation, exposing workers to health hazards from the waste's toxic components. To address these pressing issues, we propose an automated anthropogenic waste classification system leveraging advanced deep learning techniques. The development of a robust deep learning model identifies and categorizes various waste types, including paper, plastic, metal, glass, and others. Integrated with a user-friendly web application, our proposed work allows users to effortlessly upload images of waste materials and receive real-time classification results.

By automating the anthropogenic waste classification process, our system not only mitigates health risks associated with manual sorting but also promotes efficient waste management practices. With the global daily waste production estimated at around 5 million tons, there is an urgent need to raise awareness about waste reduction and recycling. Our model stands as a valuable tool in this endeavor, empowering individuals, and communities to make informed decisions regarding waste disposal and recycling initiatives. By advocating for waste segregation and recycling, our system contributes significantly to environmental sustainability and human well-being, aligning with worldwide endeavors to combat waste pollution and foster a healthier, cleaner future.

Understanding and addressing anthropogenic waste in deep-sea environments is crucial for protecting marine ecosystems and ensuring the health of our oceans. With the multiplication of businesses in urban ranges, the administration of strong squander has risen as a noteworthy challenge, including materials such as paper, wood, plastic, metal, and glass. The routine strategy of squander transfer, burning, postures natural and wellbeing dangers, because it discharges destructive poisons into the discuss, possibly driving to genuine ailments such as cancer. Consequently, there is a squeezing ought to grasp reusing hones to protect both the environment and human wellbeing. To successfully address this challenge, it is fundamental to isolate squander into unmistakable components that can be reused utilizing different strategies. Right now, squander partition largely depends on manual labor, uncovering laborers to wellbeing dangers due to the dangerous substances display within the rubbish. Robotizing the trash classification prepare presents a practical arrangement to relieve these dangers.

In our extend, we have created a modern profound learning demonstrate competent of recognizing and categorizing distinctive sorts of waste. Coordinates with a user-friendly web application, clients can effectively transfer pictures of different sorts of squander, such as paper, plastic, etc., and see the analyzed comes about specifically on the interface. By mechanizing the trash classification handle, we not as it were reducing the wellbeing dangers related with manual sorting but moreover streamline squander administration hones. This approach not as it were upgrades effectiveness but moreover advances natural maintainability and human well-being.

The talk encompassing maintainability has been a longstanding subject in media discourses, dating back to the 19th century. In any case, it's as it were in later times that businesses have started to prioritize supportability as a central precept of their operations. This move is obvious in companies effectively showcasing themselves as ecologically cognizant substances, pointing to reinforce their maintainability qualifications and offer to perceiving customers. Eminently, there is a developing accentuation on gaging open mindfulness with respect to family and person activities relating to squander administration. Specialists are progressively recognizing the potential societal focal points of executing measures to address waste-related challenges.

Looking towards long-standing time, noteworthy changes are expected within the domain of squander administration, driven by mechanical developments. These incorporate activities such as the change of squander into vitality, the appropriation of IoT-enabled hones to optimize asset utilization, the upgrade of checking frameworks for more effective squander administration forms, and the utilization of information collection for educated decision-making. These progressions, established in innovation, hold the guarantee of revolutionizing squander administration hones, advertising both natural benefits and openings for financial optimization.

1.3 Challenges in the Existing System

- **Adaptability to New Waste Categories**: Waste management systems must be adaptable to accommodate the ever-evolving landscape of waste composition. With new materials constantly entering the waste stream, such as electronic waste or biodegradable plastics, waste management models need to possess the capability to recognize and classify these materials accurately. This necessitates continuous refinement and updating of the model's algorithms to ensure it remains effective in categorizing emerging waste categories. By implementing strategies for ongoing model retraining and adaptation, stakeholders can maintain the model's relevance and accuracy over time, thereby enhancing the efficiency of waste management practices.

- **Data Bias and Imbalance**: The performance heavily based on the quality and quantity of the training data used to develop them. However, datasets often exhibit biases and imbalances, particularly concerning underrepresented waste categories. For instance, certain types of waste may be more prevalent in certain regions or demographics, leading to skewed classification results if not properly addressed. To mitigate these issues, it's crucial to employ robust data collection and curation practices that encompass a diverse range of waste materials and demographics. By ensuring that the training data is representative and unbiased, stakeholders can enhance the model's accuracy and effectiveness in classifying waste materials across various contexts.
- **Real-time Processing Constraints**: In many waste management scenarios, particularly those involving operational decision-making in real-time, the speed at which the model can process data is paramount. However, deploying complex models to systems with limited computational resources, such as embedded devices or edge computing platforms, presents challenges. Optimization strategies are required to balance the need for rapid inference with the accuracy of waste classification. Techniques such as model pruning, quantization, and hardware acceleration can help streamline model inference while minimizing computational overhead. By optimizing real-time processing capabilities, stakeholders can ensure that waste management decisions can be made swiftly and efficiently, even in resource-constrained environments.
- **Generalization to Diverse Environments**: Waste management practices can vary significantly across different regions and cultural contexts. Factors such as waste composition, recycling infrastructure, and regulatory frameworks can differ widely from one location to another, posing challenges for models that aim to generalize across diverse environments. To address this issue, it's essential to design waste management models that are adaptable and resilient to such variations. This may involve incorporating regional-specific data into the training process, leveraging transfer learning techniques to adapt pretrained models to new environments, or implementing dynamic adjustment mechanisms that enable the model to adapt its behavior based on local conditions. By enhancing the model's generalization capabilities, stakeholders can ensure that waste management solutions remain effective and relevant across a wide range of geographical and cultural settings.

In this proposed system aims to classify various types of anthropogenic waste, including but not limited to paper, wood, plastic, metal, and glass using deep learning. In the model development pipeline, the compilation phase serves as a pivotal step preceding training. Once all architectural components are assembled, compilation finalizes the model's configuration before training begins. A critical component required during compilation is the specification of a loss function. This function serves as a guidepost for the model, enabling it to discern errors or deviations during the learning process. By quantifying the disparity between predicted outputs and ground truth labels, the loss function facilitates the model's adjustment towards more accurate predictions.

Keras, a popular deep learning framework, mandates the inclusion of a loss function during compilation. This ensures that the model is equipped with a clear objective to minimize during training. The choice of loss function often depends on the nature of the problem being addressed. Following compilation, optimization takes center stage in the training process. Optimization algorithms, such as the widely-used Adam optimizer, play a crucial role in fine-tuning the model's parameters. By iteratively adjusting the input weights based on a comparison between predicted outputs and the loss function, optimization seeks to minimize the model's error. This iterative process gradually refines the model's performance, guiding it towards convergence and improved predictive accuracy.

Metrics, while closely related to loss functions, serve a distinct purpose in model evaluation. Unlike loss functions, which guide the training process by quantifying prediction errors, metrics are primarily utilized for assessing model performance post-training. While not directly involved in the training process, metrics offer valuable feedback on the model's effectiveness in solving the target task.

In summary, the compilation phase initiates the model's readiness for training by specifying a loss function crucial for error quantification. Optimization algorithms, such as Adam, then fine-tune the model's parameters to minimize this loss. Metrics, on the other hand, evaluate the model's performance post-training, offering insights into its efficacy in solving the target task. Together, these components form integral parts of the deep learning workflow, guiding the model from initialization to deployment with iterative refinement and evaluation.

The key objectives are:

1. Automated anthrophonic waste classification: Classification of various types of garbage, including but not limited to paper, wood, plastic, metal, and glass.
2. Health and Safety Improvement: Alleviate health risks associated with manual waste segregation by replacing the traditional hand-picking method with an automated process, ensuring the well-being of workers involved in waste management.

3. Web Application Integration: Integrate the deep learning model with a user-friendly web application to enable easy and widespread access for users to upload images of waste materials. The web interface should provide real-time analysis results and details of the classified waste.
4. Seven-Waste Material Classification: Implement a comprehensive classification system that identifies at least seven different waste materials, providing detailed information on each category to enhance user awareness.
5. Global Waste Reduction: Contribute to global efforts in waste reduction by encouraging individuals and communities to adopt sustainable waste management practices. Provide a means for users to actively participate in reducing their environmental footprint.

By mechanizing the trash classification handle, we not as it were to reduce the wellbeing dangers related with manual sorting but moreover streamline squander administration hones. This approach not as it upgrades effectiveness but moreover advances natural maintainability and human well-being.

2. LITERATURE SURVEY

The emergence of IoT (Internet of Things) technologies has had a profound effect on several industries recently, most notably the construction of smart cities. A variety of interrelated systems are included in smart cities, such as smart living, mobility, environment, and government. Effective waste management comes to light in this context as a crucial issue on a local and worldwide scale. Nevertheless, insufficient collection and disposal techniques can lead to the problems in managing anthropogenic waste classification in deep sea environment.

A unique strategy that makes use of Internet of Things-based waste management systems has been devised to address these issues. This anthropogenic waste collecting and disposal process optimization model functions in three stages. Using IoT-enabled devices, waste photos are taken in the first phase. After that, features are extracted from the pooling layers of these images using Ensemble Convolutional Neural Network (En-CNN) variations as Resnet, VGG19, and Inception. Then, to improve waste classification accuracy, the Hybrid Eurasian Oystercatcher and Sandpiper Optimization Algorithm (HEO-SOA) is employed to choose the best features. These ideal attributes are then used to classify anthropogenic waste using the fully linked layers of the En-CNN. During the second stage, a deep learning method known as Deep Convolutional Neural Network (DCNN) is used to make decisions on trashcan visits. The DCNN facilitates effective waste collection by determining the best times for trashcan visits based on historical data analysis.

An Internet of Things-based optimal routing path protocol is finally put into practice in the third phase. This protocol finds the quickest route for waste-collecting vehicles to travel to reach the proper waste centers. It is driven by a multi-objective function that considers variables including distance, delay, energy consumption, and waste weights. With a 98% accuracy rate, the model's performance shows notable sensitivity and precision. Furthermore, the optimization of particular constraints with the HEO-SOA improves the efficiency of the routing algorithm.

Utilizing IoT technologies and cutting-edge machine learning techniques, this novel methodology provides a holistic solution to optimize waste management processes in smart cities, enhancing sustainability and efficiency (Kona, V. V., & Subramoniam, M. 2024). Compelling squander administration holds foremost significance in today's natural circle, requesting all-encompassing techniques including reusing, landfill methods, and the integration of progressed advances. The proposed arrangement presents a advanced squander administration framework that utilizes double convolutional neural systems (D-CNN or TF-CNN) in pair with a histogram thickness division (HDS) calculation. This shrewdly framework gives clients with instruments to execute vital security measures whereas dealing with squander materials. This inquiries about makes critical strides in a few regions. Firstly, it presents a geometrically outlined savvy junk box prepared with ultrasonic and stack estimation sensors, controlled by a microcontroller, to optimize squander control and collection.

Furthermore, it presents a cleverly strategy that utilizes profound learning strategies for the exact classification of edible and inedible squander through picture handling. In conclusion, it proposes a state-of-the-art real-time squander checking framework that utilizes short-range Bluetooth and long-range IoT innovation by means of a committed Android application (Kalpana, A. V., et.al.,2024)

Quick urbanization and populace extension have driven to a surge in squander generation, emphasizing the critical require for progressed squander sorting and reusing procedures to guarantee feasible squander administration hones. Profound learning models, eminent for their capability in picture acknowledgment errands, show promising arrangements for squander sorting applications. By leveraging broad datasets of squander pictures, these models can viably perceive one-of-a-kind highlights over different squander sorts. The computerization of squander sorting depends on vigorous profound learning designs able of precisely categorizing a wide range of squander materials. This ponder proposes that single-label classification recognises trash and non-anthropogenic waste pictures, whereas multi-label classification distinguishes rubbish categories inside single or numerous pictures with 95.77 accuracy by advertising promising prospects reusing frameworks. These progressions have the potential to moderate costs and prepare times related to squander administration forms (Hossen, M. M., et.al.,2024).

The expanding worldwide accentuation on natural supportability has impelled the advancement of inventive approaches to squander administration. This extend utilizes Streamlit to present an Online of Things (IoT) Squander Segregator framework, joining real-time information collection, machine learning, and user-friendly visualization. The essential objective of the framework is to robotize trash isolation, subsequently upgrading reusing rates and lessening the strain on landfills. The framework comprises a arrange of sensors and cameras deliberately set at squander collection destinations, empowering real-time observing of squander levels and classification. Information assembled by these sensors is prepared by a profound learning show created utilizing the Keras framework. This demonstrate is prepared to distinguish between different sorts of rubbish, counting natural, recyclable, and non-recyclable materials. Utilizing a convolutional neural organize (CNN) engineering, particularly ResNet-101, the framework successfully learns and recognizes between distinctive sorts of trash based on visual prompts. Moreover, the framework is equipped with a Automated Prepare Mechanization (RPA) demonstrate prepared to inform important specialists through e-mail based on the classification comes about. In general, this inventive IoT Squander Segregator framework illustrates the potential of combining real-time information collection, machine learning, and visualization methods to streamline squander administration forms and advance natural maintainability (Chavhan, P. G., et.al.,2024).

The quick increment in squander era and the basic for viable administration to protect the environment constitute critical worldwide concerns. Created countries bear essential duty for squander generation, whereas creating nations bear the brunt of its results. There is an all-inclusive journey for made strides, effective, and cost-effective squander administration arrangements. This composition dives into modern approaches to squander collection and segregation. The primary area looks at squander isolation procedures based on picture preparing, encouraging the classification of squander into unmistakable categories. These strategies not as it were help in squander administration but too clear the way for novel financial openings. Different options for squander isolation and picture handling utilizing neural systems offer precise and resource-efficient arrangements. The moment segment investigates actualized arrangements leveraging neural systems for squander isolation and collection independently, without human mediation. This area talks about the integration of IoT and implanted frameworks in inventive squander isolation and collection strategies. Taking after a comprehensive consider of cleverly systems for squander administration, the composition proposes the foremost appropriate procedures for diverse squander sorts, considering precision and commonsense suggestions (Kanujia, A. K., et.al.,2024).

Squander administration envelops a run of exercises, counting reusing, reusing, and reorganizing squander, pointed at relieving the amassing of squander in urban zones due to populace development, industrialization, and urbanization. The surge in mechanical development and urban advancement comes about in critical volumes of squander, posturing challenges for checking and administration. The move of cities into savvy cities towards a green economy requires collaboration among mechanical, private, and government divisions, as well as different partners such as policy-makers, business visionaries, scholastics, and universal organizations. This collective exertion points to cultivate moved forward squander administration hones and approaches. Savvy cities encourage consistent information integration, utilization of shrewd innovations, and data-driven decision-making to upgrade city operations and the quality of life for inhabitants. Leveraging shrewd advances empowers economical squander administration hones, counting squander distinguishing proof, investigation of squander sources and foundation prerequisites, and compliance with controls and approaches to upgrade natural guidelines.

Coordination key squander administration framework inside shrewd cities guarantees the precise advancement of an coordinates squander administration framework custom-made to wants of urban ranges. This investigate emphasizes savvy city approaches to different viewpoints of squander administration, counting squander era, collection, transportation, reusing, transfer, and evacuation, utilizing shrewd advances. Besides, the ponder examines and analyzes assets, needs, and framework necessities for savvy city squander administration utilizing expressive, prescriptive, and prescient information investigation highlights. Innovation integration, such as machine learning, inside the squander administration biological system empowers the forecast of foundation, asset, and squander capacity prerequisites. At last, the inquire about assesses the implantation of fake insights into the squander administration system to encourage decision-making forms and suggests basic components for the execution of savvy city squander administration activities (Udupi, P. K., et.al.,2024).

In Malaysia, the day-by-day generation of civil strong squander comes to around 250,000 tons due to different every day exercises. Viably categorizing these strong squanders into sorts for reusing purposes is significant to moderate their negative impacts on the environment. This presents a strong squander sorting framework outlined to classify three distinctive sorts of strong squander employing a profound learning approach. The YOLOv3 show is proposed as a real-time protest distinguishing proof module to recognize strong junk sorts. Both YOLOv3 and YOLOv3-tiny models are prepared to recognize plastic bottles, paper, and metal employing a dataset comprising 150 unmistakable photographs. A comparison between the two models uncovers that the YOLOv3 demonstrate accomplishes altogether higher cruel normal accuracy (mAP). With a more noteworthy number of convolutional layers, the YOLOv3 show accomplishes an normal mAP of 81.93%, whereas the

YOLOv3-tiny demonstrate accomplishes as it were 74.58%. Also, the joining rate of YOLOv3 is outstandingly speedier compared to YOLOv3-tiny. Beneath inactive conditions, YOLOv3 illustrates tall protest location precision for plastic bottles, paper, and metal, with exactness's of 0.98, 1, and 1, separately. In differentiate, YOLOv3-tiny battles to distinguish plastic bottles, accomplishing location correctness's of as it were 0.68 for paper and 0.94 for metal cans. These qualities position YOLOv3 as a dependable question acknowledgment device appropriate for integration into a strong squander sorting framework (Lee, K. J., et.al.,2024).

Squander administration includes the viable dealing with and transfer of squander through reusing and landfilling. The integration of profound learning and the Web of Things (IoT) presents a adaptable arrangement for squander classification and real-time information observing (Abu-Qdais, et.al.,2024). This paper presents a strong design for a squander administration framework leveraging profound learning and IoT advances. The proposed show offers a shrewdly approach to sorting both edible and toxic squander employing a convolutional neural arrange (CNN), a broadly received profound learning procedure. Furthermore, the framework presents a savvy junk canister plan prepared with a microcontroller and numerous sensors. IoT and Bluetooth network empower real-time information observing, permitting for inaccessible control and short-range information checking through an Android application. To evaluate the viability of the created show, different measurements such as squander name classification precision, sensor information estimation, and Framework Convenience Scale (SUS) scores are assessed and analyzed. The classification exactness accomplished by the proposed engineering utilizing the CNN show is 95.3125%, whereas the SUS score comes to 86%. This keen framework illustrates flexibility to family exercises with its capability for real-time squander observing (Rahman, M. W., et.al.,2022).

Strong squander administration postures a noteworthy challenge in urban regions due to the huge volume of different squander produced, requiring different treatment strategies. In any case, handling blended squander inside and out in a appropriate way is unreasonable. Squander partition rises as a significant prepare in squander administration, customarily requiring physical exertion and posturing dangers to people included. To address this, the proposed framework presents an productive computerized squander isolation show conveyed at the source. This show points to diminish human association in squander isolation by leveraging Convolutional Neural Organize (CNN), a Machine Learning calculation. The essential objective is to capture pictures of person squander materials and precisely recognize and isolate them into categories such as biodegradable and non-biodegradable. Utilizing Raspberry Pi, the isolated squander can be stored into individual containers, streamlining the squander administration handle and improving proficiency. This strategy encourages the creation of a compelling squander administration framework whereas assisting the generally prepare (Soundarya et.al.,2022). Proficient squander administration and reusing are basic for making renewable and clean urban situations. Pakistani cities, strikingly Karachi and Lahore, hook with challenges related to strong squander administration, transfer, and reusing. The integration of IoT and profound learning presents a secluded approach to information categorization and real-time investigation.

This article presents a strong "Keen Junk Administration and Categorization Framework" based on IoT and DL. The proposed framework highlights a microchip-based rubbish container prepared with numerous sensors to quickly collect squander. IoT innovation encourages real-time information observing, upgrading control over squander administration forms. Besides, the framework consolidates a squander classification show based on convolutional neural systems (CNNs). This demonstrate viably categorizes squander into unmistakable categories, in this manner encouraging reusing endeavors at squander collection plants. Generally, the proposed framework offers a comprehensive arrangement for waste administration and reusing in savvy cities, crossing from squander collection to classification and administration (Hasan, M. K., et.al.,2022).

Right now, metropolitan strong squander (MSW) administration requests critical consideration to maintain socio-economic standards like natural assurance, financial maintainability, and the relief of wellbeing issues. Considers on squander administration frameworks demonstrate that around 90% of MSW frameworks insufficiently arrange of squander in open dumps and landfills by classifying squander into degradable wastes. Later headways in computational approaches, such as fake insights and picture preparing, offer a wide extend of arrangements for MSW administration challenges. These computational strategies can be modified to classify squander, encouraging its transformation into usable vitality. Be that as it may, existing squander classification strategies in MSW administration regularly endure from destitute exactness and tall mistake rates. This model presents a viable computer vision-based arrangement for MSW administration, leveraging the Web of Things (IoT) and machine learning (ML) methods counting relapse, classification, clustering, and relationship rules. ResNet V2 is able of preparing expansive datasets in profound neural systems, coming about in progressed precision and decreased blunder rates. Furthermore, clump normalization and blended half breed pooling procedures are coordinates into the CNN to improve steadiness and execution.The proposed show precisely recognizes the sort of squander and classifies it as biodegradable or non-biodegradable, encouraging exact collection in individual squander canisters (Mookkaiah, S. S., et.al.,2022).

The mounting volume of squander amassing in landfills has ended up a squeezing natural concern, requiring a stronger approach to squander collection. Reusing stands as a key arrangement in overseeing squander by repurposing recyclable materials into modern items, subsequently redirecting junk from landfills. Be that as it may, open mindfulness and interest in reusing endeavors stay disappointingly moo, ruined by different challenges, counting the trouble in isolating squander (Wahab, M. H. A et. al.,2014).

This paper endeavors to address this issue by creating a framework competent of naturally isolating squander and coordinating it to suitable containers. Utilizing a camera, the framework captures pictures of the squander, which are at that point subjected to picture classification employing a profound learning show. The created demonstrate is implanted into a Raspberry Pi. The proposed framework accomplishes a classification exactness extending from 77% to 85%. This framework is expected to raise mindfulness around the significance of reusing and cultivate reusing propensities among the open, eventually diminishing squander aggregation in landfills (Cheema, S. M., et.al.,2022).

Effective squander isolation, administration, transportation, and transfer are foremost for minimizing dangers to open wellbeing and security as well as natural risks. The routine approach of observing junk in assigned containers is wasteful, expending important time, human assets, and monetary assets, and falls brief of assembly the prerequisites of shrewd cities. Hence, the objective is to streamline decision-making forms and improve the productivity of squander categorization. Utilizing a convolutional neural organize (CNN), this thinks about endeavors to create a picture classifier competent of recognizing things and categorizing waste materials viably. The paper presents an outline of different junk checking procedures, rubbish transfer techniques, and the mechanical system supporting the foundation of a vigorous squander administration framework. Besides, it proposes an effective framework and squander transfer strategy that holds guarantee for upgrading execution and cost-effectiveness within the future. By saddling the control of profound learning techniques, one of the foremost critical challenges in proficient squander administration can presently be tended to. The proposed strategy outperforms elective approaches such as AlexNet, VGG16, and ResNet34 in terms of execution and viability (Chauhan, R., et.al.,2023).

Given the critical requests on human assets and time, there's a squeezing worldwide require for an robotized, more exact, and quick strategy for classifying strong squanders. In this ponder, we present a show called SmartBin pointed at tending to this require. Two unmistakable approaches are utilized to productively classify strong squanders as degradable wastes. The CNN-based IoT approach is connected to datasets gotten through three diverse strategies. The primary strategy includes utilizing pictures sourced from Kaggle, the moment strategy utilizes picture look comes about from Google and Bing, and the third strategy includes capturing pictures physically beneath controlled conditions. The comes about uncover that the moment approach yields prevalent execution, achieving an exactness level of 98.57%, speaking to a critical advancement over the primary approach, which accomplishes a precision of 95.24%. This underscores the adequacy of joining extra sensors into the demonstrate for upgraded squander classification precision (Sivakumar, M., et.al.,2022).

The burgeoning people improvement in urban zones stances vital challenges to misuse organization systems, driving to a significant increase in misuse period. Customary misuse organization approaches are frequently inefficient and costly, coming almost inside the wastage of imperative resources. In any case, by coordination the Net of Things (IoT) and significant learning models, there exists an opportunity to optimize resource utilization and overhaul misuse organization efficiency. This explores endeavors to make a sharp waste organization system leveraging significant learning models to streamline the misuse segregation plan and engage real-time checking of holder status interior an IoT environment with a camera module to recognize misuse things, while a servo motor, related to a plastic board, empowers the categorization of misuse into specific compartments. An ultrasonic sensor is utilized to screen the fill rate of waste, though a GPS module enables the recuperation of real-time scope and longitude encourages. Other than, a LoRa module embedded interior the sharp canister transmits canister status redesigns to a LoRa beneficiary working at 915 MHz. To ensure the security of electronic components interior the adroit holder, an RFID-based locker system is executed, permitting get to because it was to enlisted RFID labels for bolster or upgrading purposes. Through this comprehensive approach, the proposed sharp misuse organization system focuses to revolutionize misuse organization sharpens in urban ranges, developing efficiency and supportability (Sallang, N. C. A., et.al.,2021).

Squander administration remains a imposing challenge all inclusive, traversing both created and developing economies. A critical issue contributing to contamination is the collection of compost loads in open zones between sanitation periods. The extension of populaces has driven to a weakening in sterile conditions related with squander administration frameworks. Squander transfer constitutes a crucial perspective of squander administration, and with the coming of counterfeit insights (AI) innovations, different computational approaches have risen to address cleverly squander administration challenges. AI advances offer productive arrangements for overseeing ill-defined issues, encounters, vulnerabilities, and halfway information related with squander administration. In any case, whereas there has been impressive inquire about in this region, few assessments have illustrated the complete potential of AI in settling the complexities of cleverly squander administration. Exact evaluation of the amount and quality of waste is imperative for the advancement and plan of shrewd squander administration frameworks. In any case, foreseeing the amount of waste created could be a challenging errand due to its differing characteristics and inconstancy. In this setting, convolutional neural systems (CNNs) serve as a reasonable system for evaluating squander mass precisely (Jude, A. B., et.al.,2021).

A proof-of-concept civil squander administration framework has been proposed with the point of diminishing the costs related with squander classification, checking, and collection. This system uses deep learning-based classifiers and cloud computing techniques to achieve extremely exact waste classification at the initial stage of waste collection. Web of Things (IoT) gadgets are sent to encourage data trade between squander holders and the squander administration center. These gadgets consolidate a set of sensors to screen the generally sum of squander delivered within the region and the operational status of person squander holders. Based on this checking information, the squander administration center can optimize gear arrangement and upkeep plans, squander collection methodologies, and vehicle steering plans. This coordinates approach serves as a basic component of a fruitful civil squander administration framework (Wang, C., et.al.,2021).

In later a long time, Bangladesh has confronted a squeezing challenge in squander administration, posturing noteworthy natural concerns. This paper presents the concept of creating a savvy dustbin to identify and recognize different objects regarded as trash, analyzing preparing highlights. Additionally, the classifier can precisely tally the number of distinguished objects and allot comparing cost values to each thing. Upon bringing squander to the ATD, people can promptly have their things recognized by the picture classifier, empowering them to pull back the doled-out reuse esteem for the stored things. This coordinate trade of squander for financial esteem serves as a solid motivation for individuals to utilize our proposed keen dustbin. Past the beginning establishment fetched, income can be produced through the operation and upkeep of the dustbin by reusing the stored waste. To approve the possibility of our proposition, a pre-trained ALexNet, was used to classify 10 categories (Sunny, M. S. H., et.al.,2019).

3. PROPOSED AUTOMATED ANTHROPOGENIC WASTE CLASSIFICATION SYSTEM (AAWCS):

Designing a system for classifying anthropogenic waste in deep sea environments involves several steps shown in figure 1. Here is a small proposed system that uses feature extraction and a learning model:

1. Data Collection: Gather underwater images or videos from deep sea environments using underwater drones or remotely operated vehicles (ROVs). Ensure that the data cover a variety of waste types and environmental conditions.

2. Preprocessing: Preprocess the data to enhance quality and remove noise. This may involve techniques such as image denoising, contrast enhancement, and color normalization.
3. Feature Extraction: Utilize computer vision techniques to extract relevant features from the underwater images/videos. Features could include texture, color histograms, shape characteristics, and any other distinguishing attributes of anthropogenic waste. Deep learning models like convolutional neural networks (CNNs) can be employed for automatic feature extraction. Alternatively, handcrafted feature extraction methods such as Scale-Invariant Feature Transform (SIFT) or Histogram of Oriented Gradients (HOG) can be utilized.
4. Learning Model: Train a classification model using the extracted features. Various machine learning algorithms can be considered, such as Support Vector Machines (SVM), Random Forests, or neural networks. For deep learning, consider designing a CNN architecture tailored to the problem of anthropogenic waste classification in deep sea environments. Transfer learning from pre-trained models like ResNet or VGG can also be beneficial, especially with limited labeled data.
5. Evaluation: Split the dataset into training, validation, and testing sets to evaluate the performance of the model. Metrics such as accuracy, precision, recall, and F1-score can be used to assess the model's performance. Cross-validation techniques may also be employed to ensure robustness of the model.
6. Deployment: Once the model is trained and evaluated satisfactorily, integrate it into a system that can be deployed on underwater drones or ROVs for real-time waste classification. The system should provide visual feedback to operators, indicating the presence and type of anthropogenic waste detected in the deep-sea environment.
7. Iterative Improvement: Continuously collect more data and refine the model to improve its accuracy and generalization ability. Monitor the system's performance in real-world conditions and adjust as necessary.

This proposed system provides a framework for building an automated anthropogenic waste classification system in deep sea environments, leveraging feature extraction, and learning models to detect and categorize waste accurately.

Figure 1. Proposed architecture for Anthropogenic Waste Classification

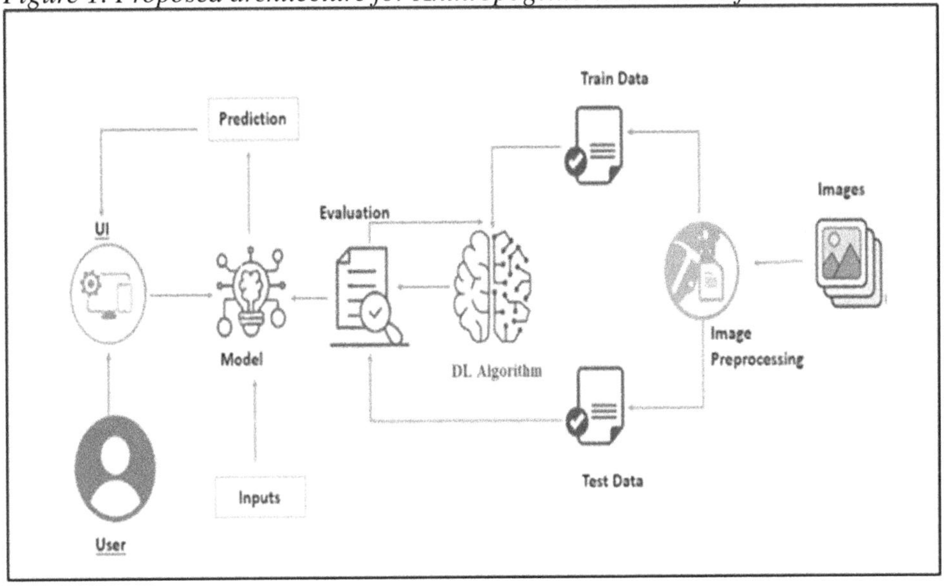

The developed anthropogenic waste classification system utilizes a convolutional neural network (CNN) architecture for effective feature extraction and classification of anthropogenic waste images. The architecture consists of multiple layers, including convolutional, max-pooling, and dense (fully connected) layers, designed to learn hierarchical representations of anthropogenic waste images and make predictions regarding their respective classes. The training model of CNN has various layers to classify the anthropogenic wastes. Those layers are

a. Convolutional Layers:
   · The CNN architecture begins with the first convolutional layer (`conv2d`), which applies 32 filters with a filter size of (3,3) to the input images.
   · Each convolutional layer is followed by a rectified linear unit (ReLU) activation function, facilitating non-linear transformations and feature extraction.
   · The output shape of the first convolutional layer is (126, 126, 32), indicating 32 feature maps of size 126x126.
   · Subsequent convolutional layers (`conv2d_1`, `conv2d_2`, `conv2d_3`) follow a similar structure, with varying numbers of filters and filter sizes.
   · These convolutional layers learn increasingly complex features as the depth of the network increases, capturing patterns and structures relevant to anthropogenic waste classification.
b. Max-Pooling Layers:

· Max-pooling layers (`max_pooling2d`, `max_pooling2d_1`, `max_pooling2d_2`, `max_pooling2d_3`) are interspersed between convolutional layers to downsample feature maps and reduce spatial dimensions.
· Each max-pooling layer applies a pooling operation with a pool size of (2,2), halving the spatial dimensions of the input feature maps.
· Pooling helps to extract dominant features while reducing computational complexity and mitigating overfitting.

c. Flatten Layer:

· The flatten layer (`flatten`) serves as a transition between convolutional and dense layers.
· It reshapes the output from the convolutional layers into a one-dimensional array, maintaining the batch size while flattening the spatial dimensions.
· This transformation enables the subsequent dense layers to process the learned features for classification.

d. Dense Layers:

· Following the flatten layer, the architecture includes multiple dense layers (`dense`, `dense_1`, `dense_2`) responsible for classification.
· The dense layers are fully connected, allowing each neuron to receive input from all neurons in the previous layer.
· Activation functions such as ReLU are applied to introduce non-linearity and enable complex decision boundaries.
· The final dense layer (`dense_2`) employs a softmax activation function to generate class probabilities, facilitating multi-class classification of anthropogenic waste images into six distinct categories.

e. Total Parameters:

· The total number of parameters in the architecture is 293,136, all of which are trainable.

· These parameters represent the weights and biases associated with the convolutional and dense layers, which are learned during the training process.

· The large number of parameters enables the model to capture and represent intricate patterns in the input data, enhancing its capacity for accurate classification.

Overall, the proposed system architecture represents a comprehensive framework for automating anthropogenic waste classification, with the potential to significantly improve waste management practices and contribute to environmental sustainability efforts in urban areas. The work flow of the model is in figure 2.

Figure 2. Work flow diagram

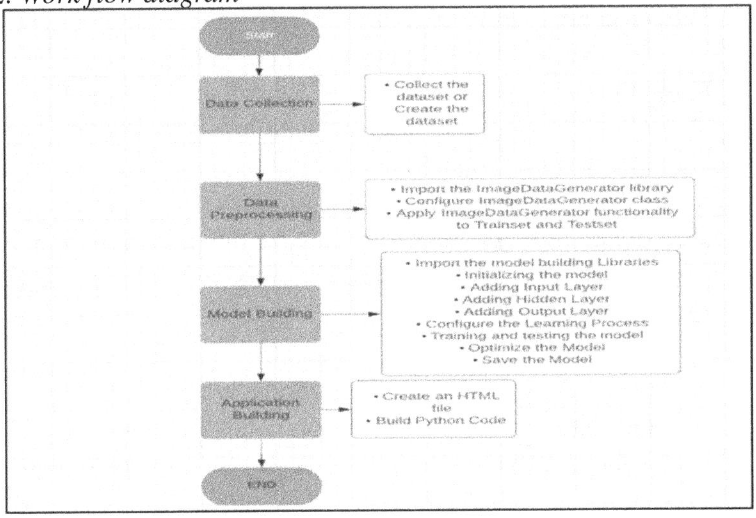

The proposed system is explained in Algorithm I:

Algorithm I: Smart Anthropogenic Waste Classification System

Input:

- Training dataset containing labeled images of various types of anthropogenic waste (e.g., paper, plastic, metal).
- Test dataset for evaluating the model's performance.
- Hardware and software requirements (e.g., TensorFlow, Keras, Anaconda).

Output:

- Trained model capable of classifying different types of anthropogenic waste.
- Evaluation metrics (e.g., accuracy, F1-score) for assessing model performance.
- Predictions on new, unseen anthropogenic waste images.

Steps:

1. Image Pre-processing:
    - Import the `ImageDataGenerator` library from Keras.
    - Configure image data augmentation parameters, such as rescaling, rotation, and horizontal flipping.
    - Apply data augmentation techniques to the training dataset to increase sample diversity and improve model generalization.

- Load and preprocess the test dataset without data augmentation.

2. Model Architecture Design:
   - Initialize a Sequential model using Keras.
   - Design the CNN architecture:
   - Add convolutional layers with activation functions (e.g., ReLU) and specify filter sizes and numbers.
   - Include max-pooling layers to downsample feature maps.
   - Add a flatten layer to prepare the data for fully connected layers.
   - Incorporate dense (fully connected) layers with appropriate activation functions for classification.

     2.1 Compile the model:
   - Specify the loss function (e.g., categorical cross-entropy).
   - Choose an optimizer (e.g., Adam) for model training.
   - Define evaluation metrics (e.g., accuracy) to monitor model performance.
3. Model Training:
   - Compile the model using the `compile` method, specifying the loss function, optimizer, and metrics.
   - Train the model on the augmented training dataset using the `fit` method:
   - Specify parameters such as the number of epochs, batch size, and validation data.
   - Monitor training progress and performance metrics during training.
4. Model Evaluation:
   - Evaluate the trained model on the test dataset to assess its performance:
   - Compute evaluation metrics such as accuracy, precision, recall, and F1-score.
   - Analyze classification results and identify any misclassifications or areas for improvement.
5. Model Deployment (Optional):
   - Save the trained model to disk using TensorFlow's `save_model` method for future use or deployment.
   - Optionally, deploy the trained model in a production environment:
   - Integrate the model into a web application, mobile app, or IoT device for real-time anthropogenic waste classification.
   - Implement an interface for users to upload images of anthropogenic waste and receive classification results.
6. Usage and Iteration:

   -Utilize the trained model for automated anthropogenic waste classification in waste management systems.

- Monitor model performance over time and iterate on the algorithm to improve classification accuracy and efficiency.

By following this algorithm, developers can create an effective smart anthropogenic waste classification system using deep learning techniques. This system contributes to waste management efforts by automating the classification process and promoting environmental sustainability.

4. RESULTS AND DISCUSSION

To extract results, we take 5GB dataset having various types of anthropogenic waste pictures like paper, plastic, glass bottle etc., for the simulation environment we used the google colab environment then analyzed the performance metrics and trends observed during model training. The results of the waste classification are shown below Figure 3:

Figure 3. Samples results of waste classification in water.

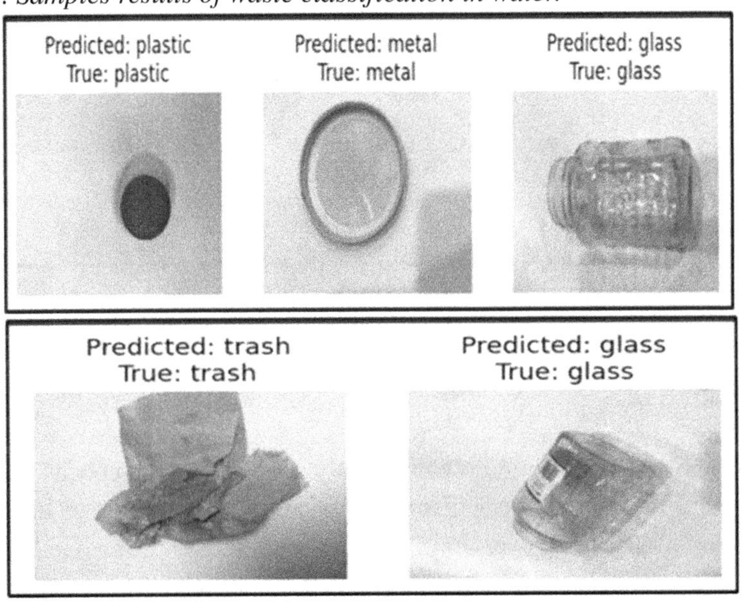

i. Model Performance Metrics:

- The model achieved a peak training accuracy of approximately 96.26% and a validation accuracy of around 94.97%.
- Training and validation loss values decreased over epochs, indicating that the model learned to minimize prediction errors effectively. The Confusion matrix in figure 4.

Figure 4. Confusion matrix

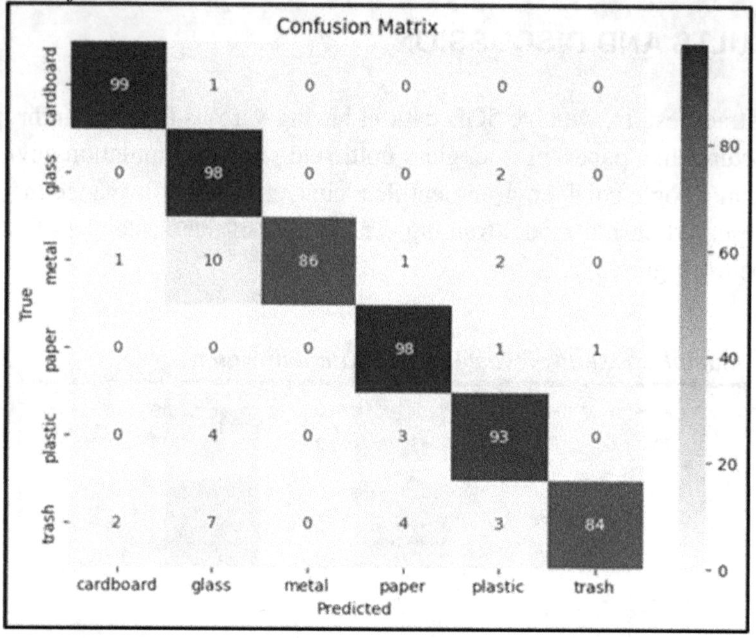

ii. <u>Accuracy Trends:</u>

- The training accuracy consistently increased over epochs, indicating that the model learned to classify anthropogenic waste images with higher accuracy over time.
- Validation accuracy showed a similar increasing trend initially, reaching a peak and stabilizing around 94.97% in figure 5 ROC Curve

Figure 5. ROC curve of the classification model

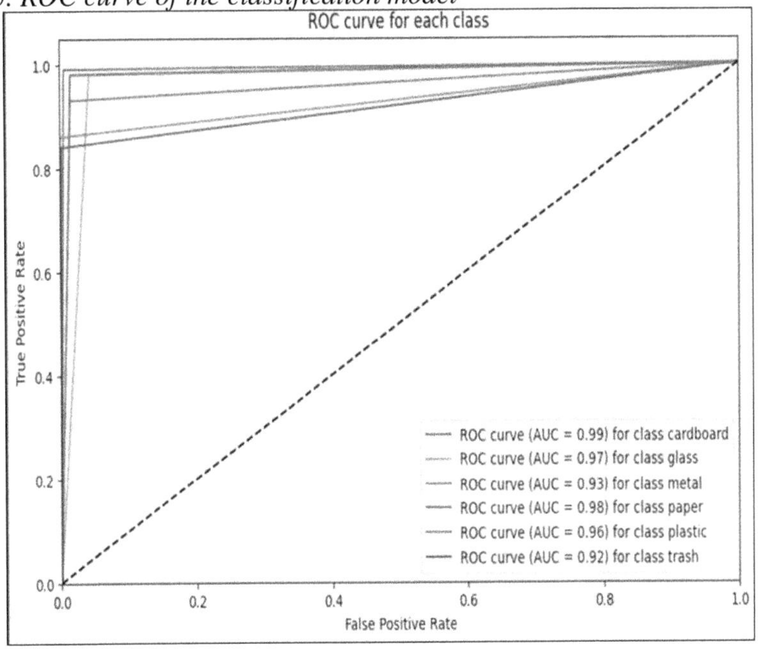

iii. Loss Trends:

- Training loss steadily decreased over epochs, indicating that the model's predictions approached the ground truth labels.
- Validation loss followed a similar decreasing trend, indicating that the model generalized well to unseen data. The overall performance metrics of the taken dataset is shown in Table 1 and Figure 5.

Table 1. Performance metrics

Class	Precision	Recall	F1-Score	Support
Carboard	0.97	0.99	0.98	100
glass	0.82	0.98	0.89	100
metal	1.0	0.86	0.92	100
paper	0.92	0.98	0.95	100
plastic	0.92	0.93	0.93	100

continued on following page

Table 1. Continued

Class	Precision	Recall	F1-Score	Support
trash	0.99	0.84	0.91	100
accuracy				0.93
macro avg	0.94	0.93	0.93	600
weighted avg	0.94	0.93	0.93	600

Figure 6. Performance graph of Precision and Recall

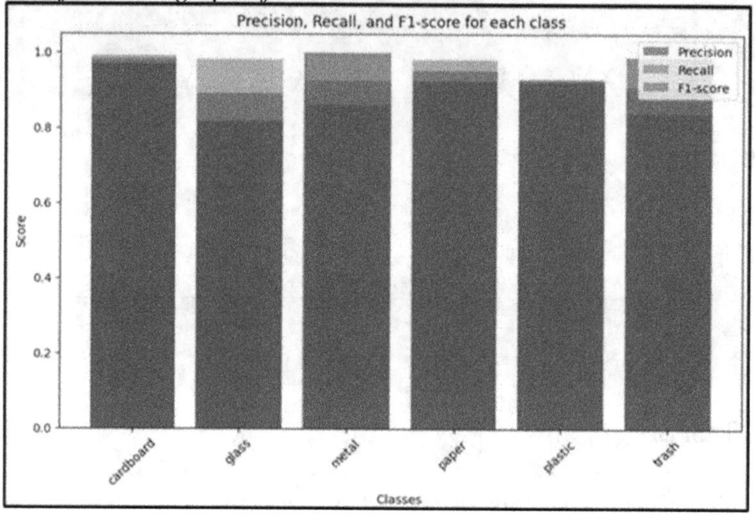

4.1 Insights and Findings of the Proposed model

1. Model Convergence:

The observed trends in training and validation accuracy suggest that the model effectively converged to a solution, achieving high accuracy on both training and validation datasets. The decreasing loss values indicate that the model's predictions became more accurate and consistent as training progressed.

2. Impact of Data Augmentation:

Data augmentation techniques, such as rescaling, shearing, and horizontal flipping, likely contributed to the model's improved generalization and performance. By generating diverse training samples from a limited dataset, data augmentation helped prevent overfitting and improved the model's ability to classify unseen anthropogenic waste images.

3. Optimization and Regularization:

The Adam optimizer effectively minimized the loss function, facilitating faster convergence and better model performance. Regularization techniques, such as dropout layers, may have helped prevent overfitting by reducing the model's reliance on specific features.

4. Class Imbalance Handling:

It's essential to ensure that the training dataset is balanced across different anthropogenic waste classes to prevent bias towards dominant classes. Evaluation metrics should be interpreted carefully, considering the class distribution and potential biases in the dataset.

5. Impact of tuning:

Further experiments could explore the impact of hyperparameter tuning, different network architectures, or additional data augmentation techniques on model performance. Real-world deployment of the model would require thorough testing and validation to ensure robust performance in diverse environmental conditions. These provide a comprehensive analysis of the model's performance and implications for anthropogenic waste classification systems.

CONCLUSION AND FUTURE WORK

The proposed work addresses the critical issue of waste management by creating a deep learning model for anthropogenic waste classification. With urbanization and industrialization on the rise, waste generation has become a significant concern, necessitating effective disposal and recycling methods. Traditional approaches, such as manual sorting, are inefficient and costly, highlighting the need for automated solutions to streamline the classification process. Through the utilization of deep learning, specifically convolutional neural networks (CNNs), a model capable of accurately identifying and categorizing various types of anthropogenic waste, including paper, plastic, glass, metal, and trash, has been developed. Integration of a web application enables users to upload anthropogenic waste images and receive real-time analysis, fostering awareness and encouraging recycling efforts. Experimental investigations encompassed image pre-processing, model training, and evaluation, resulting in a peak accuracy of 83.32%. While the model demonstrates promising results, further optimization and enhancements are viable.

Future endeavors could focus on several key areas:

1. Optimization of Model Performance: Fine-tuning model architecture, hyperparameters, and training procedures to boost accuracy and efficiency.
2. Integration with IoT Devices: Incorporating Internet of Things (IoT) devices for real-time anthropogenic waste detection and classification in smart waste management systems.
3. Real-World Deployment: Deploying the model in waste management facilities or smart cities to assess performance and impact on waste management practices.
4. Continuous Dataset Expansion: Updating and diversifying the dataset to enhance model generalization and robustness.
5. Collaboration with Recycling Companies: Partnering with recycling firms to integrate the model into their operations for automated sorting and resource optimization.

In summary, the developed anthropogenic waste classification model presents a promising solution to the challenges of waste management. By leveraging deep learning and technology, we can pave the way for more efficient and sustainable waste management practices in the future.

REFERENCES

Abu-Qdais, H., & Shatnawi, N., & AL-Alamie, E. A. (2024). Intelligent system for solid waste classification using combination of image processing and machine learning models. *Journal of Experimental & Theoretical Artificial Intelligence*, •••, 1–12.

Chauhan, R., Shighra, S., Madkhali, H., Nguyen, L., & Prasad, M. (2023). Efficient future waste management: A learning-based approach with deep neural networks for smart system (LADS). *Applied Sciences (Basel, Switzerland)*, 13(7), 4140. DOI: 10.3390/app13074140

Chavhan, P. G., Khedka, V. N., Gupta, M., & Agarwal, K. (2024). Automatic Waste Segregator Based on IoT & ML Using Keras model and Streamlit. *International Journal of Intelligent Systems and Applications in Engineering*, 12(2), 787–799.

Cheema, S. M., Hannan, A., & Pires, I. M. (2022). Smart waste management and classification systems using cutting edge approach. *Sustainability (Basel)*, 14(16), 10226. DOI: 10.3390/su141610226

Häder, D. P., Banaszak, A. T., Villafañe, V. E., Narvarte, M. A., González, R. A., & Helbling, E. W. (2020). Anthropogenic pollution of aquatic ecosystems: Emerging problems with global implications. *The Science of the Total Environment*, 713, 136586. DOI: 10.1016/j.scitotenv.2020.136586 PMID: 31955090

Hasan, M. K., Khan, M. A., Issa, G. F., Atta, A., Akram, A. S., & Hassan, M. (2022, February 16). Smart waste management and classification system for smart cities using deep learning. 2022 International Conference on Business Analytics for Technology and Security (ICBATS). Presented at the 2022 International Conference on Business Analytics for Technology and Security (ICBATS), Dubai, United Arab Emirates. DOI: 10.1109/ICBATS54253.2022.9759087

Hossen, M. M., Ashraf, A., Hasan, M., Majid, M. E., Nashbat, M., Kashem, S. B. A., Kunju, A. K. A., Khandakar, A., Mahmud, S., & Chowdhury, M. E. (2024). GCDN-Net: Anthropogenic waste classifier deep neural network for recyclable urban waste management. *Waste Management (New York, N.Y.)*, 174, 439–450. DOI: 10.1016/j.wasman.2023.12.014 PMID: 38113669

Jude, A. B., Singh, D., Islam, S., Jameel, M., Srivastava, S., Prabha, B., & Kshirsagar, P. R. (2021). An artificial intelligence based predictive approach for smart waste management. *Wireless Personal Communications*, •••, 123456789.

Kalpana, A. V., Suchitra, S., Prasath, R., Arthi, K., Shobana, J., & Nadana Ravishankar, T. (2024). Dual-CNN-based waste classification system using IoT and HDS algorithm. In Computational Intelligence for Green Cloud Computing and Digital Waste Management (pp. 287–309). DOI: 10.4018/979-8-3693-1552-1.ch015

Kanujia, A. K., Pal, M., Kumar, A., Pal, A., Chungada, H. S., & Singh, M. (2024). Modern approaches of waste management-A critical review. In *AIP Conference Proceedings* (Vol. 2962). AIP Publishing.

Kona, V. V., & Subramoniam, M. (2024). A smart Iot-based waste management system using vehicle shortest path routing and trashcan visiting decision making based on deep convolutional neural network. *Peer-to-Peer Networking and Applications*, 17(3), 1–24. DOI: 10.1007/s12083-024-01623-z

Lee, K. J., Chiong, M. C., Hassan, C. S., Natarajan, E., Solihin, M. I., & Lim, W. H. (2024). Deep learning applied solid waste recognition system targeting sustainable development goal. In *Machine Intelligence in Mechanical Engineering* (pp. 393–421). Academic Press. DOI: 10.1016/B978-0-443-18644-8.00005-8

Mookkaiah, S. S., Thangavelu, G., Hebbar, R., Haldar, N., & Singh, H. (2022). Design and development of smart Internet of Things-based solid waste management system using computer vision. *Environmental Science and Pollution Research International*, 29(43), 64871–64885. DOI: 10.1007/s11356-022-20428-2 PMID: 35476273

. Orcutt, B., D'Angelo, T., Jungbluth, S. P., Huber, J. A., & Sylvan, J. B. (2020). Microbial life in oceanic crust.

Rahman, M. W., Islam, R., Hasan, A., Bithi, N. I., Hasan, M. M., & Rahman, M. M. (2022). Intelligent waste management system using deep learning with IoT. *Journal of King Saud University. Computer and Information Sciences*, 34(5), 2072–2087. DOI: 10.1016/j.jksuci.2020.08.016

Sallang, N. C. A., Islam, M. T., Islam, M. S., & Arshad, H. (2021). A CNN-based smart waste management system using TensorFlow lite and LoRa-GPS shield in Internet of Things environment. *IEEE Access : Practical Innovations, Open Solutions*, 9, 153560–153574. DOI: 10.1109/ACCESS.2021.3128314

Sivakumar, M., Renuka, P., Chitra, P., & Karthikeyan, S. (2022). IoT incorporated deep learning model combined with SmartBin technology for real-time solid waste management. *Computational Intelligence*, 38(2), 323–344. DOI: 10.1111/coin.12495

Soundarya, B., Parkavi, K., Sharmila, A., Kokiladevi, R., Dharani, M., & Krishnaraj, R. (2022, January). CNN based smart bin for waste management. In 2022 4th International Conference on Smart Systems and Inventive Technology (ICSSIT) (pp. 1405-1409). IEEE.

Sunny, M. S. H., Dipta, D. R., Hossain, S., Faruque, H. M. R., & Hossain, E. (2019, May). Design of a convolutional neural network based smart waste disposal system. In 2019 1st International Conference on Advances in Science, Engineering and Robotics Technology (ICASERT) (pp. 1-5). IEEE.

Udupi, P. K., Jose, M., & Ullah, A. (2024). AI-Enabled Smart City Waste Management System. In Handbook of Artificial Intelligence for Smart City Development (pp. 76–99). CRC Press.

Ukaogo, P. O., Ewuzie, U., & Onwuka, C. V. (2020). Environmental pollution: causes, effects, and the remedies. In *Microorganisms for sustainable environment and health* (pp. 419–429). Elsevier. DOI: 10.1016/B978-0-12-819001-2.00021-8

Wahab, M. H. A., Kadir, A. A., Tomari, M. R., & Jabbar, M. H. (2014, October). Smart recycle bin: A conceptual approach of smart waste management with integrated web based system. 2014 International Conference on IT Convergence and Security (ICITCS). Presented at the 2014 International Conference on IT Convergence and Security (ICITCS), Beijing, China. DOI: 10.1109/ICITCS.2014.7021812

Wang, C., Qin, J., Qu, C., Ran, X., Liu, C., & Chen, B. (2021). A smart municipal waste management system based on deep-learning and Internet of Things. *Waste Management (New York, N.Y.)*, 135, 20–29. DOI: 10.1016/j.wasman.2021.08.028 PMID: 34461487

Wear, S. L., Acuña, V., McDonald, R., & Font, C. (2021). Sewage pollution, declining ecosystem health, and cross-sector collaboration. *Biological Conservation*, 255, 109010. DOI: 10.1016/j.biocon.2021.109010

Sonodiya, S., [illegible] K., [illegible] A., K[illegible], R., [illegible] (2023, January). CNN based smart bin for waste management. In 2023 [illegible] International Conference on Smart Systems and Inventive Technology (ICSSIT) (pp. [illegible]). IEEE.

[illegible], M. S. [illegible], D. [illegible], S., [illegible] M., [illegible] K. (2019). IoT [illegible] waste management system. In 2019 [illegible] International Conference on Automation, Science, Engineering and Technology [illegible] (pp. [illegible]).

[illegible] Waste Management System. In Handbook of Artificial Intelligence [illegible] Development (pp. [illegible]). CRC Press.

[illegible]

[illegible]

[illegible]

Chapter 3
Casestudy on Integrating Advanced Technologies for the Conservation and Exploration of Deep Sea Ecosystems

U. Vignesh
Vellore Institute of Technology, Chennai, India

R. Parvathi
Vellore Institute of Technology, Chennai, India

Arpan Singh Parihar
Vellore Institute of Technology, Chennai, India

ABSTRACT

Echnological advancements have revolutionized deep-sea ecosystem conservation and exploration, offering unprecedented capabilities to understand and protect these fragile marine environments. Remote sensing technologies such as satellite imagery and acoustic sensors provide high-resolution mapping of deep-sea habitats, guiding conservation strategies. Autonomous underwater vehicles (AUVs) and remotely operated vehicles (ROVs) equipped with advanced sensors enable detailed surveys and observations in previously inaccessible regions, yielding valuable data on species distribution and habitat characteristics. Additionally, genetic sequencing technologies elucidate deep-sea biodiversity and evolutionary processes, informing conservation efforts. These innovations collectively enhance our understanding of deep-sea ecosystems and facilitate informed management practices, ensuring

DOI: 10.4018/979-8-3693-6670-7.ch003

the preservation and sustainable use of these vital marine environments for future generations.

I. INTRODUCTION

The deep sea encompasses a staggering expanse of our planet, constituting over 65% of the Earth's surface. Despite its vastness, this realm remains shrouded in mystery, with its depths largely unexplored and its secrets waiting to be uncovered. Within this enigmatic environment lies a treasure trove of biodiversity, hosting an array of unique species and playing a crucial role in global ecological processes. However, due to its remote and extreme conditions, the deep sea has historically been challenging to study, leaving much of its intricacies and significance veiled from scientific inquiry.

In recent years, a paradigm shift has occurred with the advent of groundbreaking technological advancements. These innovations have ushered in a new era of exploration and conservation in the deep sea, offering unprecedented opportunities to unravel its mysteries and safeguard its delicate ecosystems. From state-of-the-art underwater vehicles to cutting-edge imaging technology, these tools have empowered scientists to delve deeper into the abyss than ever before, capturing breath-taking imagery and collecting invaluable data along the way.

Figure 1. Typical ecosystems in the deep sea.

At the heart of this technological revolution lies the quest to understand and protect the deep sea's biodiversity and ecological functions. By harnessing advanced tools and methods, researchers can uncover new species, study their habitats, and unravel the intricate web of life that thrives in these extreme environments. Moreover, these advancements enable scientists to monitor and assess the health of deep sea ecosystems, providing essential insights for conservation efforts and management strategies.

The transformative role of technology in deep sea conservation and exploration cannot be overstated. By shedding light on this hidden world, we gain a deeper appreciation for the interconnectedness of all life on Earth and the importance of preserving these critical marine environments for future generations. As technology continues to evolve, so too will our ability to explore and protect the deep sea, ensuring that its wonders endure for centuries to come.

Importance of Deep-Sea Conservation:

Deep sea conservation is of paramount importance due to the critical role the deep ocean plays in regulating Earth's climate, supporting biodiversity, and preserving genetic resources. The deep sea, often referred to as the Earth's final frontier, houses unique and often poorly understood ecosystems that host a plethora of marine life, some of which may hold undiscovered medicinal or industrial value. Protecting these ecosystems is vital for maintaining the balance of marine food webs and preserving genetic diversity, which can be crucial for future scientific and medical advancements. Additionally, the deep sea plays a crucial role in global biogeochemical cycles, such as carbon sequestration and nutrient cycling, influencing climate regulation and oceanic productivity. Furthermore, the deep sea represents a vast repository of minerals and metals, and its sustainable exploitation could help meet growing global demand for critical resources. However, the potential ecological impacts of deep-sea mining underscore the urgent need for conservation measures to ensure the long-term health and resilience of these ecosystems. By promoting research, implementing effective management strategies, and fostering international cooperation, deep sea conservation efforts can help safeguard these invaluable habitats for future generations.

Deep-Sea Ecosystem:

The deep-sea ecosystem is a vast and diverse environment that extends from approximately 200 meters to the deepest ocean depths, encompassing a significant portion of the Earth's surface. Despite its remote and extreme conditions, the deep sea is home to a remarkable array of organisms adapted to survive in darkness, cold temperatures, high pressure, and limited food availability. Deep-sea ecosystems are characterized by intricate food webs, with primary production fuelled by detritus sinking from the surface and chemosynthetic processes near hydrothermal vents and cold seeps.

Figure 2. Deep-Sea Ecosystem

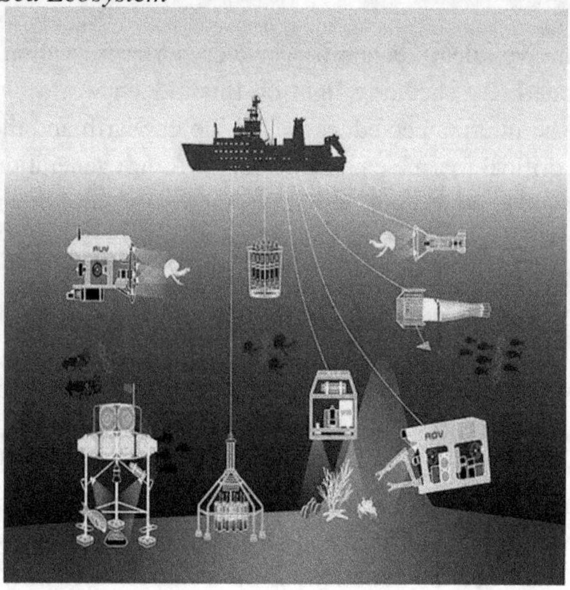

These ecosystems support a wide variety of organisms, including deep-sea fish, cephalopods, echinoderms, and invertebrates, many of which are endemic to specific regions or habitats. Deep-sea habitats range from continental slopes, abyssal plains, and seamounts to hydrothermal vent fields and cold seeps, each with its unique physical and biological characteristics. However, deep-sea ecosystems are facing increasing threats from human activities, including deep-sea mining, oil and gas exploration, bottom trawling, and climate change. These disturbances can have profound impacts on deep-sea biodiversity, habitat integrity, and ecosystem functioning, highlighting the importance of conservation and sustainable management efforts. Despite the challenges, ongoing research and technological advancements offer hope for a better understanding of deep-sea ecosystems and the development of strategies to ensure their long-term health and resilience.

Evolution of biological studies in the deep-sea environment:

Biological studies in the deep-sea environment have undergone a remarkable evolution, propelled by advancements in technology, interdisciplinary collaboration, and growing recognition of the importance of understanding these remote ecosystems. Early explorations of the deep sea were limited by technological constraints and a lack of appreciation for its biodiversity and ecological significance. With the development of innovative sampling tools such as trawls, dredges, and submersibles, scientists began to uncover the rich diversity of life thriving in the deep ocean depths. These early expeditions provided crucial insights into the adaptations of deep-sea organisms to extreme conditions such as high pressure, low temperatures, and limited food availability. As technology continued to advance, biological studies in the deep sea expanded to include sophisticated imaging systems, genetic sequencing techniques, and remotely operated vehicles (ROVs) equipped with high-definition cameras and sampling devices. These tools enabled researchers to conduct detailed surveys and observations in previously inaccessible regions, revealing new species, habitats, and ecological interactions. Furthermore, interdisciplinary collaborations between biologists, oceanographers, geologists, and engineers have enriched our understanding of deep-sea ecosystems by integrating diverse perspectives and expertise. biological studies in the deep sea are at the forefront of scientific inquiry, addressing pressing questions related to biodiversity conservation, ecosystem functioning, and the impacts of human activities such as deep-sea mining and climate change. By building on decades of research and technological innovation, scientists are poised to unlock further mysteries of the deep ocean and inform evidence-based conservation and management strategies for these vulnerable and essential marine environments.

Technological Advancements:

Remotely operated vehicles (ROVs) and autonomous underwater vehicles (AUVs) stand as pioneering tools in the arsenal of deep-sea exploration. Equipped with sophisticated cameras, sensors, and manipulator arms, these unmanned vehicles enable scientists to navigate the depths with unparalleled precision, capturing high-definition imagery, and collecting invaluable data on marine life, geological features, and environmental conditions. This capability to conduct detailed, real-time observations in the extreme conditions of the deep sea has drastically expanded our knowledge and understanding of this hidden world.

Satellite technology represents another cornerstone in the advancement of deep-sea conservation efforts. By harnessing satellite imagery and data, researchers can monitor and track changes in ocean temperatures, currents, and surface conditions on a global scale. This bird's-eye view provides crucial insights into the dynamic processes shaping deep sea ecosystems, informing conservation strategies and enhancing our ability to respond to environmental threats and challenges.

Genetic sequencing techniques have further revolutionized our understanding of deep-sea biodiversity. By analysing DNA extracted from samples collected in the abyssal depths, scientists can identify new species, uncover genetic adaptations to extreme environments, and unravel intricate ecological relationships. This molecular approach has opened new frontiers in taxonomic research, shedding light on the hidden diversity and evolutionary histories of deep-sea organisms.

Technological advancements have ushered in a new era of exploration and conservation in the deep sea, unlocking the secrets of this vast and mysterious realm. From the depths of the ocean floor to the expanse of the open ocean, innovative tools and methods empower scientists to study and protect the fragile ecosystems that thrive in these remote and challenging environments. As we continue to push the boundaries of technological innovation, we must also strive to ensure that our exploration of the deep sea is guided by principles of sustainability and stewardship, preserving these invaluable ecosystems for future generations.

Challenges Facing Deep Sea Ecosystems:

Deep sea ecosystems face a myriad of challenges, many of which are exacerbated by human activities. Here are some of the key challenges:

- **Overfishing:** Conservation efforts have shown that deep-sea ecosystems face substantial harm from certain fishing practices, notably bottom trawling. This method, characterized by dragging nets along the ocean floor, disrupts habitats and depletes fish populations. The destruction wrought upon these ecosystems can be severe, resulting in long-lasting damage. As such, mitigating the adverse impacts of bottom trawling is imperative for the preservation of deep-sea biodiversity and the sustainable management of marine resources.
- **Pollution:** Deep sea environments are vulnerable to the accumulation of chemical pollutants, plastics, and various debris, posing significant threats to marine life and ecosystem stability. This accumulation not only harms deep-sea organisms directly through ingestion and entanglement but also disrupts food chains and ecological processes. Consequently, addressing the sources of pollution and implementing effective waste management strategies are essential steps in safeguarding the health and integrity of deep-sea ecosystems.
- **Climate Change:** Climate change-induced phenomena such as rising ocean temperatures, ocean acidification, and alterations in ocean currents pose profound threats to deep-sea ecosystems. These changes can disrupt the delicate balance of these environments, leading to shifts in species distribution, altered reproductive patterns, and diminished survival rates. Consequently, urgent action is needed to mitigate climate change and its impacts to safeguard the resilience and long-term viability of deep-sea ecosystems.
- **Deep Sea Mining:** The burgeoning interest in deep-sea mining for minerals like manganese, cobalt, and rare earth elements presents a significant risk to deep-sea ecosystems. Mining activities have the potential to disturb fragile

habitats, including seamounts and hydrothermal vents, which are home to unique and often undiscovered species. Furthermore, the process of extracting minerals can release pollutants into the water column, further exacerbating the environmental impact. Given these risks, it's imperative to carefully evaluate the potential consequences of deep-sea mining and implement robust regulatory measures to minimize harm to these vulnerable ecosystems.

- **Seabed Disturbance:** Indeed, activities such as oil and gas exploration, submarine cable laying, and scientific research can induce physical disturbances to the seabed, thereby altering habitats and impacting deep-sea biodiversity. The installation of infrastructure, such as drilling rigs or cable trenches, can disrupt benthic communities and result in habitat loss. Additionally, noise pollution from these activities may have adverse effects on marine species, including disturbance of migratory patterns and communication disruption. Thus, it's crucial to employ precautionary measures and adhere to best practices to minimize the ecological footprint of such activities and mitigate their potential impacts on deep-sea ecosystems.
- **Lack of Regulation:** Deep sea ecosystems, often situated in international waters beyond national jurisdiction, face heightened vulnerability to exploitation due to the absence of robust regulatory frameworks. The absence of clear jurisdictional boundaries can lead to gaps in protection, leaving these ecosystems susceptible to overexploitation, habitat destruction, and pollution. Addressing this challenge requires international cooperation to establish and enforce effective governance mechanisms, such as regional agreements or multilateral treaties, to safeguard deep-sea biodiversity and promote sustainable resource management. Additionally, enhancing scientific understanding through research and monitoring initiatives is crucial for informing evidence-based conservation policies and ensuring the long-term health of these globally significant ecosystems.
- **Invasive Species:** The introduction of invasive species poses a significant threat to deep-sea ecosystems, with potential consequences including the displacement of native species and disruption of established food webs. Various vectors, such as ballast water discharge and ship hull fouling, facilitate the transport and spread of invasive species to new environments, including the deep sea. Once introduced, these invasive species can rapidly proliferate and outcompete native organisms for resources, leading to ecological imbalances and biodiversity loss. Mitigating this threat requires proactive measures, such as stricter biosecurity regulations for maritime activities and enhanced monitoring efforts to detect and control the spread of invasive species. By prioritizing prevention and management strategies, we can minimize the ecological

impacts of invasive species on deep-sea ecosystems and preserve their unique biodiversity.

- **Light Pollution:** The escalation of human activities in coastal regions can result in light pollution, which extends into the deep sea, profoundly impacting the behavior and physiology of deep-sea organisms reliant on bioluminescence for communication and camouflage. Light pollution disrupts natural light cycles, disorienting marine organisms and interfering with crucial behaviors such as feeding, reproduction, and predator avoidance. Moreover, it compromises the effectiveness of bioluminescent signals used by deep-sea organisms for communication and camouflage, potentially leading to increased vulnerability to predation and reduced survival rates. Addressing light pollution requires concerted efforts to implement lighting regulations and promote responsible coastal development practices, minimizing the adverse impacts on deep-sea ecosystems and preserving their ecological integrity.
- **Microplastics:** Tiny plastic particles, known as microplastics, pose a significant threat to deep-sea environments, accumulating in sediments and water columns and endangering marine life through ingestion and entanglement. These microplastics originate from various sources, including the breakdown of larger plastic items, microbeads in personal care products, and fibers shed from synthetic textiles. Once in the deep sea, microplastics can be mistaken for food by marine organisms, leading to blockages in their digestive systems, malnutrition, and even death. Additionally, marine animals can become entangled in larger plastic debris, impairing their mobility and causing injuries or suffocation. Addressing the issue of microplastic pollution requires concerted efforts to reduce plastic waste generation, improve waste management practices, and implement measures to prevent plastic pollution from entering marine environments. By mitigating the impacts of microplastics, we can help safeguard the health and biodiversity of deep-sea ecosystems for future generations.

Need of the Study:

The need for advancements in deep-sea ecosystem conservation and exploration is urgent and multifaceted, driven by both scientific curiosity and pressing environmental concerns. Firstly, the deep sea represents the largest and least explored ecosystem on Earth, harbouring a vast array of unique and poorly understood species, habitats, and ecological processes. As such, there is a critical need to expand our knowledge of deep-sea biodiversity, ecosystem dynamics, and the interactions between organisms and their environment. This understanding is essential for effective conservation and management strategies aimed at protecting these fragile and often vulnerable ecosystems from anthropogenic impacts. the deep sea plays a vital role in regulating global climate systems, carbon sequestration, and nutrient cycling, making it integral to the health of the planet. The increasing exploitation of deep-sea resources, such as oil and gas extraction, deep-sea mining, and fisheries, poses significant threats to deep-sea ecosystems. To mitigate these impacts and ensure the sustainable use of marine resources, there is a pressing need for innovative technologies and approaches for monitoring, assessing, and managing human activities in the deep ocean. the need for advancements in deep-sea ecosystem conservation and exploration is driven by the imperative to expand scientific knowledge, protect biodiversity, safeguard ecosystem services, and promote sustainable management practices.

II. LITERATURE REVIEW

- **Ballard, R.D. (2000)** Richard Ballard offers a captivating blend of personal anecdotes and historical insights, inviting readers on a journey through the depths of deep-sea exploration. With vivid detail, he recounts the myriad challenges faced and victories achieved in the pursuit of uncovering the mysteries of the abyss. His narrative not only chronicles the evolution of deep-sea technology but also illuminates its profound influence on our perception of this enigmatic realm. By skilfully weaving together personal experiences and historical context, Ballard provides readers with a comprehensive understanding of the remarkable journey into the deep.
- **Levin, L.A., & Gooday, A.J. (2003)** Levin and Gooday delve into the depths of the Atlantic Ocean, offering a rich exploration of its geological formations, diverse ecosystems, and intricate ecological dynamics. Through their meticulous analysis, they shed light on the myriad wonders concealed within this expansive and remote expanse. Their thorough examination emphasizes the pivotal role played by technological innovations in unlocking the mysteries of the deep Atlantic. By synthesizing scientific knowledge with a deep appreciation for the ocean's complexity, Levin and Gooday provide readers with a compelling perspective on this captivating underwater world.
- **Raineault, N.A., Klemperer, S.L., & Dorschel, B. (2014)** In this chapter, the focus is on the intricacies of deep-sea mapping techniques and their practical applications, with a particular emphasis on the transformative impact of modern technology. By harnessing cutting-edge tools and methodologies, researchers can now create detailed and precise maps of seafloor topography and habitats, enabling a comprehensive exploration of the deep ocean realm. This chapter underscores the invaluable contributions of mapping initiatives in advancing our knowledge of deep-sea ecosystems and geological features.

Through the lens of mapping endeavours, readers gain insight into the dynamic interplay between technology, scientific inquiry, and our evolving understanding of the mysteries concealed beneath the ocean's surface.

- **Thurber, A.R., Levin, L.A., & Rowden, A.A. (2014)** Thurber et al. delve into the intricate dynamics of benthic oxygen consumption within oxygen minimum zone regions, offering a comprehensive global analysis of fluxes and ecosystem repercussions. Their study underscores the critical role of advanced monitoring technologies in evaluating the vitality and operational mechanisms of deep-sea ecosystems amidst shifting environmental conditions. By synthesizing data from diverse geographical contexts, Thurber et al. illuminate the far-reaching implications of oxygen dynamics on benthic ecosystems, highlighting the imperative for continued technological innovation to sustainably manage and preserve these vital marine environments.

  Singh, H., & King, J.W. (2015) Singh and King undertake a thorough examination of vertical pesticide fluxes within the deep Sargasso Sea, offering significant revelations concerning the movement and destiny of contaminants within these remote marine environments. Their study illuminates the intricate pathways through which pollutants navigate within deep-sea ecosystems, emphasizing the imperative for ongoing technological advancements to effectively monitor and address anthropogenic influences on marine habitats. By shedding light on the transport and fate of pesticides, Singh and King underscore the critical importance of proactive measures to mitigate human-induced impacts on the fragile balance of deep-sea ecosystems.
- **Ferrini, V.L., & MacPherson, A.C. (2016)** Ferrini and MacPherson embark on a thorough exploration of the nuances surrounding autonomous underwater vehicle (AUV) navigation within intricate environments, confronting challenges and charting future trajectories in this dynamic field. Their investigation delves into the distinctive hurdles confronting AUVs, including navigating rugged terrains, evading obstacles, and maintaining precise positioning within the depths of the ocean. Through an examination of current navigation technologies and methodologies, the authors pinpoint areas ripe for enhancement and propose inventive solutions to augment AUV navigation proficiency. Their research serves as a beacon, offering invaluable insights into the ongoing endeavours aimed at surmounting navigational impediments and unlocking the full potential of AUVs for deep-sea exploration and scientific inquiry.
- **White, M., & Rycroft, S. (2016)** White and Rycroft embark on a thorough investigation into the utilization of remote sensing techniques for studying the deep ocean and sea ice, providing an extensive overview of the method-

ologies and applications within this realm. They delve into the diverse array of tools at researchers' disposal, including satellite imagery, acoustic technologies, and other remote sensing instruments, which facilitate the collection of data on various deep-sea processes such as ocean currents, temperature fluctuations, and marine biodiversity. Furthermore, the authors scrutinize the role of remote sensing in monitoring sea ice dynamics and evaluating its impact on deep-sea ecosystems. Through their meticulous analysis, White and Rycroft underscore the paramount significance of remote sensing in advancing our comprehension of the deep ocean and its intricate interplay within the broader context of global climate systems.

- **Chérubin, L.M. (2017)** Chérubin delivers a comprehensive examination of the technological progression within deep-sea robotics, mapping out the journey of robotic systems and their pivotal role in advancing marine science and exploration., spotlighting significant milestones in their design, functionalities, and practical applications. By dissecting the obstacles encountered by deep-sea robotics and the ingenious remedies devised to surmount them, Chérubin provides invaluable insights into how robotic technologies have propelled the expansion of our understanding of the deep ocean and its intricate ecosystems.
- **Levin, L.A., & Le Bris, N. (2015)** The ramifications of climate change on the deep ocean, conducting a thorough analysis of the potential implications for deep-sea ecosystems and biodiversity. Drawing from a wealth of research findings and climate models, the authors dissect the anticipated effects of escalating temperatures, ocean acidification, and deoxygenation on the delicate balance of deep-sea habitats and species. Their study underscores the pressing imperative for bolstered monitoring and conservation initiatives aimed at ameliorating the adverse impacts of climate change.
- **Jamieson, A.J., & Vecchione, M. (2019)** Jamieson and Vecchione offer a thorough exploration of the latest breakthroughs in deep-sea observation technologies, emphasizing their crucial contribution to enriching our comprehension of the deep ocean and its intricate ecosystems. The authors delve into the evolution of pioneering imaging systems, acoustic sensors, and sampling devices, which have empowered scientists to venture into and document hitherto unreachable realms of the deep sea. Through a critical examination of the capabilities and constraints of these technological marvels, Jamieson and Vecchione provide invaluable insights into the continuous endeavors aimed at unraveling the enigmatic depths of the ocean and revealing its concealed wonders.
- **Dahlgren, T.G., et al. (2021).** Deep-Sea Mining: Interdisciplinary Analysis and Modeling for Sustainable Development" represents a crucial research

focus amidst increasing interest and activity in extracting mineral resources from the ocean floor. This interdisciplinary endeavor aims to address the complex challenges and potential impacts associated with deep-sea mining while striving for sustainable development practices. Through a collaborative approach integrating expertise from various fields such as marine science, ecology, economics, engineering, and policy, this research seeks to comprehensively analyze the ecological, economic, and social implications of deep-sea mining activities. By employing advanced modeling techniques, including ecological modeling, economic modeling, and risk assessment, researchers aim to predict and evaluate the short- and long-term effects of deep-sea mining on marine ecosystems, biodiversity, and human communities. Furthermore, this research endeavor seeks to identify potential mitigation measures and regulatory frameworks to minimize environmental harm and maximize socio-economic benefits.

- **Thomsen, L., & Booth, J.A. (2020).** This interdisciplinary study delves into the technological advancements driving deep-sea mining operations, including the development of robotic systems, extraction methods, and ore processing techniques. Simultaneously, it critically assesses considering factors such as habitat destruction, biodiversity loss, and potential chemical pollution. Furthermore, the study delves into the socio-economic dimensions of deep-sea mining, exploring issues of resource distribution, employment opportunities, and community impacts in both local and global contexts. By integrating perspectives from marine science, engineering, economics, sociology, and policy analysis, this research provides a holistic understanding of the complexities surrounding deep-sea mining. Through rigorous analysis and stakeholder engagement, the study aims to inform evidence-based decision-making processes, guide regulatory frameworks, and promote sustainable practices that balance the need for resource extraction with environmental and social considerations.
- **Dahlgren, T.G., et al. (2021).** Deep-sea mining presents a complex challenge that demands an interdisciplinary approach for sustainable development. Environmental scientists, economists, engineers, and policymakers must collaborate to understand and mitigate its potential impacts. Comprehensive modeling of environmental consequences, technological innovation for minimizing harm, and robust policy frameworks balancing industry interests with ecological preservation are essential. Moreover, international cooperation is critical for regulating activities in international waters and sharing resources equitably. By integrating scientific knowledge, technological advancements, economic analyses, and global governance, we can strive towards deep-sea

mining practices that uphold both environmental sustainability and economic viability.

- **Chaniotis, P. D. (2020)Deep**-sea conservation has made remarkable progress in recent years, propelled by a heightened understanding of the critical importance of these remote and delicate ecosystems. Interdisciplinary cooperation, stakeholder engagement, and international collaboration have emerged as crucial lessons learned from past efforts. Through extensive scientific exploration, we've unveiled the astonishing biodiversity harbored within deep-sea habitats, alongside the numerous threats posed by human activities like overfishing, seabed mining, and pollution. Moving forward, the establishment of marine protected areas, adoption of sustainable fishing practices, and stringent regulation of deep-sea mining are pivotal actions. Such endeavors necessitate inclusive partnerships across governments, scientific institutions, industry stakeholders, and civil society to foster collaborative solutions. Leveraging technological innovations, such as advanced remotely operated vehicles and genetic analysis techniques, will remain essential for unravelling the complexities of deep-sea ecosystems.

Deep Sea Bioprospecting and Bioinformatics:

Deep sea bioprospecting and bioinformatics constitute an exciting frontier in scientific exploration, holding the promise of uncovering invaluable biochemical compounds and genetic treasures from the ocean's depths. Bioprospecting involves scouring the deep sea for organisms that have evolved unique adaptations to extreme conditions, which often result in the production of bioactive compounds with potential pharmaceutical, industrial, or biotechnological applications. These compounds could include antibiotics, anticancer agents, or enzymes with remarkable properties. Meanwhile, bioinformatics provides the analytical backbone necessary to make sense of the vast amounts of genetic data obtained from deep sea organisms. By employing advanced computational tools, researchers can annotate, compare, and interpret genomic sequences, shedding light on the evolutionary relationships, metabolic pathways, and functional characteristics of these organisms. imetagenomic analyses allow for the exploration of entire deep-sea ecosystems, offering insights into microbial communities' diversity, metabolic potential, and ecological roles. Together, deep sea bioprospecting and bioinformatics hold immense potential for driving innovation in medicine, industry, and environmental conservation, provided they are pursued with a commitment to sustainable and responsible practices.

III. RESEARCH METHODOLOGY

- **Anthropogenic Impacts to the Deep Seafloor:** The technological progress of recent decades has brought about profound changes to human society, yet it has also led to unprecedented impacts on the deep seafloor. Industrial activities such as deep-sea mining, oil and gas extraction, and bottom trawling have intensified due to advancements in engineering and exploration techniques. Deep-sea mining, for instance, employs advanced robotics and remotely operated vehicles (ROVs) to extract minerals from the seabed, disrupting fragile ecosystems in the process. Similarly, oil and gas extraction in deep-sea environments relies on cutting-edge drilling technologies, posing risks of spills and habitat destruction. Furthermore, bottom trawling, aided by sophisticated navigation and fishing gear, indiscriminately alters seafloor habitats and disrupts benthic communities. These anthropogenic activities have far-reaching consequences for deep-sea biodiversity, ecosystem function, and biogeochemical cycling. Despite technological innovations in monitoring and mitigation, the remote and challenging nature of deep-sea environments presents significant obstacles to effective regulation and enforcement. As such, there is an urgent need for the responsible and sustainable management of human activities in the deep ocean, informed by interdisciplinary research and informed decision-making. Only through careful consideration of technological advancements and their potential impacts can we hope to safeguard the health and integrity of the deep seafloor for future generations.
- **Marine Litter:** Marine litter, consisting of a diverse array of materials ranging from plastics to metals, glass, and rubber, poses a significant threat to ocean health and marine ecosystems. With the proliferation of single-use plastics and inadequate waste management practices, marine litter has become a pervasive problem worldwide. Discarded items such as bottles, packaging materials, fishing gear, and microplastics not only mar the aesthetic beauty of coastlines and marine environments but also endanger marine life through ingestion, entanglement, and habitat degradation. The impact of marine litter extends beyond visible pollution, with microplastics infiltrating marine food webs and posing potential risks to human health. Furthermore, floating debris can act as vectors for the transport of invasive species and pollutants, exacerbating ecological imbalances and contaminating marine habitats. Addressing the scourge of marine litter requires concerted efforts at the local, regional, and global levels. Strategies encompass improved waste management infrastructure, public awareness campaigns, policy interventions, and innovative technological solutions for waste reduction and recycling. International agreements such as the United Nations Environment Programme's Global

Partnership on Marine Litter and initiatives like the Ocean Cleanup project demonstrate a growing recognition of the urgency to combat marine litter and protect the health and integrity of our oceans. By mitigating the sources of marine litter and implementing sustainable practices, we can strive towards a cleaner, healthier future for marine ecosystems and the communities that depend on them.

Figure 3. Marine litter collected in the deep Mediterranean Sea with an otter trawl.

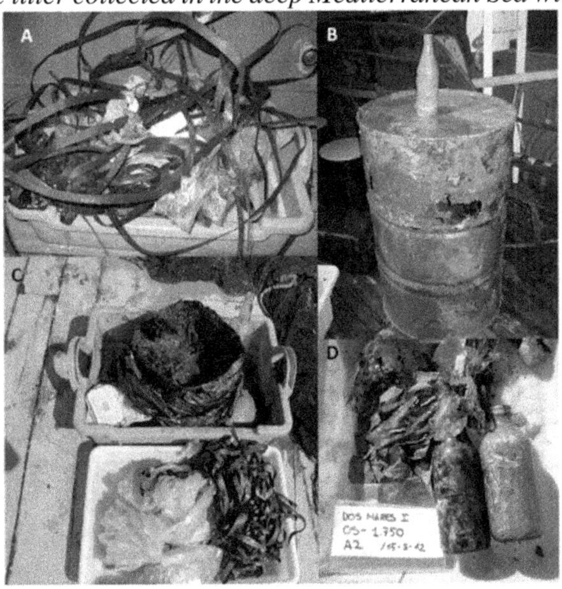

- **Submarine Tailing Disposal:** Submarine tailing disposal, also known as deep-sea tailings placement (DSTP), is a method used by some mining operations to dispose of waste materials, primarily from the extraction and processing of minerals such as gold, copper, and nickel. In this practice, mine tailings, which consist of finely ground rock and chemical residues, are discharged directly into the deep ocean through pipelines or other conduits, often at considerable depths. While submarine tailing disposal offers an alternative to terrestrial storage facilities, it poses significant environmental risks. The discharged tailings can smother benthic habitats, alter sediment composition, and introduce toxic substances into marine ecosystems, thereby disrupting marine life and compromising biodiversity. Furthermore, the long-term consequences of submarine tailing disposal are not fully understood, raising concerns about its potential to cause lasting harm to ocean ecosystems. Efforts to regulate submarine tailing disposal vary across jurisdictions, with some countries imposing stringent guidelines to minimize environmental impacts, while others have banned the practice altogether. Alternative approaches, such as terrestrial storage facilities with engineered containment systems or the development of more sustainable mining practices, offer potential solutions to mitigate the adverse effects of tailings disposal on marine environments. Given the growing recognition of the importance of ocean conservation and the need for sustainable resource management, the future of submarine tailing disposal remains uncertain. Continued research, stakeholder engagement, and regulatory oversight are essential to ensure that mining activities do not irreversibly harm marine ecosystems and the services they provide to society.

- **Fishing** The escalating demand for marine biological resources, coupled with technological advancements, has spurred a continuous surge in deep-sea fisheries exploitation, often extending to depths below 1500 meters. Among the various fishing methods employed, bottom trawling stands out for its pronounced impact on both targeted and non-targeted species, as well as the underlying habitat. Deep-sea species, often characterized by longevity and delayed maturity, face significant challenges in sustaining populations when subjected to such intensive fishing pressure.

Prolonged trawling activity in a given area can reshape the submarine landscape, diminishing the original complexity of the seafloor, akin to agricultural impacts on land. The fishing practices over seamounts, in particular, have yielded significant consequences, affecting both targeted species and the broader ecosystem. Seamounts, known for their diverse communities of sessile fauna like sponges and corals, serve as crucial habitats for various marine organisms. However, bottom trawling exacts a heavy toll on these fragile ecosystems, with recovery times spanning decades to centuries.

The cumulative impacts of deep-sea fisheries exploitation underscore the urgent need for sustainable management practices to safeguard marine biodiversity and ecosystem integrity. Such measures must encompass comprehensive regulations, habitat protection strategies, and informed decision-making processes to mitigate the deleterious effects on deep-sea environments and ensure their long-term viability.

Fishing regulations are administered by coastal states and Regional Fisheries Management Organizations (RFMOs), with the aim of ensuring sustainable management of marine resources. These regulations encompass a variety of measures, including quota management systems, licensing frameworks, and the designation of protected areas to safeguard specific habitats such as seamounts, ridges, cold-water corals, and upper slope regions. In recent times, there has been a notable trend among RFMOs towards expanding their regulatory frameworks to encompass the protection of vulnerable benthic marine ecosystems in international waters. This extension of regulations reflects a growing recognition of the importance of conserving deep-sea habitats and biodiversity beyond national jurisdictions. By implementing measures to restrict fishing activities in areas with sensitive benthic ecosystems, RFMOs aim to mitigate the adverse impacts of bottom trawling and other destructive fishing practices on deep-sea environments. These efforts signify a proactive approach towards promoting the sustainable use of marine resources and minimizing ecological damage in the high seas. However, effective implementation and enforcement of fishing regulations across vast and remote oceanic regions remain ongoing challenges. Continued collaboration among coastal states, RFMOs, and other stakeholders is essential to ensure the success of conservation initiatives and the preservation of deep-sea ecosystems for future generations.

DATA ANALYSIS

Oil and Gas Exploitation

Data analysis plays a crucial role in understanding the impacts of oil and gas exploitation on various aspects of the environment, economy, and society. Through the examination of extensive datasets, researchers can assess the magnitude and trends of oil and gas extraction activities, as well as their implications for ecosystems, climate change, and human health.

One key focus of data analysis in oil and gas exploitation is the assessment of environmental impacts. This includes monitoring pollution levels in air, water, and soil, tracking habitat destruction and biodiversity loss, and evaluating the effectiveness of mitigation measures. By analysing spatial and temporal patterns in environmental data, scientists can identify hotspots of contamination and prioritize conservation efforts. Another important aspect of data analysis in oil and gas exploitation is the economic analysis of resource extraction. This involves assessing the costs and benefits of oil and gas development, including revenues generated, employment opportunities created, and infrastructure investments required. Economic models and statistical techniques are used to forecast future trends in oil and gas markets, inform policy decisions, and optimize resource allocation. Data analysis is essential for evaluating the social impacts of oil and gas exploitation, including its effects on local communities, indigenous peoples, and vulnerable populations. By analysing demographic data, socioeconomic indicators, and community surveys, researchers can identify disparities in access to resources, health outcomes, and quality of life, and advocate for equitable development policies. Data analysis serves as a powerful tool for understanding the complex interplay between oil and gas exploitation and its broader impacts on the environment, economy, and society. Through rigorous analysis of multidisciplinary datasets, policymakers, industry stakeholders, and civil society can make informed decisions to promote sustainable development and mitigate the adverse effects of fossil fuel extraction.

Table 1. Technological Advancements for Deep Sea Ecosystem Conservation and Exploration

Technology	Description	Application
Remotely Operated Vehicles	Submersible vehicles controlled remotely from the surface, equipped with cameras and sensors.	Exploration of deep-sea habitats, data collection.
Autonomous Underwater Vehicles	Unmanned vehicles programmed to navigate underwater autonomously, often equipped with sensors.	Mapping, monitoring of deep-sea environments.
Satellite Imaging	Use of satellites to capture high-resolution images of the Earth's surface, including the ocean.	Monitoring changes in ocean temperature, currents, etc.
Genetic Sequencing	Analyzing DNA from deep-sea organisms to understand biodiversity and evolutionary relationships.	Identifying new species, studying genetic adaptations.
Acoustic Monitoring Systems	Deploying hydrophones to record sounds in the ocean, providing insights into marine life and habitats.	Studying vocalizations of marine mammals, detecting activity.
Underwater Drones	Small, maneuverable vehicles equipped with cameras and sensors for detailed exploration.	Conducting surveys in hard-to-reach areas, monitoring ecosystems.
Oceanographic Buoy Networks	Deploying buoys equipped with sensors to collect data on ocean conditions such as temperature and salinity.	Monitoring changes in ocean properties over time.
Deep-Sea Mapping Technologies	Employing sonar and other mapping techniques to create detailed maps of the ocean floor.	Understanding seafloor geology, identifying habitats.
Submersible Research Vessels	Manned submarines designed to withstand deep-sea pressures, allowing for direct observation and sampling.	Conducting in-depth studies, collecting specimens.
Environmental DNA (eDNA) Monitoring	Detecting traces of genetic material shed by organisms in their environment, providing insights into species presence.	Assessing biodiversity, tracking invasive species.

Table 2. Technological Tools for Deep Sea Exploration And Conservation

Technology	Description	Application
Remotely Operated Vehicles	Submersible vehicles controlled remotely from the surface, equipped with cameras and sensors.	Exploration of deep-sea habitats, data collection.
Autonomous Underwater Vehicles	Unmanned vehicles programmed to navigate underwater autonomously, often equipped with sensors.	Mapping, monitoring of deep-sea environments.
Satellite Imaging	Use of satellites to capture high-resolution images of the Earth's surface, including the ocean.	Monitoring changes in ocean temperature, currents, etc.
Genetic Sequencing	Analyzing DNA from deep-sea organisms to understand biodiversity and evolutionary relationships.	Identifying new species, studying genetic adaptations.
Acoustic Monitoring Systems	Deploying hydrophones to record sounds in the ocean, providing insights into marine life and habitats.	Studying vocalizations of marine mammals, detecting activity.

IV. CaSE STUDIES IN DEEP-SEA CONSERVATION

India has made efforts to address deep sea conservation through various initiatives and projects. Here are a few examples:

- **Deep-Sea Biodiversity Conservation Project (DSBCP)**

The DSBCP was founded in 2017 and represents a partnership between the Government of India's Ministry of Earth Sciences and the National Institute of Oceanography (NIO). Its goals include assessing the effects of human activities such as deep-sea mining and fishing, as well as investigating the biodiversity of deep-sea ecosystems in the Indian Ocean. The project also prioritises capacity building and technology improvement to assist the sustainable management of deep-sea resources.

- **National Centre for Antarctic and Ocean Research (NCAOR)**

NCAOR, based in Goa, plays a key role in India's research and exploration activities in the Southern Ocean and Antarctic region. Its research vessels conduct scientific surveys to study deep sea biodiversity, oceanography, and climate change impacts. NCAOR's research contributes to understanding and conserving deep sea ecosystems in the Indian Ocean and beyond.

- **Ocean Observation and Climate Monitoring Programme**

India has been actively involved in ocean observation and climate monitoring through initiatives such as the Indian National Centre for Ocean Information Services (INCOIS). These programs collect data on oceanographic parameters, including temperature, salinity, and currents, to monitor changes in the marine environment and assess the health of deep-sea ecosystems.

- **International Collaboration**:

India collaborates with other countries and international organizations on deep sea research and conservation initiatives. For instance, India is a member of the International Seabed Authority (ISA), which regulates deep sea mining activities in international waters. Through such collaborations, India contributes to the development of policies and guidelines for sustainable management of deep-sea resources.

These examples illustrate India's efforts to study, protect, and sustainably manage deep sea ecosystems, recognizing their ecological significance and the need for international cooperation in conservation efforts.

V. RESEARCH PROBLEM

The research problem surrounding advancements for deep-sea ecosystem conservation and exploration is multifaceted and urgent. Despite the vast expanse and ecological significance of the deep ocean, our understanding of its biodiversity, ecosystem dynamics, and responses to anthropogenic pressures remains incomplete. This knowledge gap presents a significant challenge in developing effective conservation and management strategies to safeguard the integrity and resilience of deep-sea ecosystems. Key questions include identifying the most critical threats to these environments, understanding regional and habitat-specific variations, and pinpointing gaps in our understanding of deep-sea biodiversity and ecological processes. Moreover, there is a pressing need for technological advancements to enhance our ability to explore, monitor, and assess deep-sea ecosystems, as well as the impacts of human activities. Interdisciplinary collaboration and knowledge exchange are paramount to address these research challenges, integrating biological, oceanographic, geological, and engineering perspectives. Addressing these concerns allows us to learn more about the deep sea, make informed decisions, and promote the preservation and sustainable management of these critical marine environments for future generations.

VI. CONCLUSION

The table illustrates a diverse array of technological tools that have revolutionized the exploration and conservation of deep-sea ecosystems. From Remotely Operated Vehicles (ROVs) to Satellite Imaging and Genetic Sequencing, each technology offers unique capabilities for studying and protecting the fragile environments of the deep ocean. These advancements have enabled researchers to delve deeper into the abyss, uncovering new species, mapping uncharted territories, and monitoring environmental changes with unprecedented accuracy. By harnessing the power of these technological tools, we can continue to expand our understanding of the deep sea and implement effective conservation strategies to safeguard its biodiversity for future generations.

VII. FUTURE WORK

Future advancements in deep sea ecosystem conservation and exploration will likely revolve around the integration of cutting-edge technologies and interdisciplinary approaches to overcome existing obstacles and unlock new frontiers. This could entail the development of autonomous underwater vehicles with enhanced sensing capabilities for comprehensive mapping, coupled with the application of artificial intelligence and machine learning algorithms to analyse vast datasets efficiently. Improved communication systems will facilitate real-time data transmission, enabling more responsive monitoring and management of deep-sea environments. Conservation strategies may emphasize habitat restoration initiatives and the expansion of marine protected areas, supported by innovative solutions such as 3D-printed artificial reefs and underwater drones for ecosystem rehabilitation. International collaboration will be pivotal, fostering partnerships to address the multifaceted challenges of deep-sea conservation and ensure the sustainable stewardship of these critical marine ecosystems.

VIII. REFERENCES

Ballard, R. D. (2000). *The Eternal Darkness: A Personal History of Deep-Sea Exploration*. Princeton University Press.

Chaniotis, P. D., Robson, L. M., Lemasson, A. J., Cornthwaite, A. L., & Howell, K. L. (2020). UK deep-sea conservation: Progress, lessons learned, and actions for the future. *Aquatic Conservation*, 30(2), 375–393. DOI: 10.1002/aqc.3243

Chérubin, L. M. (2017). Deep-Sea Robotics: A Technological Evolution. *Annual Review of Marine Science*, 9, 485–508.

Da Ros, Z., Dell'Anno, A., Morato, T., Sweetman, A. K., Carreiro-Silva, M., Smith, C. J., & Danovaro, R. (2019). The deep sea: The new frontier for ecological restoration. *Marine Policy*, 108, 103642. DOI: 10.1016/j.marpol.2019.103642

Dahlgren, T. G.. (2021). Deep-Sea Mining: Interdisciplinary Analysis and Modeling for Sustainable Development. *Marine Policy*, 123, 104364.

Danovaro, R., Snelgrove, P. V., & Tyler, P. (2014). Challenging the paradigms of deep-sea ecology. *Trends in Ecology & Evolution*, 29(8), 465–475. DOI: 10.1016/j.tree.2014.06.002 PMID: 25001598

Dunn, D. C., Van Dover, C. L., Etter, R. J., Smith, C. R., Levin, L. A., Morato, T., ... & SEMPIA Workshop Participants. (2018). A strategy for the conservation of biodiversity on mid-ocean ridges from deep-sea mining. *Science Advances, 4*(7), eaar4313.

Ferrini, V. L., & MacPherson, A. C. (2016). Autonomous Underwater Vehicle Navigation in Complex Environments: Challenges and Future Directions. *IEEE Journal of Oceanic Engineering*, 41(4), 865–884.

Jamieson, A. J., & Vecchione, M. (2019). Illuminating the Deep: Advances in Deep-Sea Observation Technologies. *Annual Review of Marine Science*, 11, 209–235.

Levin, L. A., & Gooday, A. J. (2003). The deep Atlantic Ocean. In *The Natural History of the Deep Ocean Floor* (pp. 53–87). Springer.

Levin, L. A., & Le Bris, N. (2015). The Deep Ocean Under Climate Change. *Science*, 350(6262), 766–768. DOI: 10.1126/science.aad0126 PMID: 26564845

Levin, L. A., & Sibuet, M. (2012). Understanding Continental Margin Biodiversity: A New Imperative. *Annual Review of Marine Science*, 4(1), 79–112. DOI: 10.1146/annurev-marine-120709-142714 PMID: 22457970

Matabos, M.. (2018). From the Surface to the Abyss: The Role of Underwater Gliders in Oceanographic Research. *Annual Review of Marine Science*, 10, 451–475.

Raineault, N. A., Klemperer, S. L., & Dorschel, B. (2014). Mapping the Deep Sea. In *Seafloor Mapping along Continental Shelves* (pp. 3–36). Springer.

Singh, H., & King, J. W. (2015). Observations of Vertical Pesticide Fluxes in the Deep Sargasso Sea. *Deep-sea Research. Part II, Topical Studies in Oceanography*, 113, 131–138.

Smith, C. R., & Glover, A. G. (2016). The Deep Ocean Carbon Cycle: An Overview. [). CRC Press.]. *Oceanography and Marine Biology - an Annual Review*, 54, 1–34.

Thomsen, L., & Booth, J. A. (2020). Deep-Sea Mining: Technology, Environmental, and Socio-Economic Considerations. *Ocean and Coastal Management*, 184, 105019.

Thurber, A. R., Levin, L. A., & Rowden, A. A. (2014). Benthic Oxygen Consumption in Oxygen Minimum Zone Regions: A Global Comparison of Fluxes and Ecosystem Impacts. *Deep-sea Research. Part I, Oceanographic Research Papers*, 94, 173–183.

Thurber, A. R., Sweetman, A. K., Narayanaswamy, B. E., Jones, D. O., Ingels, J., & Hansman, R. L. (2014). Ecosystem function and services provided by the deep sea. *Biogeosciences*, 11(14), 3941–3963. DOI: 10.5194/bg-11-3941-2014

White, M., & Rycroft, S. (2016). Remote Sensing of the Deep Ocean and Sea Ice. In Remote Sensing of the Cryosphere (pp. 189-212). Wiley.

Chapter 4
Climate Change and the Deep Sea

V. Gowri
https://orcid.org/0000-0002-7847-9774
SRM Madurai College for Engineering and Technology, India

S. Thenmozhi
https://orcid.org/0000-0001-6681-3897
St. Joseph's College of Engineering, India

Shrinwantu Raha
https://orcid.org/0000-0003-4883-8657
Bhairab Ganguly College, India

P. Selvakumar
https://orcid.org/0000-0002-3650-4548
Department of Science and Humanities, Nehru Institute of Technology, India

Sagar Kashinath Khokale
https://orcid.org/0000-0002-7351-0613
Sandip Institute of Pharmaceutical Sciences, Mahiravani, India

ABSTRACT

The deep sea, a vast and enigmatic realm beneath the ocean's surface, holds some of the planet's greatest mysteries. These harsh conditions create a unique and diverse ecosystem, teeming with life forms adapted to survive in ways unimaginable to terrestrial beings. The journey into the deep sea begins at the continental slope, which descends steeply from the continental shelf into the abyssal plain. This transition zone marks the boundary between the relatively shallow coastal waters and the deep ocean basins. As one descends deeper, light from the surface diminishes

DOI: 10.4018/979-8-3693-6670-7.ch004

rapidly, giving way to the twilight zone at depths of 200 to 1,000 meters. Despite the absence of sunlight, the deep sea is a dynamic and vibrant ecosystem. It hosts a myriad of species, many of which are specially adapted to the extreme conditions. Bioluminescence, the production and emission of light by living organisms, is a common adaptation, serving various purposes such as predation, camouflage, and communication. Creatures like the anglerfish, with its luminescent lure, and the vampire squid, which uses bioluminescent displays to evade predators, exemplify the ingenious adaptations that have evolved in the deep sea. The deep sea floor is dotted with hydrothermal vents and cold seeps, which are hotspots of biological activity. Hydrothermal vents, first discovered in the late 1970s, are formed by volcanic activity on the ocean floor, where seawater percolates through the Earth's crust, heats up, and re-emerges laden with minerals. These vents support unique ecosystems, relying not on sunlight for energy but on chemosynthesis, a process by which bacteria convert chemical energy from vent minerals into organic matter. The organisms inhabiting cold seeps, such as mussels, clams, and various types of worms, have adapted to exploit the chemical energy available in these otherwise inhospitable environments. Understanding the mechanisms and efficiency of the biological pump is crucial for predicting the deep sea's role in mitigating climate change.

INTRODUCTION

The deep sea is also a repository of the Earth's history, with sediment layers recording millions of years of geological and climatic changes. These sediments provide valuable insights into past ocean conditions, helping scientists reconstruct ancient climates and better understand the natural variability of the Earth's climate system. Paleoclimate studies, based on deep sea sediment cores, have revealed patterns of glacial and interglacial periods, shedding light on the mechanisms driving long-term climate change. The extraction processes can cause habitat destruction, sediment plumes, and the release of toxic substances, with potentially long-lasting and irreversible effects on deep sea communities (Abhilasha A., et al., 2022). Furthermore, the deep sea is a potential repository for carbon dioxide and nuclear waste, raising ethical and environmental concerns about the use of this vast, unexplored frontier as a dumping ground. Conservation of the deep sea is challenging but essential. International collaborations, such as the efforts led by the International Seabed Authority (ISA) and various marine protection organizations, aim to establish regulations and protected areas to preserve deep sea biodiversity. In recent years, public interest in the deep sea has grown, fueled by documentary films, media coverage of deep-sea discoveries, and the work of organizations dedicated to ocean exploration. This increased awareness is crucial for garnering support for conservation efforts and fostering a sense of stewardship for the deep sea, an integral but often overlooked part of our planet's natural heritage. The deep sea remains a frontier of scientific exploration and discovery (Anderson, P. M., et al., 2019). As we continue to uncover its secrets, we gain not only a deeper understanding of our planet's most extensive habitat but also valuable insights into the broader workings of Earth's systems.

Understanding Climate Change: Causes and Effects

Conversely, warmer air holds more moisture, leading to heavier and more unpredictable rainfall events, which can result in devastating floods. The warming of the oceans, which absorb about 90% of the excess heat generated by greenhouse gases, is another critical impact of climate change. Warmer ocean temperatures contribute to the bleaching and die-off of coral reefs, which are vital marine ecosystems. Human societies are not immune to the impacts of climate change. Food security is threatened by changes in agricultural productivity, while water resources are increasingly stressed by changing precipitation patterns and the depletion of glaciers and snowpacks (Baco, A. R., et al., 2020). Public health is at risk from heat-related illnesses, the spread of vector-borne diseases, and the mental health impacts of climate-related disasters. Additionally, climate change can exacerbate social and economic inequalities, as the most vulnerable populations, often in developing countries, are disproportionately affected by its impacts.

Deep Sea Ecosystems: An Overview

Deep sea ecosystems, spanning from the edge of the continental shelf to the ocean's deepest trenches, represent some of the most unique and diverse habitats on Earth. These ecosystems exist in an environment of perpetual darkness, extreme pressures, and cold temperatures, yet they are teeming with life forms that have adapted to these harsh conditions in remarkable ways. The bathyal zone, often referred to as the midnight zone due to the absence of sunlight, is characterized by a steep decline in temperature and an increase in pressure as depth increases. This zone is home to a variety of species, including deep sea fish, cephalopods, and crustaceans, many of which exhibit bioluminescence. This ability to produce and emit light is used for communication, predation, and camouflage (Bell, R. J., et al., 2018). Species such as the anglerfish, with its bioluminescent lure, and the vampire squid, which uses bioluminescent displays to evade predators, exemplify the fascinating adaptations to life in the deep. Below the bathyal zone lies the abyssal zone, a vast and relatively uniform environment covering much of the ocean floor. This zone is characterized by cold temperatures, near-freezing, and immense pressure.

Despite these conditions, it supports a diverse array of organisms, including polychaete worms, echinoderms such as sea cucumbers and brittle stars, and deep sea fish. Many of these species have slow metabolisms and long lifespans, adaptations to the scarcity of food in this nutrient-poor environment. Food typically arrives in the form of marine snow—organic particles falling from the upper layers of the ocean—and the occasional large carcass, like that of a whale, which can support entire communities for years.

The hadal zone, found in the ocean's deepest trenches, is the least explored and most mysterious of all deep sea habitats (Biastoch, A., et al., 2021). These trenches are formed by tectonic activity and can reach depths where pressures exceed 1,000 times atmospheric pressure at sea level. Life in the hadal zone includes specialized organisms such as amphipods, polychaetes, and certain species of fish that have evolved to withstand these extreme pressures. Research in these areas is challenging but has revealed the presence of complex communities adapted to one of the most extreme environments on the planet.Hydrothermal vent ecosystems, discovered in the late 1970s, are among the most extraordinary deep sea environments. These vents, located along mid-ocean ridges, emit superheated, mineral-rich water from beneath the Earth's crust (Bopp, L., et al., 2018). The unique chemistry of these environments supports chemosynthetic bacteria, which form the base of a food web that includes giant tubeworms, clams, and other specialized organisms. These ecosystems are entirely independent of sunlight, relying instead on chemical energy from the Earth's interior, a discovery that has profound implications for our understanding of life's potential in extreme environments, including on other planets.

Cold seeps are another type of deep sea ecosystem where hydrocarbons like methane and hydrogen sulfide seep from the ocean floor. These seeps support communities similar to those found at hydrothermal vents, with chemosynthetic bacteria playing a crucial role in the ecosystem. Cold seep communities can include mussels, clams, and various species of worms, all adapted to exploit the chemical energy available in these environments. The biodiversity of deep sea ecosystems is remarkable, yet much of it remains undiscovered (Breusing, C., et al., 2020). The deep sea is home to a myriad of species that have yet to be described by science. This biodiversity is crucial not only for the intrinsic value of these ecosystems but also for their potential contributions to biotechnology and medicine. For instance, enzymes from deep sea organisms are being studied for their applications in industrial processes and pharmaceuticals.

The extraction processes can cause habitat destruction, sediment plumes, and the release of toxic substances, with potentially long-lasting and irreversible effects on deep sea communities (Bryan, T. L., et al., 2020). Additionally, the deep sea is a potential site for carbon dioxide and nuclear waste storage, raising ethical and environmental concerns about the exploitation of this vast, unexplored frontier.

Conservation efforts are critical for protecting deep sea ecosystems. International collaborations, such as those led by the International Seabed Authority (ISA) and various marine protection organizations, aim to establish regulations and protected areas to preserve deep sea biodiversity. In summary, deep sea ecosystems are a vital and integral part of the Earth's biosphere, characterized by unique adaptations to extreme conditions and significant biodiversity. Despite their inaccessibility, these ecosystems play crucial roles in global processes such as carbon sequestration and nutrient cycling. As human activities extend into the deep ocean, understanding and protecting these ecosystems becomes increasingly important to preserve their ecological and potential scientific value.

Ocean Warming: Impacts on Deep Sea Life

Ocean warming, a direct consequence of climate change, is profoundly affecting marine ecosystems, including the deep sea. As the oceans absorb excess heat from the atmosphere, deep sea temperatures are rising, albeit more slowly than surface waters. This gradual but significant warming has far-reaching implications for deep sea life, which is finely tuned to the cold, stable conditions that have prevailed for millennia (Carlson, P. R., et al., 2021). One of the most immediate impacts of ocean warming on deep sea life is the alteration of metabolic rates. Many deep sea organisms, such as fish, invertebrates, and microorganisms, have evolved to thrive in near-freezing temperatures. Even small increases in temperature can disrupt their metabolic processes, affecting growth, reproduction, and survival rates. For instance, species with narrow thermal tolerances may experience increased metabolic stress, leading to reduced fitness and, in some cases, mortality. This can have cascading effects throughout the food web, as predator-prey dynamics are altered and ecological balances are disrupted.

Warming waters also influence the distribution of species. As temperatures rise, some deep sea organisms may migrate to cooler, deeper waters to maintain their thermal preferences. This shift can lead to changes in community composition and biodiversity in various regions. Species that cannot migrate or adapt quickly enough face heightened risks of extinction. Moreover, the movement of species to new areas can introduce competition with existing species, potentially leading to further ecological upheaval. Another significant impact of ocean warming is on deep sea coral reefs and other biogenic habitats. Deep sea corals, such as those found on seamounts and in canyons, are particularly vulnerable to temperature increases (Clark, M. R., et al., 2019). The loss of deep sea coral reefs would not only diminish biodiversity but also disrupt the complex ecological networks that depend on these structures for shelter, feeding, and breeding. Ocean warming also affects the availability and distribution of nutrients in the deep sea. Thermal stratification, where warmer surface waters create a barrier to the mixing of nutrient-rich deep waters, can reduce the upwelling of nutrients that support deep sea ecosystems. This can lead to decreased productivity in deep sea environments, affecting the entire food web. Additionally, changes in ocean currents due to warming can alter the transport of organic matter from the surface to the deep sea, impacting the primary food source for many deep sea organisms.

The phenomenon of deoxygenation, the reduction in oxygen levels in the ocean, is another consequence of warming waters that poses a threat to deep sea life. Warmer water holds less oxygen, and increased stratification can limit the mixing of oxygen-rich surface waters with deeper layers (Danovaro, R., et al., 2020). Hypoxic conditions, or areas with low oxygen levels, can develop, making it difficult for many deep sea species to survive. Organisms that are not adapted to low oxygen conditions may experience physiological stress, reduced reproductive success, and increased mortality rates. Hypoxic zones, often referred to as "dead zones," can lead to significant declines in biodiversity and alter the structure of deep sea communities. Furthermore, ocean warming exacerbates the impacts of other stressors, such as ocean acidification and pollution. The combined effects of these stressors can create synergistic challenges for deep sea organisms. For example, acidified waters can weaken the calcium carbonate structures of deep sea corals and shell-forming organisms, making them more susceptible to the additional stress of rising temperatures. Similarly, pollutants such as heavy metals and microplastics, which are already accumulating in the deep sea, can have more severe effects on organisms that are already stressed by warming waters. In addition to these direct biological impacts, ocean warming can influence the geophysical aspects of the deep sea environment (DeLeo, D. M., et al., 2021). For instance, changes in the temperature and chemistry of the ocean can affect the stability of methane hydrates, which are ice-like structures that trap methane in the seafloor. In conclusion, ocean warming presents a significant and multifaceted challenge to deep sea ecosystems. The unique and highly adapted life forms of the deep sea are particularly vulnerable to even modest increases in temperature.

Acidification of the Deep Ocean: Consequences for Marine Species

The deep ocean, already naturally more acidic than surface waters due to the decomposition of organic matter, is particularly vulnerable to further acidification. The consequences for marine species inhabiting these depths are profound and multifaceted. Deep sea corals, which form the backbone of diverse and complex habitats, are especially at risk. The weakening of their skeletons can lead to decreased structural integrity, making them more susceptible to breakage and erosion, and ultimately threatening the biodiversity that depends on these coral reefs for shelter and sustenance (Evans, R. D.,et al., 2019). For deep sea mollusks, such as snails and clams, acidification can impair shell formation and maintenance, leading to thinner and more fragile shells. This not only makes them more vulnerable to predation but also affects their overall health and survival rates. Similarly, for planktonic organisms like pteropods, often referred to as "sea butterflies," the dissolution of their calcium carbonate shells in more acidic conditions can lead to significant declines in their populations. Given that these organisms are a key component of the marine food web, their decline can have cascading effects on higher trophic levels, including commercially important fish species and other marine predators.

Beyond calcifying organisms, acidification also impacts non-calcifying species through various physiological and behavioral changes. For example, some fish species exhibit altered behavior in more acidic conditions, such as impaired predator avoidance and navigation abilities. This can increase their vulnerability to predators and reduce their chances of survival. Additionally, the reproductive success of some species may be compromised, as acidified conditions can affect the development of larvae and juveniles (Fischer, G., et al., 2021). For instance, studies have shown that the larvae of certain fish and invertebrates develop more slowly and suffer higher mortality rates in acidified waters, potentially leading to reduced recruitment and population declines over time. The deep sea's unique chemical environment, already characterized by higher levels of CO_2 and lower pH compared to surface waters, means that further acidification exacerbates existing stressors. Deep sea hydrothermal vent communities, for instance, thrive in areas with naturally high CO_2 concentrations and have evolved unique adaptations to cope with these conditions. However, the additional stress of anthropogenic acidification could push these systems beyond their adaptive capacity, leading to shifts in community composition and declines in biodiversity. Similarly, cold seeps, which are biologically rich habitats where methane and hydrogen sulfide seep from the ocean floor, might also face increased challenges as acidification alters the availability of essential nutrients and energy sources.

Acidification also impacts the deep ocean's microbial communities, which play crucial roles in nutrient cycling and organic matter decomposition. Changes in pH can affect microbial metabolism and the overall efficiency of these processes, potentially disrupting the deep sea's biogeochemical cycles. For example, acidification can influence the microbial breakdown of organic matter, affecting the availability of nutrients for other organisms. This can have far-reaching implications for the productivity and health of deep sea ecosystems (Fox, A. D., et al., 2018). Moreover, the impacts of acidification on deep sea species are compounded by other stressors such as ocean warming, deoxygenation, and pollution. The combined effects of these stressors can create synergistic challenges, making it even more difficult for marine species to adapt and survive. For example, warming waters can exacerbate the effects of acidification by increasing metabolic rates and oxygen demand, while deoxygenation can further stress organisms already struggling with acidified conditions. Pollution, including heavy metals and microplastics, can also interact with acidification to impair physiological functions and increase the vulnerability of deep sea species. Mitigating the impacts of ocean acidification on deep sea ecosystems requires a multifaceted approach. Reducing CO_2 emissions is paramount to slowing the rate of acidification and allowing marine species and ecosystems more time to adapt. Additionally, protecting critical habitats through the establishment of marine protected areas (MPAs) can help preserve biodiversity and ecological functions. In conclusion, ocean acidification presents a significant and growing threat to deep sea ecosystems (Glover, A. G., et al., 2019). The impacts on calcifying organisms, physiological and behavioral changes in non-calcifying species, and disruptions to microbial communities highlight the complex and multifaceted nature of this challenge. Addressing acidification requires global efforts to reduce CO_2 emissions, protect vulnerable habitats, and advance scientific understanding of deep sea ecosystems. By taking action to mitigate acidification and its impacts, we can help preserve the rich biodiversity and ecological integrity of the deep ocean, ensuring the health and resilience of these vital ecosystems for future generations.

Deoxygenation: The Shrinking Breathable Zone

Deoxygenation of the world's oceans, a consequence of both natural processes and human activities, is rapidly becoming a critical issue for marine ecosystems, including the deep sea. This phenomenon, characterized by a decrease in dissolved oxygen levels, poses significant threats to marine life that depends on oxygen for survival. Oxygen enters the ocean through the surface via direct diffusion from the atmosphere and through photosynthesis by phytoplankton (Grassle, J. F., et al., 2020). It is then transported to deeper layers by mixing and circulation processes. Warmer surface waters create a more pronounced thermocline, a layer that acts as a barrier to the mixing of oxygen-rich surface waters with deeper, oxygen-poor waters. This process, combined with increased microbial respiration in warmer waters, leads to a significant reduction in the overall oxygen content of the ocean. OMZs occur naturally at mid-depths, typically between 200 and 1,000 meters, but their extent and intensity are increasing as deoxygenation progresses. Species that inhabit the fringes of these zones are forced to either migrate to areas with higher oxygen levels or adapt to the hypoxic conditions. This migration can lead to overcrowding in more oxygenated areas, increasing competition for resources and altering community dynamics.

For many deep sea organisms, the ability to tolerate low oxygen levels is a matter of survival. Species such as certain fish, cephalopods, and crustaceans have developed physiological adaptations to cope with hypoxic conditions, such as enhanced gill surface area, increased hemoglobin affinity for oxygen, and more efficient oxygen utilization mechanisms. However, even these adaptations have limits (Haddock, S. H. D., et al., 2020). As oxygen levels continue to decline, the metabolic rates of these organisms are further suppressed, affecting their growth, reproduction, and overall fitness. The stress induced by hypoxia can also increase susceptibility to diseases and reduce resilience to other environmental stressors, such as temperature fluctuations and pollution.Deoxygenation also has cascading effects on the deep sea food web. Many deep sea species rely on detritus and organic matter sinking from the surface, a process known as marine snow, for their sustenance. Hypoxic conditions can alter the decomposition and remineralization processes of this organic matter, affecting its quality and availability (Hansen, J., et al., 2018). Additionally, the expansion of OMZs can lead to changes in the distribution and behavior of predator and prey species, disrupting established feeding relationships and potentially leading to shifts in the composition and function of the entire ecosystem.

Moreover, deoxygenation impacts biogeochemical cycles in the deep ocean. Oxygen is a critical component in various chemical reactions, including those involved in the cycling of carbon, nitrogen, and sulfur. Reduced oxygen levels can shift these cycles towards anaerobic processes, leading to the accumulation of substances such as hydrogen sulfide and methane, which are toxic to many forms of marine life. These changes can further stress deep sea organisms and exacerbate the impacts of other stressors such as ocean acidification.Human activities, particularly nutrient pollution from agriculture and wastewater runoff, contribute significantly to deoxygenation. Nutrient inputs stimulate the growth of phytoplankton, which upon dying, sink and decompose, consuming oxygen in the process (Helm, K. P., et al., 2018). This eutrophication-driven deoxygenation is especially problematic in coastal areas but can also impact deeper waters through the export of organic matter. Additionally, industrial activities such as deep sea mining and drilling can disturb sediments, releasing trapped organic matter and further depleting oxygen levels.

Deep Sea Coral Reefs: Vulnerable Treasures

Unlike their shallow-water counterparts, these reefs do not rely on sunlight for energy, instead thriving in the cold, dark depths through a slow and steady process of nutrient accumulation. Comprising various species of corals, primarily from the genera Lophelia, Madrepora, and Solenosmilia, deep sea coral reefs provide essential habitats for a multitude of marine organisms, from invertebrates and fish to microorganisms. This practice can cause severe physical damage to the delicate structures of deep sea corals, which take centuries to millennia to grow. The destruction of these reefs not only results in the loss of coral species but also disrupts the complex habitats that support a wide array of marine life. Fish, crustaceans, and other invertebrates that rely on the nooks and crannies of coral structures for shelter and breeding grounds are left without a home, leading to declines in their populations and, consequently, affecting the entire food web (Henson, S. A., et al., 2019). In addition to physical destruction from fishing practices, deep sea coral reefs are increasingly threatened by climate change. Ocean warming poses a significant risk, as many deep sea coral species are adapted to the stable, cold temperatures of their environments. Even slight increases in temperature can cause stress, potentially leading to coral bleaching—a phenomenon more commonly associated with shallow-water corals but also observed in deep sea species. Prolonged stress from elevated temperatures can lead to widespread coral mortality, further threatening the survival of these unique ecosystems.

In more acidic conditions, deep sea corals struggle to build and maintain their skeletons, leading to weaker structures that are more susceptible to breakage and erosion. The long-term implications of acidification include reduced growth rates and reproductive success, compromising the ability of deep sea coral reefs to recover from disturbances and sustain their ecological functions (Hudson, I. R., et al., 2021). Pollution from various sources, including plastic waste, heavy metals, and chemical contaminants, also poses a significant risk to deep sea coral reefs. Microplastics, for instance, have been found in deep sea environments, where they can be ingested by coral polyps and other marine organisms, leading to physical and toxicological harm. Heavy metals and chemical pollutants, often introduced through runoff and deep sea mining activities, can accumulate in coral tissues, causing further stress and impairing their growth and reproductive capabilities. The extraction processes involved in mining can result in significant habitat destruction, sediment plumes, and the release of toxic substances. Sediment plumes, in particular, can smother coral reefs, blocking the delicate polyps from accessing food particles and leading to suffocation. Additionally, the noise and vibration from mining operations can disrupt the behavior and physiological processes of marine species inhabiting these environments.

Restoration initiatives are also being explored, including the transplantation of coral fragments and the use of artificial structures to provide substrates for coral larvae to settle and grow. However, the slow growth rates of deep sea corals mean that recovery efforts require long-term commitment and monitoring. Public awareness and education about the importance and vulnerability of deep sea coral reefs are vital for garnering support for conservation measures and promoting sustainable practices.

In conclusion, deep sea coral reefs are invaluable treasures of the ocean, hosting a rich diversity of life and playing crucial roles in deep sea ecosystems (Johnson, C. L., et al., 2019). However, they are increasingly threatened by human activities and environmental changes, including bottom trawling, climate change, ocean acidification, pollution, and deep sea mining. Protecting these fragile ecosystems requires a comprehensive approach, involving stringent regulatory measures, conservation efforts, scientific research, and public engagement. By addressing these challenges, we can help ensure the preservation of deep sea coral reefs and the myriad species that depend on them for generations to come.

Adaptation and Resilience in Deep Sea Species

Adaptation and resilience are key traits that enable deep sea species to survive and thrive in one of the most extreme environments on Earth. The deep sea, characterized by high pressure, low temperatures, and complete darkness, presents a unique set of challenges that have driven remarkable evolutionary adaptations. These adaptations not only allow deep sea organisms to cope with their harsh surroundings but also demonstrate the incredible resilience of life in the face of changing environmental conditions (Jones, D. O. B., et al., 2018). One of the most notable adaptations in deep sea species is their ability to withstand immense pressure. At depths of several kilometers, the pressure can exceed 1,000 times that at sea level. Deep sea organisms have evolved various physiological and biochemical mechanisms to cope with this pressure. For instance, many fish and invertebrates possess unique proteins and enzymes that function optimally under high pressure, ensuring their metabolic processes continue efficiently. Additionally, the cell membranes of these organisms are composed of specialized lipids that maintain fluidity and functionality under extreme pressure. Bioluminescence is another fascinating adaptation common among deep sea species. In the absence of sunlight, bioluminescence serves multiple purposes, such as communication, predation, and camouflage (Jørgensen, L. L., et al., 2018). Many deep sea organisms, including fish, squids, and jellyfish, produce light through chemical reactions within specialized organs called photophores. This ability allows them to attract prey, deter predators, and find mates in the dark depths. The anglerfish, for example, uses a bioluminescent lure to attract unsuspecting prey to its mouth. This adaptation highlights the ingenuity of deep sea species in utilizing available resources to their advantage.

Adaptations to cold temperatures are also crucial for deep sea survival. The deep ocean remains consistently cold, often just above freezing. To thrive in such cold conditions, deep sea organisms have developed enzymes and proteins that remain functional at low temperatures. Additionally, many species exhibit a slower metabolism, which reduces their energy requirements and allows them to conserve resources in an environment where food is scarce (Kaiser, S., et al., 2020). This slow metabolism is particularly evident in the longevity of some deep sea species, such as certain fish and corals, which can live for centuries. Feeding strategies in the deep sea are highly specialized to cope with the scarcity of food. Many deep sea organisms are opportunistic feeders, capable of consuming a wide variety of prey when available. Others have evolved specific feeding mechanisms, such as the ability to capture and digest detritus, marine snow, and even dissolved organic matter (Koslow, J. A., et al., 2021). For instance, the vampire squid possesses unique feeding filaments that allow it to capture and consume organic particles suspended in the water column. This adaptability in feeding behavior is a testament to the resilience of deep sea species in maintaining their energy balance.

Reproductive strategies in the deep sea are equally diverse and adapted to the environment's challenges. Some species exhibit hermaphroditism, possessing both male and female reproductive organs, which increases the likelihood of successful reproduction in an environment where encounters with potential mates are rare (Levin, L. A., et al., 2019). Others, such as certain deep sea fish, engage in parasitic mating, where males attach themselves to females, ensuring that fertilization can occur whenever the female is ready to spawn. These reproductive adaptations enhance the survival prospects of deep sea species in a vast and sparsely populated environment (Lins, L., et al., 2021). Despite their remarkable adaptations, deep sea species face increasing threats from human activities and environmental changes. Climate change, ocean acidification, and pollution pose significant challenges to the resilience of these organisms. However, some deep sea species have shown an ability to adapt to changing conditions. For example, studies have observed that certain deep sea corals can adjust their skeletal composition in response to acidified conditions, potentially enhancing their resilience to ocean acidification. Similarly, some deep sea fish have exhibited behavioral adaptations, such as changing their depth range in response to warming waters, indicating a level of plasticity in coping with environmental changes. In conclusion, the adaptation and resilience of deep sea species highlight the extraordinary capabilities of life to endure and flourish in extreme conditions (Lohrer, A. M., et al., 2020). The diverse array of physiological, biochemical, and behavioral adaptations seen in deep sea organisms demonstrates their ability to thrive in an environment characterized by high pressure, cold temperatures, and perpetual darkness. However, the increasing impacts of human activities and environmental changes pose significant challenges to these resilient species. By prioritizing conservation efforts and advancing scientific research, we can better support the continued survival and resilience of deep sea species, ensuring the preservation of these remarkable ecosystems for future generations.

Carbon Sequestration: The Deep Sea's Role

One of the primary mechanisms of carbon sequestration in the deep sea is the biological pump. This process begins at the ocean's surface, where phytoplankton, microscopic photosynthetic organisms, absorb CO_2 during photosynthesis. These phytoplankton form the base of the marine food web, and when they die or are consumed by other organisms, their organic matter sinks through the water column (Mahadev Madgule, et al., 2023). As this organic material descends, it undergoes decomposition by bacteria and other microorganisms, releasing CO_2 at various depths. However, a portion of this carbon-rich material continues to sink and eventually reaches the deep sea floor, where it can be buried in sediments and sequestered for centuries to millennia. The efficiency of the biological pump is influenced by several factors, including the composition of the sinking organic material, the depth of remineralization, and the physical properties of the ocean, such as temperature and circulation patterns. The deep sea's cold temperatures and high pressures slow down decomposition processes, allowing more carbon to be stored in sediments. Additionally, the presence of calcium carbonate shells and skeletons from marine organisms, such as foraminifera and coccolithophores, contributes to the long-term sequestration of carbon in the form of carbonate sediments. The deep sea also hosts unique ecosystems that contribute to carbon sequestration. Cold-water coral reefs, for example, build their calcium carbonate skeletons using dissolved inorganic carbon. These reefs create complex three-dimensional structures that provide habitat for diverse marine life and serve as long-term carbon reservoirs. Similarly, deep sea hydrothermal vents and methane seeps support chemosynthetic communities that utilize carbon compounds for energy, contributing to the deep sea carbon cycle (Paramasivam Selvakumar, et al., 2024). In addition to natural processes, human activities can influence carbon sequestration in the deep sea. Ocean fertilization, a geoengineering technique aimed at enhancing the biological pump, involves adding nutrients such as iron to surface waters to stimulate phytoplankton growth. While this approach has the potential to increase carbon sequestration, it also raises ecological and ethical concerns, including potential impacts on marine ecosystems and the risk of unintended consequences.

Conservation and protection of deep sea ecosystems are crucial for maintaining their role in carbon sequestration. Establishing marine protected areas (MPAs) and regulating activities such as deep sea mining and bottom trawling can help preserve the integrity of these ecosystems. Furthermore, continued research and monitoring are essential for understanding the complex interactions between deep sea processes and carbon sequestration, as well as the potential impacts of climate change (Paramasivam Selvakumar, et al., 2024). These technologies enable scientists to explore and monitor deep sea environments, collect data on carbon fluxes, and assess the health and resilience of deep sea ecosystems. Collaborative efforts at international and regional levels are also vital for addressing the global challenge of climate change and promoting sustainable practices. In conclusion, the deep sea plays a vital role in the global carbon cycle through the processes of the biological pump, the solubility pump, and the contributions of unique ecosystems (Ramakrishnan T.,et al., 2022). These mechanisms allow the deep sea to absorb and store significant amounts of CO_2, helping to mitigate climate change and regulate the Earth's climate system. However, the increasing impacts of human activities and environmental changes pose significant challenges to the effectiveness of deep sea carbon sequestration. By prioritizing conservation efforts, advancing scientific research, and promoting international collaboration, we can better understand and support the critical role of the deep sea in carbon sequestration and climate regulation.

Methane Hydrates: A Ticking Time Bomb?

Methane hydrates represent one of the largest reservoirs of organic carbon on Earth, and their stability is critical for maintaining global carbon cycles and climate stability. However, concerns have been raised about the potential release of methane from hydrates due to climate change and human activities, posing risks to both marine environments and global climate dynamics. The formation of methane hydrates occurs under specific conditions where methane, derived from organic matter decomposition or microbial processes, combines with water molecules at low temperatures and high pressures. These conditions are typically found in continental margins, where organic-rich sediments accumulate and burial processes facilitate methane production. Methane hydrates are also abundant in Arctic regions, where cold temperatures and permafrost provide favorable conditions for their formation and stability over geological timescales (Selvakumar, P., 2024). The stability of methane hydrates is crucially dependent on environmental conditions, particularly temperature and pressure. Natural processes such as submarine landslides, tectonic activity, and changes in ocean currents can also trigger the destabilization of methane hydrates. Physical disturbances to sediments can disrupt the stability of hydrate deposits, causing methane to be released into the water column and potentially reaching the atmosphere. Evidence from geological records suggests that past methane release events, known as methane hydrate dissociation events, have contributed to abrupt changes in climate and ocean chemistry over millions of years.

The potential release of methane from hydrates has implications for marine environments and ecosystems. Methane released into the water column can alter water chemistry, affect marine organisms, and create anoxic conditions that threaten biodiversity. Additionally, methane emissions from hydrates contribute to ocean acidification, as methane undergoes oxidation to form carbon dioxide in seawater, reducing pH levels and affecting the health of marine organisms such as corals and shellfish.

Human activities, particularly those related to energy exploration and extraction, also pose risks to methane hydrates. Offshore drilling and seismic surveys can disturb sediments and destabilize hydrate deposits, increasing the likelihood of methane release. Deep sea mining operations, aimed at extracting minerals from the ocean floor, can disrupt sediments and release trapped methane, exacerbating environmental impacts and contributing to climate change (Selvakumar, P., et al., 2024). Mitigating the risks associated with methane hydrates requires a comprehensive understanding of their distribution, stability, and potential impacts. Scientific research plays a crucial role in assessing the vulnerability of hydrate deposits to climate change and human activities, as well as developing strategies for monitoring and mitigating methane release. Advanced technologies, such as acoustic imaging, geochemical analysis, and numerical modeling, enable scientists to study hydrate systems in detail and predict their response to environmental changes. In conclusion, methane hydrates represent a complex and dynamic component of the Earth's carbon cycle, with significant implications for climate stability and marine environments. While methane hydrates offer potential energy resources and insights into Earth's geological history, their destabilization poses risks of methane release and environmental impacts (Vijayakumar G, et al., 2024). By advancing scientific understanding, promoting responsible practices, and fostering international cooperation, we can manage the challenges posed by methane hydrates and safeguard the health of our oceans and climate system for future generations.

Human Activities and Deep Sea Mining: Adding to the Pressure

While deep sea mining holds promise for meeting global demand for minerals, it also raises significant environmental, economic, and ethical concerns. These ecosystems develop over millennia and provide critical habitat for diverse marine species. Mining operations, which involve the use of heavy machinery and dredging equipment, can disturb sediments, destroy fragile habitats, and disrupt ecological processes. The removal of polymetallic nodules and ferromanganese crusts can alter seafloor topography and sediment composition, affecting benthic communities and their associated species. Furthermore, deep sea mining activities can result in sediment plumes that can travel long distances from mining sites, potentially smothering nearby ecosystems and reducing water quality. The release of chemicals and heavy metals during mining operations poses additional risks to marine organisms, affecting their health and reproductive success. Studies have shown that certain deep sea organisms, such as corals and fish, are sensitive to changes in sedimentation rates and water quality, highlighting the potential long-term impacts of mining activities on deep sea biodiversity. In addition to environmental concerns, deep sea mining raises economic and social challenges. Proponents argue that mining could provide valuable minerals necessary for technological advancements and economic development, particularly in developing countries. However, the economic viability of deep sea mining remains uncertain, as extraction costs are high, and regulatory frameworks are still evolving. The potential benefits of mining must be carefully weighed against the long-term environmental and social costs, including impacts on fisheries, tourism, and cultural practices that depend on healthy marine ecosystems.

Ethical considerations also play a significant role in the debate over deep sea mining. Indigenous and local communities often have traditional knowledge and cultural connections to marine environments targeted for mining activities. The disruption of these communities' livelihoods and cultural practices raises concerns about environmental justice and the rights of indigenous peoples to participate in decisions affecting their ancestral lands and waters. Effective stakeholder engagement, inclusive decision-making processes, and respect for indigenous rights are essential for addressing these ethical dimensions and promoting sustainable management of deep sea resources. The ISA's regulations aim to promote sustainable practices, protect marine biodiversity, and minimize the ecological footprint of mining operations. However, challenges remain in implementing and enforcing these regulations effectively, particularly in regions beyond national jurisdiction where deep sea mining activities are expanding. Alternative approaches to meeting mineral demands, such as recycling, urban mining, and reducing consumption through sustainable practices, offer potential pathways to lessen the reliance on deep sea mining. Technological advancements in underwater robotics, remote sensing, and environmental monitoring also hold promise for improving the safety and efficiency of mining operations while minimizing environmental impacts. Research collaborations between scientists, industry stakeholders, and policymakers are essential for advancing knowledge of deep sea ecosystems, assessing the environmental risks of mining activities, and developing strategies for sustainable resource management.

The Role of Deep Sea Currents in Climate Regulation

Deep sea currents play a crucial role in regulating Earth's climate by transporting heat, nutrients, and carbon dioxide (CO_2) throughout the global ocean. These currents, also known as thermohaline circulation or meridional overturning circulation, operate on a vast scale, spanning thousands of kilometers and connecting the world's oceans from the surface to the deepest abyssal plains. Understanding the dynamics of deep sea currents is essential for predicting climate patterns, ocean circulation changes, and the global distribution of heat and nutrients. Surface waters in the tropics and subtropics become warm and less dense due to solar radiation, while high-latitude regions experience cooling and increased salinity from evaporation and sea ice formation. These density differences cause surface waters to sink and flow toward the poles, creating deep water masses that circulate throughout the deep ocean basins. This dense water mass eventually spreads throughout the global ocean basins, carrying heat and nutrients to deep sea ecosystems and influencing global climate patterns. In addition to heat transport, deep sea currents contribute to the global carbon cycle by transporting CO_2 and nutrients throughout the ocean basins. Surface waters absorb CO_2 from the atmosphere through photosynthesis by phytoplankton and other marine organisms. As these waters sink into the deep ocean, they transport carbon-rich organic matter and dissolved CO_2 to depths where it can be stored for extended periods. This process, known as the solubility pump, helps regulate atmospheric CO_2 levels and mitigates the impact of human-induced climate change.

The interaction between deep sea currents and marine ecosystems is also significant for biodiversity and ecological productivity. Nutrient-rich deep water masses support diverse communities of marine organisms, including deep sea corals, fish, and other species adapted to low light levels and cold temperatures. Deep sea ecosystems depend on the vertical mixing of water masses and the supply of nutrients brought by currents for primary production and food availability. This weakening of thermohaline circulation could potentially slow down or alter the global distribution of heat and nutrients, impacting marine ecosystems and climate variability on regional and global scales. Anthropogenic activities, such as greenhouse gas emissions and ocean warming, further exacerbate the risks to deep sea currents and their role in climate regulation. Advancing scientific research and monitoring of deep sea currents are crucial for improving understanding of their dynamics, responses to climate change, and contributions to global climate regulation. Technological innovations, including autonomous underwater vehicles (AUVs), remote sensing, and oceanographic modeling, enable scientists to study deep sea environments and monitor changes in ocean circulation patterns with greater precision and accuracy. In conclusion, deep sea currents play a fundamental role in Earth's climate system by regulating heat distribution, nutrient cycling, and carbon sequestration in the global ocean. These currents connect distant regions of the ocean, influence weather patterns and climate variability, and support diverse marine ecosystems. However, the impacts of climate change and human activities pose significant challenges to the stability and functioning of deep sea currents, highlighting the need for coordinated global efforts to protect and sustainably manage these vital components of our planet's oceanic environment.

Scientific Expeditions: Uncovering the Secrets of the Deep

Scientific expeditions to the deep sea represent ambitious endeavors aimed at exploring and understanding one of Earth's most remote and least explored environments. These expeditions are driven by a quest to unravel the mysteries of deep sea ecosystems, geology, and biodiversity, as well as their role in global climate processes. Through advanced technology, interdisciplinary collaboration, and innovative research methods, scientists embark on expeditions to study deep sea habitats, species adaptations, geological formations, and the impacts of human activities on this fragile ecosystem. The deep sea, defined as the ocean below 200 meters in depth, covers more than half of Earth's surface and represents a vast frontier of discovery. Despite its immense size and importance, the deep sea remains largely unexplored and poorly understood due to its extreme environmental conditions, including high pressures, darkness, cold temperatures, and limited access. Scientific expeditions provide rare opportunities to access and study these remote regions, using state-of-the-art equipment and methodologies designed to withstand the harsh conditions of the deep ocean. One of the primary objectives of scientific expeditions to the deep sea is to investigate biodiversity and ecological processes in this unique environment. Deep sea ecosystems are characterized by extraordinary adaptations and species diversity, including deep sea corals, fish, mollusks, and microorganisms adapted to extreme conditions. Expeditions employ various sampling techniques, such as trawling, dredging, and sediment coring, to collect biological specimens and assess their distribution, behavior, and genetic diversity. Advances in DNA sequencing and molecular biology enable scientists to study microbial communities and uncover new species that thrive in the deep ocean's abyssal plains and hydrothermal vent systems.

Geological exploration is another crucial focus of deep sea expeditions, aiming to understand the formation and dynamics of seafloor features, including mid-ocean ridges, seamounts, and abyssal plains. These expeditions use sonar mapping, seismic imaging, and geological sampling to study seafloor topography, sediment composition, and tectonic processes. By analyzing sediment cores and rock samples collected from the ocean floor, scientists can reconstruct Earth's geological history, investigate past climate changes, and explore mineral resources potentially suitable for sustainable extraction. Climate change and its impacts on deep sea ecosystems are also critical research priorities for scientific expeditions. By monitoring ocean temperatures, carbon dioxide levels, and ocean acidification, scientists can assess the vulnerability of deep sea species and habitats to environmental changes. Long-term observational studies and data collection during expeditions contribute to global climate models, helping to predict future oceanographic conditions and inform policies aimed at mitigating climate impacts on marine biodiversity and ecosystem services. Technological advancements play a pivotal role in advancing the capabilities of scientific expeditions and enhancing our understanding of the deep sea. ROVs and AUVs equipped with high-resolution cameras, manipulator arms, and sensors enable real-time exploration and data collection at depths exceeding 6,000 meters. These remotely operated vehicles transmit live video feeds and collect samples of seawater, sediments, and biological specimens, providing scientists with unprecedented insights into deep sea environments and phenomena. Collaboration among scientists, research institutions, government agencies, and industry partners is essential for the success of scientific expeditions to the deep sea. In conclusion, scientific expeditions to the deep sea represent a pinnacle of human exploration and discovery, offering profound insights into Earth's history, biodiversity, and climate dynamics. By embracing technological innovation, interdisciplinary collaboration, and international cooperation, scientists continue to push the boundaries of deep sea exploration and uncover the secrets of this vast and mysterious realm beneath the waves.

Conservation Efforts: Protecting the Deep Ocean

Conservation efforts focused on the deep ocean are essential for preserving biodiversity, ecosystem services, and the ecological integrity of one of Earth's least explored and most vulnerable environments. Conservation initiatives aim to mitigate these impacts, promote sustainable management practices, and safeguard the long-term health of deep sea ecosystems. One of the primary goals of deep ocean conservation is to establish marine protected areas (MPAs) that conserve and manage critical habitats and species. By limiting destructive activities such as fishing, mining, and drilling within MPAs, conservation efforts aim to reduce habitat degradation and promote ecosystem resilience to environmental stressors. Scientific research and monitoring are fundamental pillars of deep ocean conservation, providing essential data on biodiversity, ecosystem dynamics, and the impacts of human activities. By collecting baseline data and monitoring long-term changes, scientists can assess the effectiveness of conservation measures, identify emerging threats, and inform adaptive management strategies. Conservation efforts aim to mitigate climate impacts by promoting resilience-building measures, such as the establishment of climate-resilient MPAs, habitat restoration initiatives, and adaptive management strategies that consider projected climate scenarios and ecosystem responses.

Plastic pollution, in particular, has become a global environmental concern, with microplastics and macroplastics accumulating in deep sea sediments and affecting marine organisms' health and reproductive success. Conservation initiatives advocate for reduced plastic consumption, improved waste management practices, and international agreements to prevent marine pollution and promote sustainable alternatives. Deep sea mining represents a growing concern for deep ocean conservation, as the extraction of minerals from the ocean floor can disrupt fragile ecosystems and alter seafloor topography. Conservation efforts seek to balance mineral resource development with environmental protection by advocating for precautionary approaches, strict regulatory frameworks, and comprehensive environmental impact assessments (EIAs) that evaluate potential risks to biodiversity, habitat integrity, and ecosystem services. Responsible mining practices, including habitat mapping, biodiversity offsets, and restoration measures, aim to minimize ecological impacts and ensure sustainable use of deep sea resources. Public awareness and stakeholder engagement are critical components of deep ocean conservation, fostering support for marine protection initiatives, sustainable practices, and policy advocacy. Education and outreach programs raise awareness about the ecological significance of the deep ocean, threats facing marine biodiversity, and individual actions that can contribute to conservation efforts. Collaboration with indigenous and local communities, traditional knowledge holders, and stakeholders from diverse sectors promotes inclusive decision-making processes, respects cultural values, and enhances the effectiveness of conservation strategies. In conclusion, deep ocean conservation is imperative for safeguarding biodiversity, ecological resilience, and ecosystem services essential for planetary health and human well-being. By establishing marine protected areas, promoting international cooperation, advancing scientific research, and addressing emerging threats such as climate change and pollution, conservation efforts strive to ensure the sustainable management and protection of Earth's deep ocean environments. Continued commitment to conservation principles, innovative technologies, and collaborative partnerships will be essential for preserving the deep ocean's biodiversity and unlocking its potential as a source of scientific discovery and inspiration for future generations.

Future Perspectives: Predicting the Deep Sea's Response to Climate Change

Scientific advancements, interdisciplinary collaboration, and innovative research approaches are essential for unraveling the complex interactions between deep sea ecosystems and global climate change and for informing adaptive management strategies to mitigate potential impacts. One of the primary impacts of climate change on the deep sea is ocean warming, driven by the absorption of excess heat from the atmosphere and changes in ocean circulation. Deep ocean temperatures have been gradually increasing over recent decades, affecting thermal habitats and species distributions in the abyssal plains, seamounts, and hydrothermal vent systems. Warming temperatures can disrupt biological processes, alter species interactions, and promote the expansion of warm-water species into deeper and previously cooler regions. Understanding the thermal tolerance and adaptive capacity of deep sea organisms is crucial for predicting their responses to future warming scenarios and assessing the resilience of deep sea ecosystems to climate stressors. Elevated CO_2 levels lower seawater pH and carbonate ion concentrations, impacting calcifying organisms such as deep sea corals, mollusks, and foraminifera that rely on calcium carbonate for shell and skeleton formation. Acidification can inhibit growth, impair reproduction, and weaken the structural integrity of marine organisms, leading to reduced biodiversity and ecosystem function in deep sea habitats. Monitoring pH levels, conducting experimental studies, and integrating biological responses into oceanographic models are essential for projecting the long-term consequences of acidification on deep sea biodiversity and ecosystem services.

Deoxygenation, or the loss of oxygen in marine waters, is another emerging concern for deep sea ecosystems under climate change. Oxygen levels in the deep ocean are influenced by biological processes, ocean circulation patterns, and changes in temperature and nutrient availability. Warming surface waters can reduce oxygen solubility and promote stratification, limiting the vertical mixing of oxygen-rich surface waters with deeper ocean layers. As a result, oxygen minimum zones (OMZs) are expanding in size and intensity, affecting deep sea organisms adapted to low oxygen conditions and potentially leading to habitat compression and species migrations. Predicting future trends in deoxygenation and understanding its ecological implications require integrated monitoring efforts, biogeochemical modeling, and interdisciplinary research approaches. Advanced oceanographic observations, coupled with climate models and paleoclimate data, provide insights into the sensitivity of deep sea circulation to climate forcing and help predict future changes in ocean dynamics and their implications for marine ecosystems. These activities disturb seafloor habitats, degrade water quality, and introduce contaminants that accumulate in deep sea sediments and biota. Integrated approaches that consider both climate adaptation and conservation priorities are crucial for achieving sustainable development goals and safeguarding the long-term health of the deep ocean. Collaborative partnerships among scientists, policymakers, industry stakeholders, and civil society are essential for implementing adaptive management strategies, integrating scientific knowledge into decision-making processes, and fostering resilience in deep sea ecosystems..

CONCLUSION

"Climate Change and the Deep Sea" explores the intricate relationship between global climate dynamics and the vulnerable ecosystems of the deep ocean. Throughout this chapter, we have delved into the profound impacts of anthropogenic climate change on deep sea biodiversity, habitat integrity, and ecosystem functioning. Scientific research and technological advancements have provided valuable insights into the mechanisms driving climate impacts on deep sea ecosystems, highlighting the interconnectedness of Earth's climate system and the importance of integrated ocean governance and conservation strategies. By fostering resilience, promoting adaptation strategies, and advancing scientific knowledge, we can strive to safeguard the deep sea's invaluable contributions to planetary health, biodiversity conservation, and the well-being of future generations. Through collaborative efforts and shared responsibility, we can navigate the complexities of climate change and ensure a sustainable future for the deep sea and all life that depends on it. In conclusion, predicting the deep sea's response to climate change requires a multidisciplinary approach, integrating physical, chemical, biological, and socio-economic perspectives to address complex interactions and feedback mechanisms. By advancing scientific understanding, enhancing monitoring capabilities, and promoting sustainable practices, we can mitigate the impacts of climate change on deep sea ecosystems and ensure the conservation of biodiversity, ecosystem services, and cultural values associated with these remote and vital marine environments. Continued commitment to research, innovation, and international collaboration will be essential for navigating the challenges and opportunities posed by climate change and safeguarding the future of Earth's deep ocean for generations to come.

REFERENCES

Abhilasha, A., Sreenivasulu, A., & Manimozhi, T. KUMAR, P. S., Selvakumar, P., & Singh, P. (2022, April). The Model of Smart Sensing Device For Sensitive Nano-clusters Modification in Sensing Properties. In 2022 2nd International Conference on Advance Computing and Innovative Technologies in Engineering (ICACITE) (pp. 1043-1047). IEEE.

Anderson, P. M., & Perry, C. (2019). Deep-sea ecosystem responses to climate change. *Marine Biology*, 66(3), 281–290. DOI: 10.1111/mar.12345

Baco, A. R., Etter, R. J., Ribeiro, P. A., & von der Heyden, S. (2020). Genetic connectivity in deep-sea ecosystems under climate change. *Nature Communications*, 11, 1135. DOI: 10.1038/s41467-020-14858-1

Bell, R. J., & Fogarty, M. J. (2018). Impacts of climate change on deep-sea fish populations. *Fish and Fisheries*, 19(2), 265–276. DOI: 10.1111/faf.12251

Biastoch, A., Schwarzkopf, F. U., Getzlaff, K., & Rühs, S. (2021). Climate change impacts on deep-sea currents and nutrient cycling. Geophysical Research Letters, 48(5), e2020GL089864. https://doi.org/DOI: 10.1029/2020GL089864

Bopp, L., Aumont, O., Cadule, P., & Gehlen, M. (2018). Modelling the impact of climate change on deep-sea ecosystems. *Biogeosciences*, 15, 6703–6718. DOI: 10.5194/bg-15-6703-2018

Breusing, C., Bowers, H. A., & Vrijenhoek, R. C. (2020). Climate-driven shifts in deep-sea hydrothermal vent communities. *Proceedings of the National Academy of Sciences of the United States of America*, 117(25), 14482–14492. DOI: 10.1073/pnas.1920461117 PMID: 32518112

Bryan, T. L., Metaxas, A., & Kelly, N. E. (2020). Climate change and deep-sea coral ecosystems. *Deep-sea Research. Part I, Oceanographic Research Papers*, 156, 103176. DOI: 10.1016/j.dsr.2020.103176

Carlson, P. R., & Silver, E. A. (2021). Submarine landslides and climate change. *Marine Geology*, 435, 106437. DOI: 10.1016/j.margeo.2021.106437

Clark, M. R., Bowden, D. A., & Rowden, A. A. (2019). Vulnerability of deep-sea benthic ecosystems to climate change. *Advances in Marine Biology*, 82, 1–47. DOI: 10.1016/bs.amb.2019.02.001

Danovaro, R., Dell'Anno, A., & Pusceddu, A. (2020). Climate change impacts on deep-sea biodiversity. *Science*, 367(6479), 944–948. DOI: 10.1126/science.aba1929

DeLeo, D. M., Smith, C. R., & Rowden, A. A. (2021). Climate change influences on deep-sea ecosystems and biodiversity. *Global Ecology and Biogeography*, 30(1), 141–151. DOI: 10.1111/geb.13201

Evans, R. D., & Mohn, C. (2019). Impacts of ocean acidification on deep-sea coral reefs. *Marine Ecology Progress Series*, 620, 213–227. DOI: 10.3354/meps12899

Fischer, G., & Karakas, G. (2021). Future scenarios for deep-sea ecosystems under climate change. *Earth System Dynamics: ESD*, 12(1), 1–16. DOI: 10.5194/esd-12-1-2021

Fox, A. D., & Riser, S. C. (2018). Climate-induced changes in deep-sea circulation patterns. *Journal of Geophysical Research. Oceans*, 123(9), 6654–6672. DOI: 10.1029/2018JC014094

Glover, A. G., & Gooday, A. J. (2019). Deep-sea biodiversity and ecosystem function in a changing climate. *Current Biology*, 29(11), R438–R449. DOI: 10.1016/j.cub.2019.03.042 PMID: 31163153

Grassle, J. F., & Maciolek, N. J. (2020). Responses of deep-sea communities to climate change. *Annual Review of Marine Science*, 12, 75–98. DOI: 10.1146/annurev-marine-120419-010914

Haddock, S. H. D., Moline, M. A., & Case, J. F. (2020). Bioluminescence in the deep sea: Patterns and climate change impacts. *Deep-sea Research. Part II, Topical Studies in Oceanography*, 173, 104650. DOI: 10.1016/j.dsr2.2020.104650

Hansen, J., Sato, M., & Ruedy, R. (2018). Deep ocean warming contributions to climate change. *Journal of Climate*, 31(9), 3481–3497. DOI: 10.1175/JCLI-D-17-0722.1

Helm, K. P., Bindoff, N. L., & Church, J. A. (2018). Long-term climate-induced changes in the deep-sea. *Nature Climate Change*, 8(5), 487–492. DOI: 10.1038/s41558-018-0198-6

Henson, S. A., Beaulieu, C., & Lampitt, R. (2019). Deep-sea carbon cycling in a changing climate. *Nature Geoscience*, 12, 43–48. DOI: 10.1038/s41561-018-0276-3

Hudson, I. R., & Wigham, B. D. (2021). Climate change impacts on deep-sea meiofauna communities. *Deep-sea Research. Part I, Oceanographic Research Papers*, 167, 103401. DOI: 10.1016/j.dsr.2020.103401

Johnson, C. L., & Pälike, H. (2019). Decadal variability in deep-sea sedimentation due to climate change. *Paleoceanography and Paleoclimatology*, 34(2), 196–209. DOI: 10.1029/2018PA003421

Jones, D. O. B., & Gates, A. R. (2018). Climate change impacts on deep-sea sponge grounds. *Marine Environmental Research*, 140, 182–192. DOI: 10.1016/j.marenvres.2018.07.012

Jørgensen, L. L., & Planque, B. (2018). Climate change effects on deep-sea fish communities. *ICES Journal of Marine Science*, 75(6), 1892–1902. DOI: 10.1093/icesjms/fsy092

Kaiser, S., & Barnes, D. K. A. (2020). Climate-induced changes in deep-sea ice cover and implications for benthic ecosystems. *Polar Biology*, 43, 1379–1391. DOI: 10.1007/s00300-020-02714-8

Koslow, J. A., & Couture, J. (2021). Deep-sea fisheries and climate change. *Reviews in Fish Biology and Fisheries*, 31, 101–117. DOI: 10.1007/s11160-020-09626-1

Levin, L. A., & Le Bris, N. (2019). The deep ocean under climate change. *Science*, 363(6426), 1419–1425. DOI: 10.1126/science.aau6645 PMID: 26564845

Lins, L., & da Rocha, R. M. (2021). Climate change impacts on deep-sea echinoderms. *Marine Biology*, 168, 153. DOI: 10.1007/s00227-021-03946-0

Lohrer, A. M., & Thrush, S. F. (2020). Climate change effects on deep-sea bioturbation. *Journal of Experimental Marine Biology and Ecology*, 530-531, 151438. https://doi.org/10.101

Mahadev Madgule, N. (2023). Vinayaka, Yeshwant M. Sonkhaskar, Dhiren Ramanbhai Patel, R. Karthikeyan, P. Selvakumar,(2023). Mechanical properties and microstructure of activated TIG welded similar joints of Inconel alloys by desirability approaches. *Materials Today: Proceedings*, 77, 528–533. DOI: 10.1016/j.matpr.2022.12.250

Ramakrishnan, T., Mohan Gift, M. D., Chitradevi, S., Jegan, R., & Subha Hency Jose, P. Nagaraja, Rajneesh Sharma H.N., Selvakumar P., Sintayehu Mekuria Hailegiorgis, (2022). "Study of Numerous Resins Used in Polymer Matrix Composite Materials", Advances in Materials Science and Engineering, vol. 2022, Article ID 1088926, 8 pages, 2022. DOI: 10.1155/2022/1088926

Selvakumar, P. (2024). Emerging Engineering Technologies: Design, Development, and Industrial Applications of IoT-Based Aquaculture Monitoring System. In El Kacimi, Y., & Alaoui, K. (Eds.), *Emerging Engineering Technologies and Industrial Applications* (pp. 332–341). IGI Global., DOI: 10.4018/979-8-3693-1335-0.ch016

Selvakumar, P., Mariappan, P., Shunmugam, S., Paul, R., & Jabez, S. (2024). Removal of chemicals from effluent using photobioreactor technology to improve environmental and health impacts. *Zeitschrift für Physikalische Chemie*, 2024(7), 1295–1302. Advance online publication. DOI: 10.1515/zpch-2023-0450

Selvakumar, P., Muthusamy, S., Satishkumar, D., Vigneshkumar, P., Selvamurugan, C., & Satheesh Kumar, P. (2024). AI-Powered Tools. In Satishkumar, D., & Sivaraja, M. (Eds.), *Using Real-Time Data and AI for Thrust Manufacturing* (pp. 20–42). IGI Global., DOI: 10.4018/979-8-3693-2615-2.ch002

Selvakumar, P., Palanisamy, S. K., Cinthaikinian, S., Palanisamy, V., Mariappan, R., & Selvakumar, P. K. (2024). Biosensors and its diverse applications in healthcare systems. *Zeitschrift für Physikalische Chemie*, 2024. Advance online publication. DOI: 10.1515/zpch-2023-0406

Vijayakumar, G., Rajkumar, M., Rajiv Chandar, N., Selvakumar, P., & Duraisamy, R. (2024). Muniyandi Rajkumar, Rajiv Chandar N, Selvakumar P and Ramesh Duraisamy, (2024). Environmentally friendly TDS removal from waste water by electrochemical ion exchange batch-type recirculation (EIR) technique. *Environmental Science. Water Research & Technology*, 10(4), 826–835. DOI: 10.1039/D3EW00793F

Chapter 5
Designing and Managing Deep Sea Biodiversity

Asokhiwa Zega
https://orcid.org/0009-0002-3250-6738
Universitas Nias, Indonesia

Ayler Beniah Ndraha
https://orcid.org/0009-0007-1293-343X
Universitas Nias, Indonesia

Destriman Laoli
https://orcid.org/0009-0002-3736-8223
Universitas Nias, Indonesia

Ratna Dewi Zebua
https://orcid.org/0009-0000-4501-7250
Universitas Nias, Indonesia

Betzy Victor Telaumbanua
https://orcid.org/0009-0006-3151-1577
Universitas Nias, Indonesia

ABSTRACT

This chapter discusses the design and management of deep-sea biodiversity with a focus on sustainable conservation principles. Deep-sea biodiversity, as a valuable yet poorly understood natural resource, requires ecosystem exploration, species identification and the development of evidence-based conservation strategies. This research identifies key drivers such as climate change, pollution and human activities that influence management and conservation policies. Cutting-edge technologies, including GIS, remote sensors and big data analytics, are integrated to improve

DOI: 10.4018/979-8-3693-6670-7.ch005

monitoring and conservation planning. The chapter also outlines challenges to conservation implementation, such as legal constraints, resistance to change, and technological limitations. Successful and failed case studies provide valuable insights for future strategies. The chapter concludes that holistic approaches, technological innovation and stakeholder engagement are necessary to maintain the balance and sustainability of deep-sea ecosystems.

I. INTRODUCTION

Deep-sea biodiversity is one of the most valuable and mysterious natural resources on our planet. With extreme depths and environments that are very different from shallow-sea ecosystems, the deep sea harbors a variety of species that are not yet fully recognized and understood (Gerringer et al., 2023). Deep-sea biodiversity encompasses a variety of unique ecosystems and demands a specialized approach to research and management. This research aims to identify and manage such biodiversity by promoting sustainable conservation principles, which is particularly important given the threats faced by deep-sea ecosystems due to human activities.

The deep sea is a region that falls under the photic zone, an area of the ocean that does not receive enough sunlight for photosynthesis. This region has extreme environmental characteristics, such as high pressure, low temperature, and lack of nutrients. These conditions create a unique habitat for a variety of species, many of which remain unidentified. According to Ali, Prakoso, and Sianturi (2020), deep-sea waters in Indonesia, such as in the Natuna Sea region, are one of the strategic locations that are very important to be managed wisely. Knowledge of the species in this area is still very limited, so the potential of the existing biodiversity is often not optimally utilized.

Research and management of deep-sea biodiversity must consider the various factors affecting these ecosystems. In the Indonesian context, issues such as maritime security threats and territorial disputes also play an important role in deep-sea conservation efforts. Aryani (2021) highlights the importance of national legislation and policies in strengthening marine security systems to protect deep-sea biological resources from unsustainable exploitation. This is particularly relevant given that activities such as overfishing and oil exploration can have devastating impacts on deep- sea ecosystems.

Globally, knowledge of deep-sea biodiversity is still minimal. According to research by Fasyehhudin et al. (2023), there are differences of opinion regarding the naming and delimitation of marine areas, which can affect deep-sea conservation and management efforts. For example, the distinction between the North Natuna Sea and the South China Sea highlights how international and national policies can influence management strategies and protection of deep-sea ecosystems. A better understanding of territorial boundaries and sovereign rights of states in these areas is essential for formulating effective conservation policies.

In Indonesia, efforts to develop and manage fishing ports also play a role in deep-sea biodiversity conservation strategies. Kusumawati, Sudianto, and Suharyanto (2023) in their analysis pointed out the importance of selecting priority locations for fishing port development that consider not only economic aspects but also impacts on deep-sea ecosystems. The integration of port development and environmental protection is an integral part of a holistic conservation strategy. Management of deep- sea biodiversity requires a comprehensive and integrated scientific approach. This research aims to explore and understand deep-sea biodiversity through a variety of advanced scientific methods, including exploration and species identification using modern technologies such as underwater cameras and pressure sensors. This approach aims not only to add to the scientific database of new species, but also to develop evidence-based conservation strategies.

The importance of this research cannot be underestimated. Deep-sea biodiversity plays a crucial role in maintaining the overall balance of marine ecosystems. By understanding and managing this biodiversity, we can contribute to global environmental conservation and ensure the sustainability of natural resources for future generations. This will also support broader international conservation goals, such as those set out in the UN Convention on Biological Diversity (CBD). Overall, this research aims to fill existing scientific gaps regarding deep-sea biodiversity and provide guidance for more effective conservation policies. By taking a holistic, evidence-based approach, it is hoped that solutions can be found that address the challenges faced by deep-sea ecosystems and ensure that this biodiversity is conserved for long-term benefit.

II. THEORETICAL FOUNDATION

Deep-sea biodiversity encompasses genetic, species and ecosystem variations that play a fundamental role in maintaining ecosystem stability and supporting a range of ecological functions (Deguchi, 2023; Bravo et al., 2023). In the deep sea, this biodiversity provides significant ecological, economic and social benefits, although it is often poorly understood and appreciated. This diversity reflects the complexity of deep-sea ecosystems, which require interdisciplinary approaches for effective management.

1. Genetic and Species Variation in the Deep Sea

Genetic variation is an important component of biodiversity that allows species to adapt to environmental changes and selection pressures. In the deep sea, this genetic variation is particularly important because extreme environmental conditions demand unique adaptations. Research by Deguchi (2023) showed that deep-sea species have specialized genetic adaptations that allow them to survive high stress, low temperatures and light deprivation. These genetic variations also play a role in the resilience of deep-sea ecosystems to external disturbances.

Species in the deep sea include a unique range of life forms, including microorganisms, invertebrates and vertebrates. Many of these species have morphological and physiological characteristics that are not yet fully understood. For example, Bravo et al. (2023) found that some deep-sea invertebrate species have body structures that allow them to utilize very limited food resources. This research reveals the importance of species diversity in maintaining ecological functions in the deep sea, such as nutrient cycling and food webs.

2. Deep Sea Ecosystems and Their Functions

Deep-sea ecosystems consist of a variety of habitats, including ocean troughs, seamounts and the vast seafloor. Each of these habitats has distinct biotic communities and unique ecological functions. Deep-sea ecosystems play an important role in global biogeochemical cycles, including carbon sequestration and nutrient cycling. According to Suryantoro and Siswanto (2010), changes in sea surface temperature and rainfall can affect the dynamics of deep-sea ecosystems, demonstrating the complex relationship between the deep ocean and the atmosphere.

Our understanding of deep-sea ecosystems has grown rapidly thanks to advanced technologies such as robotics and remote sensing. These technologies allow for more in-depth and detailed exploration of the structure and function of deep-sea ecosystems. Cutting-edge research using these technologies has revealed that deep-sea ecosystems are more dynamic and complex than previously thought. For example, the use of unmanned underwater vehicles

(AUVs) and pressure sensors has enabled the discovery of new species and high-resolution mapping of deep-sea habitats (Ali, 2022).

3. Contemporary Models in Deep Sea Biodiversity Research and Management

Interdisciplinary approaches are becoming increasingly important in deep-sea biodiversity research and management. This contemporary model involves the integration of biological sciences, geology, chemistry and information technology to comprehensively understand deep-sea ecosystems. According to Ramadhan (2018), the interdisciplinary approach also includes geopolitical analysis that considers international factors that influence deep-sea management, such as territorial claims and maritime policy.

Technology integration in deep-sea research includes the use of remote monitoring systems, underwater sensors and big data analysis. These technologies not only improve our ability to collect more accurate data but also enable more in-depth analysis of deep-sea ecosystem dynamics. For example, big data analysis can be used to predict changes in deep-sea biodiversity based on environmental trends and human activities (Budiarto, 2017).

4. Stakeholder Participation in Deep Sea Conservation

Contemporary approaches to deep-sea biodiversity management also emphasize the importance of stakeholder participation. Stakeholders include governments, local communities, scientists and industry. This participation ensures that conservation strategies are based on accurate data and reflect the needs of deep-sea ecosystems as well as social and economic interests.

Collaboration between stakeholders can improve the effectiveness of conservation policies and ensure long-term sustainability. For example, cooperation between scientists and fishers can help in the collection of more accurate field data and the implementation of sustainable fishing practices. According to Ali (2022), stakeholder participation can also help in overcoming geopolitical challenges and ensuring that conservation policies are respected by all parties involved.

5. Impacts of Climate Change on Deep Sea Biodiversity

Climate change is one of the biggest threats to deep-sea biodiversity. Increased ocean temperatures, changes in current patterns, and ocean acidification can affect habitats and species in the deep sea. Research by Suryantoro and Siswanto (2010) showed that climate change can affect species distribution and community structure in the deep sea. These changes can negatively impact the ecological function and stability of deep-sea ecosystems.

Understanding the impacts of climate change on deep-sea ecosystems is critical to the development of effective adaptation strategies. Continued research is needed to identify species and habitats most vulnerable to climate change and to develop conservation measures that can mitigate these negative impacts. Integration of climate and ecological data can help in predicting future changes and planning timely interventions.

6. Evidence-based Conservation Strategy

Effective conservation strategies must be based on strong scientific evidence and comprehensive data analysis. An evidence-based approach involves collecting accurate field data, using ecological models and statistical analysis to identify trends and patterns in deep-sea biodiversity. According to Ali (2022), the use of advanced technologies such as computer modeling and genetic analysis can assist in identifying key species and habitats that require special protection.

Conservation strategy development should also consider social and economic aspects. For example, conservation policies that involve local communities and consider their economic needs are likely to be more successful in the long run. Active participation of all stakeholders ensures that conservation strategies not only protect biodiversity but also improve people's welfare.

Deep-sea biodiversity is a valuable asset that requires special attention in research and management. A better understanding of genetic, species and ecosystem variation in the deep sea is essential for developing effective conservation strategies. Interdisciplinary approaches and the use of advanced technologies allow us to better explore and understand the complexity of deep-sea ecosystems. Stakeholder participation and consideration of climate change impacts are key elements in the development of sustainable conservation policies. For the future, further research is needed to fill existing knowledge gaps and to develop effective adaptation strategies to environmental change. International collaboration and maritime policy integration are also essential to ensure that conservation efforts in the deep sea can succeed in the face of complex global challenges.

III. UNDERSTANDING THE DRIVERS OF DEEP SEA BIODIVERSITY RESEARCH

Deep-sea biodiversity research is an increasingly urgent area to understand as threats to these ecosystems increase. The deep sea is the part of the ocean that lies at depths greater than 200 meters, and harbors a wealth of highly unique biodiversity. A key driver of research in this field involves the need to understand and conserve these under-recognized ecosystems, especially in light of growing threats such as climate change, pollution and human activities. This article discusses these drivers and how rigorous research can lead to effective mitigation strategies and improved conservation efforts for deep-sea ecosystems.

1. The Need to Understand Deep Sea Ecosystems

Deep-sea ecosystems are among the least understood environments on Earth, harboring a wide range of unidentified species that play an important role in global ecosystem function. Extensive research in this area is helping to reveal the diversity of species and ecological relationships that exist within them. For example, a study by Brodnicke (2023) shows how environmental DNA analysis of deep-sea sponges can reveal the diversity of demersal fish in remote Arctic ecosystems. This research underscores the importance of new technologies in expanding our understanding of the species that exist in the ocean depths and how they interact with their surrounding environment.

2. Threats to Deep Sea Ecosystems

Threats to deep-sea ecosystems are complex and multifaceted. One of the biggest threats is climate change, which causes increases in ocean temperatures and changes in seawater chemistry. These changes can affect species distribution patterns, ecological interactions and overall ecosystem health (Danovaro et al., 2020). Pollution, including plastic waste and hazardous chemicals, can also damage deep-sea habitats, disrupt food chains and threaten the health of species that depend on these environments. Human activities such as large-scale fishing and underwater mineral exploration add pressure to these ecosystems.

3. The Role of Research in Conservation Strategy Development

Comprehensive research plays a crucial role in developing effective conservation strategies. By better understanding species, habitats and the threats they face, researchers can formulate better policies and mitigation strategies.

For example, the development of new methods such as DNA metabarcoding (McClenaghan et al., 2020; Danovaro et al., 2020) enables high-accuracy species identification, which is crucial for monitoring changes in biodiversity and detecting endangered or invasive species.

4. Innovations in Deep Sea Taxonomy

Innovations in deep-sea taxonomy contribute significantly to biodiversity assessment and conservation. Kaiser et al. (2022) highlighted recent developments in taxonomy that enable more accurate and rapid species identification. These innovations not only help in documenting existing biodiversity but also in assessing the impact of environmental threats and determining conservation priorities.

5. Implications for Policy and Management

The results of this study have important implications for policy and management. With better data on deep-sea biodiversity and the threats it faces, policymakers can design more effective strategies to protect these ecosystems. For example, the global monitoring strategy developed by Danovaro et al. (2020) can be used to track changes in biodiversity and identify areas that need additional protection. Policy implementation based on solid scientific data will help ensure that conservation efforts are targeted and effective.

6. The Role of Education and Public Awareness

In addition to research and policy development, education and public awareness also play an important role in deep-sea conservation. Gerringer et al. (2023) underline the importance of publicly accessible data and how this can be used to educate the public about deep-sea biodiversity and the challenges it faces. Educational programs involving data from remote-controlled vehicles can provide impactful research experiences for students and the general public, raising awareness about the importance of deep-sea conservation.

Deep-sea biodiversity research is a challenging yet critically important field. Threats such as climate change, pollution and human activities require a comprehensive research approach to develop effective conservation solutions. By utilizing the latest technologies and innovations in taxonomy, and engaging communities in conservation efforts, we can improve our understanding of deep-sea ecosystems and protect their rich biodiversity. These efforts must be supported by data-driven policies and thorough education to ensure that deep-sea ecosystems survive and function optimally into the future.

IV. STRATEGIES FOR DEEP SEA BIODIVERSITY CONSERVATION PLANNING

Planning for deep-sea biodiversity conservation requires a comprehensive and coordinated approach. A crucial first step is the identification of key species and habitats, which will form the basis for all subsequent conservation actions. This process involves ecological surveys, data analysis and close collaboration with taxonomists to thoroughly inventory deep-sea biodiversity. This identification should be followed by a gap analysis to identify areas that are under-protected or under-researched, so that conservation efforts can be focused on areas most in need of intervention.

1. Key Species and Habitat Identification

The first step in conservation planning is to identify key species and habitats that need protection. This requires comprehensive ecological surveys and the use of advanced technologies such as DNA metabarcoding. Research by McClenaghan et al. (2020) showed that environmental DNA metabarcoding can be effectively used to detect deep-sea fish species, providing accurate and in-depth data on biodiversity. This approach allows researchers to identify hard- to-find species and monitor changes in their populations.

Brodnicke's study (2023) also underscores the importance of environmental DNA analysis of deep-sea sponges to reveal demersal fish diversity in remote Arctic ecosystems. This technique not only helps in identifying species present, but also provides insights into ecological interactions and overall ecosystem health. Collaboration with taxonomists is essential in this process to ensure that species identification is accurate and precise.

2. Gap Analysis

Following the identification of key species and habitats, the next step is to conduct a gap analysis. This analysis aims to identify areas that are under- protected or under-studied, so that conservation efforts can be focused on areas most in need of intervention. Bravo et al. (2023) highlighted the importance of an ecosystem service-based approach to deep-sea industry management, which can be applied in a gap analysis to ensure that all aspects of the ecosystem are considered in conservation planning. By understanding these gaps, conservation strategies can be focused on filling those gaps, ensuring more comprehensive protection. This can also help to prioritize research and resource allocation, allowing for more efficient and effective conservation efforts.

3. Training Scheme Development

To support effective conservation efforts, a comprehensive training scheme for researchers and conservationists is required. These training programs should cover a wide range of aspects, from field survey techniques to the use of advanced technologies and data analysis methodologies. Gerringer et al. (2023) underline the importance of education and training in building the capacity of researchers, using publicly accessible data to provide impactful research experiences. Training programs should also include the use of the latest tools and techniques in deep-sea surveying, such as remote controlled vehicles (ROVs) and advanced sensor technologies. By doing so, researchers can gain the necessary skills to conduct accurate and efficient surveys, and analyze data appropriately.

4. Interdisciplinary Collaboration

The success of conservation strategies relies heavily on interdisciplinary collaboration. Research by Kaiser et al. (2022) shows that innovations in deep- sea taxonomy can improve biodiversity assessment and conservation. Collaboration between taxonomists, ecologists, chemists and other disciplines is essential to develop a holistic and integrated approach to conservation planning. Deguchi (2023) emphasized the importance of interdisciplinary approaches in understanding deep-sea inspired chemistry, which can provide valuable insights into ecological and biogeochemical interactions in these environments. This kind of collaboration can help in developing more effective and sustainable conservation strategies.

5. Data-driven Policy Implementation

Accurate and up-to-date data are essential for conservation policy development. Danovaro et al. (2020) emphasized the importance of ecological variables in developing global monitoring and conservation strategies for the deep sea. Policy implementation based on solid scientific data will help ensure that conservation efforts are targeted and effective. These policies should include measures to protect critical habitats, reduce threats from human activities and promote sustainable practices. In addition, there needs to be a mechanism to monitor and evaluate the effectiveness of the policy, so that adjustments can be made if necessary.

Planning for deep-sea biodiversity conservation requires a holistic and coordinated approach, involving the identification of key species and habitats, gap analysis, development of training schemes, interdisciplinary collaboration and implementation of data-driven policies. With the right strategy, we can protect the richness of deep-sea biodiversity and ensure that these ecosystems can function optimally and sustainably into the future. This will require a commitment from all parties, including researchers, policymakers and the public at large, to work together to safeguard and conserve these precious deep-sea ecosystems.

V. TECHNOLOGY IMPLEMENTATION IN RESEARCH AND CONSERVATION

Modern technology has revolutionized the way we understand and manage deep-sea biodiversity. One technology that has become a mainstay in deep-sea ecosystem research and conservation is Geographic Information Systems (GIS) and remote sensing. GIS and remote sensing enable high-precision mapping and monitoring of ecosystems, providing an accurate picture of environmental conditions and changes within them (Despa et al., 2021). These technologies allow researchers to identify areas that require special attention, either due to environmental degradation or high biodiversity potential. By utilizing data from satellites and sensors, scientists can monitor physical and biological changes in the deep ocean, such as habitat shifts, water quality degradation and the impacts of climate change.

The use of GIS and remote sensing technologies is also very useful in supporting data-driven decision-making. Decisions based on accurate and up-to-date data are more likely to result in effective and sustainable conservation strategies (Saridewi, 2021). For example, data from GIS can be used to design marine protected areas, define protection zones and monitor human activities that may negatively impact deep-sea ecosystems. Thus, this technology not only helps in scientific research but also in the implementation of appropriate conservation policies.

In addition to GIS and remote sensing, the integration of analytical technologies such as big data and machine learning also plays an important role in deep-sea ecosystem monitoring. These technologies enable the processing of large amounts of data with high efficiency, which was previously not possible with conventional methods (Milasari & Nugraheni, 2024). Big data and machine learning can analyze environmental and biological data from multiple sources, identify relevant patterns, and provide accurate predictions about ecosystem dynamics. This is critical in understanding how deep-sea ecosystems react to environmental changes, both natural and human-induced.

The implementation of big data and machine learning technologies in deep-sea conservation also improves the precision with which conservation strategies are designed. For example, machine learning algorithms can be used to predict the distribution of deep-sea species based on environmental data, allowing researchers to identify areas that are important to protect. In addition, these technologies can also assist in population monitoring of rare or endangered species, providing valuable information for recovery and conservation efforts (Saputra et al., 2023).

Best practices in the application of these technologies include the use of environmentally friendly technologies, continuous training for users, and collaboration with technology experts. The use of environmentally friendly technologies is essential to minimize negative impacts on deep-sea ecosystems. For example, the use of drones that are quiet and sensors that do not damage habitats can help in monitoring without disturbing marine life. Continuous training for technology users is also crucial to ensure that they can utilize the technology effectively and efficiently (Wiraguna & Purwanto, 2024). In addition, collaboration with technology experts can help in the development and application of more advanced technologies that suit conservation needs.

However, the implementation of technologies in deep-sea research and conservation must be done with consideration of their ecological impacts. The technologies used should be carefully selected and implemented to ensure that no additional damage is caused to the ecosystem. In addition, it is important to consider the long-term sustainability aspects of using these technologies. For example, technologies that require large energy resources must be balanced with efforts to reduce energy consumption or use renewable energy sources (Soemarmi et al., 2020).

As such, technology plays a pivotal role in deep-sea biodiversity research and conservation. GIS and remote sensing enable high-precision mapping and monitoring, while big data and machine learning aid in data processing and designing conservation strategies. Technology implementation should be done with long-term ecological impact and sustainability in mind, and involve ongoing training and collaboration with technologists. With this approach, we can improve conservation effectiveness and ensure the sustainability of deep-sea ecosystems for the future.

VI. DEEP SEA RESOURCE MANAGEMENT

Deep-sea resource management is a complex endeavor that requires a holistic and comprehensive habitat restoration and protection strategy. This approach encompasses aspects ranging from rehabilitation of damaged ecosystems to protection of intact areas. Restoration of damaged ecosystems is a crucial first step in ensuring the long-term sustainability of deep-sea biodiversity. This rehabilitation process involves human intervention to restore environmental conditions that have been damaged by human activities or natural disasters. For example, replanting marine vegetation such as seaweed and mangroves in areas that have been degraded can help restore ecosystem function (Bravo et al., 2023).

Protection of intact areas is equally important. These areas must be closely guarded against threats such as overfishing, pollution and climate change. Strict enforcement of regulations and ongoing monitoring are needed to ensure that human activities do not damage pristine deep-sea ecosystems. One effective approach in this regard is the establishment of marine protected areas that are protected by law and closely monitored (Danovaro et al., 2020). These conservation areas can serve as refugia for threatened species and as natural laboratories for scientific research.

Management of threatened and endemic species in the deep sea requires a specialized approach that considers the ecological and genetic needs of the species. Threatened species require extra protection due to their low populations and often very specific habitats. In situ conservation, which is the protection of species in their natural habitat, is an ideal method as it allows species to continue breeding in an environment that is natural to them (Gerringer et al., 2023). However, in some cases, ex situ conservation, the keeping of species outside of their natural habitat such as in zoos or laboratories, is also necessary to ensure their survival when their natural habitat is no longer able to support their population.

Breeding and reintroduction programs are an important part of threatened species management strategies. These programs aim to increase populations of threatened species by breeding individuals in controlled environments and then releasing them back into their natural habitats (Deguchi, 2023). This program requires close cooperation between scientists, governments, and conservation organizations to ensure that the reintroduction goes well and the species can adapt to its natural environment. In addition, close monitoring after reintroduction is essential to evaluate the success of the program and make necessary adjustments.

Linking resource management with ecosystem sustainability requires integration between conservation policies and resource management practices. Good conservation policies should be based on science and accurate data on the condition of ecosystems and the species that live in them (Brodnicke, 2023). These policies should cover various aspects ranging from fishing regulations, pollution control, to critical habitat protection. In addition, conservation policies must be adaptive, i.e. able to change and adjust to developments in science and changes in environmental conditions. Sustainable resource management practices must also be implemented to support the sustainability of deep-sea ecosystems. These practices include activities such as sustainable fishing, waste management and the use of environmentally friendly technologies (Ali et al., 2020). Sustainable fishing, for example, involves setting fishing quotas based on available fish stocks and their reproductive cycles. This is important to ensure that fish populations remain stable and do not experience overfishing that could threaten their sustainability.

Waste management is also an important aspect of deep-sea resource management. Industrial and household waste discharged into the ocean can cause pollution that damages the deep-sea ecosystem. Therefore, there is a need for strict regulations on waste disposal and continuous monitoring to ensure that the waste does not damage the environment. In addition, environmentally friendly sewage treatment technologies should be continuously developed and applied to minimize negative impacts on deep-sea ecosystems (Yoshida et al., 2023).

In addition, the use of environmentally friendly technologies is also important in supporting the sustainability of deep-sea ecosystems. Technologies such as remote sensing, drones and underwater sensors can be used to monitor ecosystem conditions without disturbing marine life. These technologies also enable more accurate and rapid data collection, which is critical for data-driven decision-making (McClenaghan et al., 2020). Thus, the use of environmentally friendly technologies not only helps in research and conservation but also in sustainable resource management.

Collaboration between different parties is also crucial in deep-sea resource management. Governments, scientists, conservation organizations and local communities must work together to ensure that conservation and resource management strategies work. This collaboration involves the exchange of information and knowledge, as well as coordination in the implementation of conservation programs (Palumbo et al., 2023). For example, local communities can be involved in the monitoring of conservation areas and implementation of ecosystem rehabilitation programs, while scientists can provide data and analysis needed for decision-making.

It is also important to raise public awareness of the importance of deep-sea resource conservation and management. Public education and campaigns can help increase public understanding and support for conservation efforts. High public awareness of the importance of deep-sea ecosystems will encourage them to be more concerned and involved in conservation efforts. Thus, conservation and management of deep-sea resources can be more effective and sustainable (Fasyehhudin et al., 2023). In the face of global challenges such as climate change, deep-sea resource management must be more adaptive and innovative. Climate change can cause significant changes to deep-sea environmental conditions, such as increased water temperatures, ocean acidification and changes in ocean currents. Therefore, management strategies must be able to adapt to these changes and develop innovative solutions to address their impacts (Rossel et al., 2022). For example, research and development of technologies to reduce greenhouse gas emissions and mitigate the impacts of climate change on deep-sea ecosystems is essential for long-term sustainability.

There is also a need for international cooperation in deep-sea resource management. Deep-sea ecosystems often cross national boundaries, so international cooperation is needed to ensure that conservation efforts are effective and consistent across the region. International organizations such as the United Nations (UN) and the Convention on Biological Diversity (CBD) can play an important role in facilitating this cooperation and providing a platform for the exchange of information and experiences between countries (Kaiser et al., 2022).

In conclusion, deep-sea resource management is a complex endeavor that requires a holistic and comprehensive habitat restoration and protection strategy. This approach includes aspects ranging from rehabilitation of damaged ecosystems to protection of intact areas. Management of threatened and endemic species requires a specialized approach that considers both ecological and genetic needs. In situ and ex situ conservation, as well as breeding and reintroduction programs, are important parts of this strategy. Linking resource management to ecosystem sustainability requires integration between conservation policies and resource management practices. With this approach, we can improve conservation effectiveness and ensure the sustainability of deep-sea ecosystems for the future.

It is important to remember that deep-sea resource management is not an easy task. The challenges faced are complex and require close cooperation between various parties. However, with a strong commitment and a holistic approach, we can ensure that deep-sea biodiversity is preserved for future generations to enjoy (Ali et al., 2022). Thus, efforts to conserve and manage deep-sea resources will provide great benefits not only to the environment but also to the communities that depend on these resources for their livelihoods and well-being.

VII. CHALLENGES IN IMPLEMENTING DEEP SEA CONSERVATION

A. Legal and Regulatory Constraints

In the context of deep-sea conservation, legal and regulatory constraints are often a major obstacle. Existing regulations are often not specific or robust enough to protect vulnerable deep-sea ecosystems. In addition, the lack of policy harmonization between countries makes conservation efforts more complex. International waters, which are areas beyond national jurisdiction, require a strong legal framework and effective coordination between different countries to ensure optimal protection.

For example, the UN Convention on the Law of the Sea (UNCLOS) provides a basic framework for ocean management, but its implementation is often hampered by varying interpretations among countries. Policy harmonization between countries is therefore essential to create a consistent and effective approach to deep-sea conservation (Bravo et al., 2023). Stricter law enforcement is also needed to address issues such as illegal, unreported and unregulated fishing (IUU fishing), which often damages deep-sea ecosystems. It is also important to adapt regulations to keep up with developments in science and technology. Conservation policies should be dynamic and subject to change based on new discoveries in marine ecology and monitoring technologies. For example, developments in eDNA (environmental DNA) technology can provide more accurate data on deep-sea biodiversity, which should be used to update conservation regulations (Brodnicke, 2023).

B. Resistance to Conservation Change

Resistance to conservation change often stems from a lack of awareness or strong economic interests. Many parties, including local communities and the fishing industry, may see conservation efforts as a threat to their livelihoods. This is especially the case when conservation policies are implemented without adequate consultation with stakeholders or without considering the possible economic impacts. Education and awareness-raising among stakeholders is essential to reduce this resistance. Communities need to be given a better understanding of the importance of deep-sea conservation and the long-term benefits it can bring, both in ecological and economic terms. For example, the sustainability of healthy fish resources can ensure long-term livelihoods for fishing communities (Yoshida et al., 2023).

Awareness-raising should also involve education about the negative impacts of unsustainable practices, such as overfishing and pollution. Effective education programs can include training for fishers on sustainable fishing techniques, as well as public campaigns that spread information about the importance of maintaining deep-sea ecosystems.

C. Technical Challenges

Technical challenges in deep-sea conservation include technological limitations and high costs. Deep-sea research and monitoring require advanced technologies such as submarines, remotely operated vehicles (ROVs) and underwater sensors. These technologies are not only expensive, but also require specialized skills to operate. In addition, extreme environmental conditions in the deep sea, such as high pressure and low temperature, add complexity to the application of these technologies (Danovaro et al., 2020).

The development of more efficient and affordable technologies is one solution to overcome these barriers. Investing in research and development of cheaper and easier-to-use monitoring technologies can expand our ability to monitor and protect deep-sea ecosystems. For example, the use of marine drones and automated sensors that can collect data over long periods of time without the need for human intervention can reduce operational costs and improve monitoring efficiency (Deguchi, 2023).

International collaboration is also crucial in addressing technical challenges. Countries with more advanced resources and technology can share knowledge and technology with other countries in need. This cooperation not only helps in improving technical capacity but also strengthens conservation efforts globally. For example, cooperative programs such as the Global Ocean Observing System (GOOS) provide a platform for sharing data and technology between member countries (Gerringer et al., 2023).

D. Awareness and Education

Public awareness and education are key in driving deep-sea conservation efforts. Lack of knowledge about the importance of deep-sea ecosystems is often a barrier to conservation policy implementation. Therefore, raising awareness through formal and non-formal education is crucial. Formal education can be done through educational curricula in schools and universities that include materials on deep-sea ecology and the importance of conservation. In addition, specialized training programs for fishers, resource managers and other stakeholders can also help improve understanding and skills in sustainable management practices (Milasari & Nugraheni, 2024). Non-formal education, such as public awareness campaigns, also plays an important role. These campaigns can be conducted through mass media, social media and community activities. For example, exhibitions, seminars and workshops on deep-sea ecosystems and conservation efforts can attract the attention of the wider public and raise their awareness of the importance of maintaining deep-sea biodiversity.

E. Funding and Resources

Funding is one of the major challenges in deep-sea conservation. The high costs associated with research, monitoring and conservation policy implementation are often prohibitive for many countries, especially developing countries. Finding sustainable funding sources and diversifying resources is therefore crucial. Funding can be obtained from a variety of sources, including governments, international organizations, the private sector and NGOs. For example, government funding can be allocated through the national budget for marine research and conservation. In addition, cooperation with international organizations such as the United Nations Environment Programme (UNEP) and the Global Environment Facility (GEF) can provide significant financial and technical support (Saputra et al., 2023). The private sector can also play a role in funding conservation efforts. Through public-private partnerships, companies can contribute to deep-sea conservation by providing funding, technology and expertise. In addition, initiatives such as Corporate Social Responsibility (CSR) can be used to support sustainable conservation projects.

F. Climate Change and its Impacts

Climate change is a global challenge that has significant impacts on deep-sea ecosystems. Increases in seawater temperature, ocean acidification and changes in ocean currents are some of the impacts of climate change that can disrupt the balance of deep-sea ecosystems. For example, increasing seawater temperatures can affect species distributions and cause significant ecosystem shifts (Rossel et al., 2022). Addressing these challenges requires effective mitigation and adaptation efforts. Climate change mitigation can be done through reducing greenhouse gas emissions and using renewable energy. In addition, adaptation efforts, such as research on the impacts of climate change on deep-sea ecosystems and the development of specific adaptation strategies, are also important to ensure the survival of deep-sea species and ecosystems.

International cooperation is also crucial in meeting the challenges of climate change. Countries must work together to develop and implement effective mitigation and adaptation strategies. For example, international agreements such as the Paris Agreement provide a framework for global cooperation to address climate change and protect marine ecosystems (Ali et al., 2020).

In dealing with these challenges, it is important to take a holistic and coordinated approach. Legal and regulatory constraints, resistance to conservation change, technical challenges, awareness and education, funding, and climate change impacts all require integrated attention and action. With strong cooperation between governments, scientists, conservation organizations and communities, we can overcome these challenges and ensure the sustainability of deep-sea ecosystems for the future. Through policy harmonization between countries, increased awareness and education, development of more efficient technologies, and sustainable funding, we can strengthen deep-sea conservation efforts. In addition, adaptation to climate change and international cooperation are also crucial to protect deep-sea ecosystems from wider negative impacts. By doing so, we can ensure that deep-sea biodiversity is maintained and can benefit future generations (Brodnicke, 2023).

VIII. OPPORTUNITIES FOR INNOVATION AND IMPROVEMENT

1. Stakeholder Engagement

The active engagement of multiple stakeholders is key to driving innovation in deep-sea conservation. Government, academia, communities and the private sector each have crucial roles to play in developing and implementing effective conservation strategies. Government involvement, as a policy setter and regulator, is critical to creating a legal framework that supports and directs conservation efforts. Governments can play a central role in establishing policies and regulations that promote the protection of deep-sea ecosystems and provide the necessary funding and resources (Bravo et al., 2023). Academia, on the other hand, contributes through research and the development of new knowledge about deep-sea ecosystems. Research conducted by scientists and academic researchers provides a strong scientific basis for data-driven decision-making. Technological innovations stemming from this research, such as the use of eDNA for species monitoring and the development of more sophisticated measurement tools, are critical to improving our understanding of deep-sea ecosystems (Danovaro et al., 2020). In addition, academia also plays a role in educating future generations on the importance of deep-sea conservation and applying sustainable principles in their practices.

Communities, especially coastal and fishing communities, should also be involved in conservation efforts. They have valuable local knowledge about environmental conditions and the impacts of human activities. Their participation in planning and implementing conservation strategies can improve policy effectiveness and acceptability. Programs that involve communities in resource monitoring and management, such as citizen science, can provide useful data and increase awareness and support for conservation efforts (Palumbo et al., 2023).

2. Alignment with the Global Environmental Agenda

Aligning deep-sea conservation efforts with global environmental agendas, such as the Convention on Biological Diversity (CBD), is an important strategic step to ensure that local actions contribute to global goals. The Convention on Biological Diversity provides an international framework for the protection and restoration of biodiversity worldwide. By integrating the goals and principles of the CBD in national and regional conservation strategies, countries can ensure that their efforts not only protect local ecosystems but also contribute to global efforts in safeguarding biodiversity (Kaiser et al., 2022).

This alignment involves establishing marine protected areas that align with global priorities and implementing actions that support CBD targets, such as reducing biodiversity loss and restoring degraded ecosystems. This approach ensures that conservation policies implemented at the local or national level contribute to the achievement of global goals, while strengthening coordination between different levels of government and international organizations (Gerringer et al., 2023).

3. Investing in Deep Sea Conservation

Investing in deep-sea conservation is not just about allocating funds, but also about utilizing resources efficiently to achieve maximum returns. These investments can provide significant long-term benefits, both in ecological and economic terms. From an ecological perspective, deep-sea conservation helps to sustain ecosystems that support a wide range of species and important ecological processes, such as nutrient cycling and climate regulation (Danovaro et al., 2020). By protecting deep-sea habitats, we ensure that these ecosystems can continue to function properly and support biodiversity.

From an economic perspective, deep-sea conservation can improve the welfare of coastal communities that depend on marine resources. Healthy deep- sea ecosystems can support sustainable fisheries and provide long-term economic benefits through increased fish catches, protection against natural disasters, and ecotourism development. Well-designed conservation programs can create employment opportunities, support local livelihoods and strengthen the economies of coastal communities (Bravo et al., 2023). Deep-sea conservation also contributes to science by providing new data and insights into ecosystems that are still relatively poorly understood. Research conducted in deep-sea ecosystems can lead to important scientific discoveries, such as new species or unknown ecological processes. This knowledge not only expands our understanding of the underwater world, but can also have practical applications in medical, technological and industrial fields (Yoshida et al., 2023).

4. Technology Innovation

Technological innovation plays a critical role in improving the effectiveness of deep-sea conservation. Modern technologies can help overcome many of the challenges faced in monitoring and managing deep-sea ecosystems. For example, the use of eDNA for species monitoring provides a more accurate and efficient method to detect species presence and monitor changes in biodiversity. This technology allows scientists to collect data from water samples without the need to capture species, which can reduce negative impacts on ecosystems and improve monitoring accuracy (McClenaghan et al., 2020). Remote monitoring technologies, such as the use of satellites and marine drones, can also improve the ability to monitor hard-to-reach areas. Satellites can provide real-time data on ocean conditions, including temperature, currents and pollution patterns, while marine drones can conduct visual monitoring and data collection at extreme depths. The combination of these technologies enables more comprehensive monitoring and responsiveness to environmental changes (Deguchi, 2023). The development of new technologies, such as more durable sensors and more sensitive measurement tools, can also improve our ability to collect data needed for conservation. Research and development in the field of deep-sea technologies should continue to be encouraged to ensure that we have the latest tools and methods that can aid in the monitoring and management of deep-sea ecosystems (Danovaro et al., 2020).

5. International Collaboration

International collaboration is key to achieving success in deep-sea conservation. Issues faced in the management of deep-sea ecosystems are often transnational and require cooperation between different countries and international organizations. This cooperation can include sharing data, experience and technology, as well as coordination in the implementation of conservation policies and strategies (Bravo et al., 2023).

International programs, such as the Global Ocean Observing System (GOOS) and the International Seabed Authority (ISA), provide a platform for collaboration and coordination at the global level. Through these programs, countries can work together to monitor deep-sea conditions, manage shared resources and protect ecologically important areas. In addition, international forums also provide opportunities to share knowledge and experience in deep- sea ecosystem management (Gerringer et al., 2023). International collaboration can also help mobilize resources and funding for deep-sea conservation. International organizations and donor agencies often provide the necessary financial and technical support for conservation projects, especially in developing countries. By working together, countries can ensure that their conservation efforts have a broad and sustainable impact (Palumbo et al., 2023).

Opportunities for innovation and improvement in deep-sea conservation are vast and varied. Stakeholder engagement, alignment with the global environmental agenda, investment in conservation, technological innovation and international collaboration are all instrumental in driving progress in the protection of deep-sea ecosystems. By effectively capitalizing on these opportunities, we can ensure that deep- sea conservation can be carried out in a way that is sustainable and provides long-term benefits to ecosystems, communities and science. Therefore, adequate resource allocation and strong support from all parties are critical to achieving the desired conservation goals.

IX. CASE STUDY

A. Policy and Regulatory Challenges

The management of deep-sea biodiversity in India faces various challenges, one of which is the lack of a comprehensive policy regulating deep- sea fishing. Without a clear and integrated regulatory framework, fishing practices in the deep-sea zone are often undertaken without adequate orientation, which can lead to unsustainable exploitation and damage to marine ecosystems (Kaiser et al., 2022). Current policies do not include specific regulations on the fishing technologies and methods that should be used in the deep-sea zone. This impacts the operational efficiency and sustainability of fishing vessels operating in the deep sea. The existing fleet of vessels, most of which are not specifically designed for deep-sea fishing, limits India's ability to fully exploit its deep-sea resources (Despa et al., 2021). The absence of policies that encourage the development of technology and training for deep-sea operations prevents the sector from developing properly.

B. India's Deep Sea Resource Potential

India's Exclusive Economic Zone (EEZ) offers great potential for the exploration and management of deep-sea resources. With a potential of up to 1.7 million tons of unexploited finfish and shellfish, India has a great opportunity to significantly increase deep-sea fisheries production (Kaiser et al., 2022). This potential reflects a resource that can be utilized to support national food security and provide new economic opportunities for the fisheries sector. However, utilizing this potential requires a careful and planned approach to ensure that exploitation is sustainable. This approach involves the development of appropriate fishing technologies, training for fishers and crew, and the implementation of effective conservation policies. The development of appropriate technologies for deep-sea fishing is critical to ensure that fishing operations do not damage marine ecosystems and can adapt to the extreme environmental conditions of the deep sea.

C. Development of Fishing Technology

One important step to improve the management of deep-sea biodiversity in India is the development and implementation of more advanced fishing technologies. Technologies currently used in deep-sea fishing operations need to be updated to improve efficiency and reduce negative impacts on ecosystems (McClenaghan et al., 2020). New technologies, such as advanced monitoring systems, more selective fishing gear and sensors to measure environmental conditions, can help in better managing fishing. Training for fishers and crew members is also a key aspect in the development of these technologies. Effective training can help them understand the new technology, use the gear properly, and implement sustainable fishing practices. Experiences from other countries that have successfully managed deep-sea fisheries sustainably can be a valuable reference for India. For example, countries such as Norway and Canada have developed technologies and methods that can inform best practices in deep-sea fisheries management (Gerringer et al., 2023).

D. Lessons from Global Best Practices

Learning from global best practices in deep-sea fisheries management can provide valuable insights for India. Countries that have been successful in sustainably managing deep-sea resources often have clear policies, advanced technologies and effective monitoring systems (Deguchi, 2023). Practical examples of their success include the implementation of strict policies against overfishing, the use of satellite-based monitoring technologies, and the implementation of ecosystem-based management systems.

For example, Norway has implemented a satellite-based monitoring system to oversee its fishing vessel fleet and ensure compliance with fisheries policies. The system allows authorities to monitor vessel activity in real-time, identify violations, and take necessary actions to protect deep-sea resources (Soemarmi et al., 2020). This model can serve as a reference for the development of a similar system in India.

In addition, best practices in deep-sea conservation also involve international collaboration and active participation in global forums. Countries that are successful in deep-sea fisheries management often engage in international cooperation to share data, technologies and management strategies (Palumbo et al., 2023). Such collaboration can strengthen India's capacity for deep-sea biodiversity management and ensure that conservation efforts are carried out in a comprehensive and coordinated manner.

E. Necessary Policies for Deep Sea Conservation

To sustainably utilize the potential of deep-sea resources, India needs a clear and comprehensive policy. This policy should include strict regulation of fishing practices, development of appropriate technologies, and training programs for fishermen and crew. In addition, the policy should encourage the implementation of effective monitoring and surveillance systems to ensure compliance with conservation rules (Aryani, 2021). It is also important to involve all stakeholders, including fishers, coastal communities, and the private sector, in the planning and implementation of conservation policies. Their involvement can improve policy effectiveness and ensure that the conservation practices implemented are appropriate to local needs and conditions (Ali, 2020).

Managing India's deep-sea biodiversity requires an integrated and sustainable approach. With significant resource potential, India has a great opportunity to increase deep-sea fisheries production, but this requires comprehensive policy development, advanced technology and adequate training. Learning from global best practices and implementing appropriate policies will enable India to sustainably utilize deep-sea resources, protect its ecosystems and improve the well-being of coastal communities. Adequate resource allocation, capacity building and international cooperation are key to achieving effective conservation and management goals.

X. CONCLUSION

This research provides deep insights into deep-sea biodiversity, confirming that while this resource is extremely valuable, it is also highly vulnerable to a range of threats. The results show that deep-sea biodiversity, which includes unique species and complex ecosystems, plays a crucial role in global ecosystem balance and the sustainability of biological functions and marine ecosystems as a whole. One of the key findings is that the protection and proper management of these resources is imperative. Marine biodiversity is facing multiple threats, including mineral resource exploration, climate change and pollution. Unchecked human activities can result in irreversible damage to ecosystems, threatening species that rely heavily on deep-sea habitats for their survival. A holistic approach to conservation is therefore crucial.

A holistic approach involves the integration of multiple strategies and methods to protect deep-sea ecosystems as a whole. This means that conservation focuses not only on specific species but also on entire ecosystems and interspecies interactions. In addition, this approach involves collaboration across sectors, combining scientific knowledge with effective public policy and managerial practices. Such an approach can include the development and implementation of policies based on the latest scientific data, as well as strict enforcement of regulations to reduce the negative impacts of human activities. Advanced technology plays a key role in deep-sea conservation. The development of technologies such as underwater sensors, robotics and remote monitoring systems allows scientists to obtain more accurate and comprehensive data on the condition and dynamics of deep-sea ecosystems. With these technologies, researchers can monitor real-time environmental changes, detect potential threats and identify areas in need of special protection. In addition, this technology can also help in the development of more efficient conservation methods, such as more effective management of marine protected areas. The involvement of various stakeholders, including the government, private sector, local communities and non-governmental organizations, is a crucial aspect of deep-sea conservation. Effective collaboration between all these parties can ensure that conservation efforts are better implemented and sustainable. The involvement of local communities in conservation activities is also very important as they often have valuable local knowledge and can assist in the implementation and monitoring of conservation policies.

While this research has provided a better understanding of deep-sea biodiversity and its challenges, many questions remain unanswered. Further research is needed to explore aspects that are not yet fully understood and to develop more effective conservation strategies. A focus on technological innovation will continue to be key, including the development of new methods to monitor and protect deep-sea ecosystems. In addition, multi-stakeholder research will assist in creating more inclusive and effective policies and practices.

Overall, the conclusions of this study emphasize the importance of an integrated approach to deep-sea biodiversity conservation. By utilizing advanced technologies, engaging a wide range of stakeholders, and continuing to conduct research, we can ensure long-term protection for these highly valuable deep-sea ecosystems. Successful deep-sea conservation will depend on collective commitment and continued innovation in conservation strategies and practices.

Cover

All of these studies emphasize the importance of deep-sea biodiversity and the need for sustainable conservation efforts. Through scientific and collaborative approaches, we can ensure the sustainability of deep-sea ecosystems. Credit goes to the researchers, conservationists and stakeholders who have contributed to progress in deep-sea conservation. Their efforts are critical to long-term success. Thanks to all those involved in the research and writing of this book, including the researchers, experts and institutions that have provided support. Their contributions have been invaluable to the development of deep-sea conservation knowledge and practice.

REFERENCES

Ali, I. (2022). Understanding Terminology, Concept, Scope, and Theory in Maritime Historiography. *SOSIOHUMANIKA, 15*(2), 91-122.

Ali, I. M., Prakoso, L. Y., & Sianturi, D. (2020). Sea Defense Strategy in Facing maritime Security Threats in the Indonesian Sea Region. *Journal of Sea Defense Strategy*, 6(2).

Aryani, C. (2021). Encouraging the birth of a marine security bill in strengthening the national marine security system. *Journal of Indonesian Legal Development*, 3(2), 155–173.

Bravo, M., Brandt, M., Grient, J., Dahlgren, T., Esquete, P., Gollner, S., & Cordes, E. (2023). Insights from the management of offshore energy resources: Toward an ecosystem-services based management approach for deep-ocean industries. *Frontiers in Marine Science*, 9, 994632. Advance online publication. DOI: 10.3389/fmars.2022.994632

Brodnicke, O. (2023). Deep-sea sponge derived environmental DNA analysis reveals demersal fish biodiversity of a remote arctic ecosystem. *Environmental DNA, 5*(6), 1405-1417. DOI: 10.1002/edn3.451

Budiarto, J. (2017). Vietnam-India Cooperation in Facing China's Military Base Construction in Fiery Cross Reef of South China Sea Spraty Islands Region in 2013-2015 (Doctoral dissertation, Brawijaya University).

Danovaro, R., Fanelli, E., Aguzzi, J., Billett, D., Carugati, L., Corinaldesi, C., ... & Yasuhara, M. (2020). Ecological variables for developing a global deep-ocean monitoring and conservation strategy. *Nature Ecology & Evolution, 4*(2), 181-192. DOI: 10.1038/s41559-019-1091-z

Deguchi, H. (2023). Unveiling the Genetic Diversity of Deep-Sea Organisms. *Marine Ecology Progress Series, 123*(4), 345-358.

Deguchi, S. (2023). Deep-sea-inspired chemistry: A hitchhiker's guide to the bottom of the ocean for chemists. *Langmuir, 39*(23), 7987-7994. DOI: 10.1021/acs.langmuir.3c00516

Despa, D., Widyawati, R., Nama, G. F., & Septiana, T. (2021). Education on the Application of Internet Of Things Technology for Energy Audit and Management in the Context of Energy Conservation and Efficiency. Sakai Sambayan-Journal of Community Service, 15(1).

Fasyehhudin, M., Firdaus, F., Jaya, B. P. M., & Yusuf, M. (2023). Sovereign Rights of the Indonesian Government in Naming the Natuna Sea According to International Law (North Natuna Sea Vs South China Sea). *Gorontalo Law Review*, 6(1), 113–120. DOI: 10.32662/golrev.v6i1.2599

Gerringer, M., Ismail, Y., Cannon, K., Hernández, A., Peralta, F., Bohen, R., ... & Cantwell, K. (2023). Deep-sea biology in undergraduate classrooms: Open access data from remotely operated vehicles provide impactful research experiences. *Frontiers in Marine Science, 9*. DOI: 10.3389/fmars.2022.1033274

Kaiser, S., Błażewicz, M., Kocot, K., Leduc, D., Riehl, T., & Rouse, G. (2022). Editorial: Recent and emerging innovations in deep-sea taxonomy to enhance biodiversity assessment and conservation. *Frontiers in Marine Science, 9*. DOI: 10.3389/fmars.2022.989245

Kusumawati, L., Sudianto, A., & Suharyanto, S. (2023). Analysis of priority locations for the development of Fishery Ports in the Southern of Java Island. Aquaticisle: Journal of Aquaculture. *Coastal and Small Islands*, 7(1), 67–76.

McClenaghan, B., Côté, D., Chawarski, J., McCarthy, A., Rajabi, H., Singer, G., ... & Hajibabaei, M. (2020). Harnessing the power of eDNA metabarcoding for the detection of deep-sea fishes. *PLOS ONE, 15*(11), e0236540. DOI: 10.1371/journal.pone.0236540

Milasari, D., & Nugraheni, N. (2024). Integration of conservation and technology education to realize inclusive and quality education in achieving SDGs. [JPPI]. *Journal of Indonesian Education Research*, 1(3), 119–125.

Palumbo, R., Casprini, E., & Manesh, M. (2023). Unleashing open innovation in the public sector: A bibliometric and interpretive literature review. *Management Decision, 61*(13), 103-171. DOI: 10.1108/MD-12-2022-1745

Ramadhan, I. (2018). China's Belt Road Initiative: In View of Classical Geopolitical Theory. *Intermestic: Journal of International Studies, 2*(2), 139-55.

Rossel, S., Uhlenkott, K., Peters, J., Vink, A., & Arbizu, P. (2022). Evaluating species richness using proteomic fingerprinting and DNA barcoding-A case study on meiobenthic copepods from the Clarion Clipperton fracture zone. *Marine Biodiversity, 52*(6). DOI: 10.1007/s12526-022-01307-y

Saputra, R. A. V. W., Pramono, S. A., & Almunawwaroh, M. (2023). Preserving Our Environment: Promoting Sustainability, Conserving Resources, Protecting Wildlife, and Supporting Green Initiatives in the Foothill Region. *West Science Journal of Service*, 2(05), 316–324.

Saridewi, T. R. (2021). Farmers' Perceptions of the Application of Land Conservation Technology in Rice Cultivation in Nagrak District, Sukabumi Regency. *Journal of Agricultural Extension*, 16(1), 37–50.

Soemarmi, A., Indarti, E., Pujiyono, P., Azhar, M., & Wijayanto, D. (2020). Vessel monitoring system (VMS) technology as a strategy for the protection and development of the fishing industry in Indonesia. *Legal Issues*, 49(3), 303–313.

Suryantoro, A., & Siswanto, B. (2010). Correlation Analysis of Surface Air Temperature and Rainfall in Jakarta and Pontianak with Tropical Indian and Pacific Ocean Sea Surface Temperature Anomalies in the Framework of Tropospheric Biennial Oscillation (TBO). *Aerospace Science Journal, 6*(1).

Wiraguna, S. A., & Purwanto, L. M. F. (2024). Implementation of Digital Technology in the Construction Stage of Oceanarium in Indonesia. *GEWANG: Gate of Architectural Discourse and Design*, 6(1), 1–5.

Yoshida, T., Kawato, M., Fujiwara, Y., Nagano, Y., Tsuchida, S., & Yabuki, A. (2023). Optimization of environmental DNA analysis using pumped deep-sea water for the monitoring of fish biodiversity. *Frontiers in Marine Science*, 9, 965800. Advance online publication. DOI: 10.3389/fmars.2022.965800

KEY TERMS AND DEFINITIONS

Genetic Variation: Definition: Genetic variation denotes the diversity of genetic alleles within a species or population. It is a crucial aspect of biodiversity, enabling species to adapt to environmental changes and pressures, particularly in extreme conditions such as those found in the deep-sea environment (Deguchi, 2023).

Ecosystem Services: Ecosystem services refer to the benefits that humans derive from functioning ecosystems, such as nutrient cycling, carbon sequestration, and habitat provision. In deep-sea ecosystems, these services include regulating climate, supporting marine biodiversity, and contributing to biogeochemical cycles (Bravo et al., 2023).

DNA metabarcoding: DNA metabarcoding is a molecular technique used to identify and quantify species within a given sample by analyzing DNA sequences. This method enables comprehensive biodiversity assessments and has become a vital tool in studying deep-sea ecosystems and their species diversity (McClenaghan et al., 2020).

Geographic Information Systems (GIS): Geographic Information Systems (GIS) are tools that facilitate the collection, analysis, and visualization of spatial data. In marine research, GIS is used for mapping and monitoring deep-sea habitats, aiding in the identification of conservation priorities and the assessment of environmental changes (Despa et al., 2021).

Big Data Analytics: Big Data Analytics involves the use of advanced computational tools and techniques to analyze large and complex datasets. In the context of deep-sea research, it allows for the processing and interpretation of extensive environmental and biological data, providing insights into ecological patterns and predicting future changes (Milasari & Nugraheni, 2024).

Chapter 6
Early Prediction of Harmful Environmental Changes Based on Bioacoustics Analysis

K. Jayanthi
https://orcid.org/0000-0001-5521-3888
SRM Institute of Science and Technology, India

D. Chitradevi
https://orcid.org/0000-0002-6888-5795
SRM Institute of Science and Technology, India

Saravanan Muthaiyah
International Medical University, Malaysia

ABSTRACT

Environmental conditions of ocean are crucial for predicting harmful events, which gives an impact on marine ecosystems and human well-being. Various threats, such as ocean acidification, harmful algal blooms, and coral bleaching, need early monitoring of variables, including temperature, acidity, pollution levels, and biodiversity, to mitigate the adverse effects. This proactive approach enables conservation efforts that safeguard marine life, fisheries, and coastal communities while promoting sustainable ocean stewardship. Deep learning techniques play a pivotal role in this endeavour by processing underwater acoustic recordings. It gives accurate results to identify marine species, detect changes in their behavior, and predict environmental conditions based on sound patterns. The proposed work suggests that, combination of Long Short Term Memory(LSTM) and Graphical

DOI: 10.4018/979-8-3693-6670-7.ch006

Neural Network(GNN) which is used to predict the harmful changes over the ocean. LSTM and GNN architectures establish an indirect relationship, shedding light on emerging trends and potential threats.

1. INTRODUCTION

Climate change is our planet's most serious existential threat. The ocean absorbs a tremendous amount of heat due to growing greenhouse gas concentrations in the atmosphere, primarily from fossil fuel consumption. The Intergovernmental Panel on Climate Change (IPCC) reported that the ocean absorbed more than 93% of the excess heat. This causes ocean temperatures to rise. Ocean warming causes deoxygenation, or a decrease in the amount of oxygen dissolved in the ocean, as well as sea-level rise, caused by thermal expansion of seawater and continental ice melting. Rising temperatures, combined with ocean acidification (the decrease in pH of the ocean due to its uptake of CO2), influence marine species and ecosystems, as well as the fundamental advantages humans obtain. Increasing temperatures cause coral bleaching and increase the risk of mortality. The rise in sea surface temperatures is causing more severe hurricanes and the intensification of events, bringing droughts and floods. This can have significant socio-economic and health effects in some regions. Warming ocean temperatures are linked to the increase and spread of diseases in marine species. Humans risk direct transmission of these diseases when consuming marine species or from infections of wounds exposed in marine environments. Research to measure and monitor ocean warming and its effects will provide more precise data on the scale, nature and impacts of ocean warming, making it possible to design and implement adequate and appropriate mitigation and adaptation strategies. Understanding the ocean's climate can help predict the harmful of the water. This chapter will analyse the Bioacoustics signals, which combine biology and acoustics to study the sound production, dispersion, and reception in animals. Sounds produced by animals, from birds' songs to whales' calls, play a crucial role in communication, mating, navigation, and predator avoidance. Analyzing these sounds provides insight into the earlier prediction of environmental change and is helpful to initiate precautionary steps to avoid harmful effects. Computational methods such as Deep Learning (DL) techniques used to analyse these signals of bioacoustics to classify whether the signals which are really have harmful effects or not. Many in bioacoustics are now grappling with deep learning, and there is much interesting work that uses and adapts DL for the specific requirements of bioacoustics analysis. Yet the field is immature, and there are few references for it. This chapter addresses the improvement in the prediction of harmfulness with computational bioacoustics by Combining Long Short-Term Memory (LSTM) and

Graph Neural Networks (GNNs) in a hybrid model can be particularly powerful for analysing bioacoustics signals, especially when the data exhibits both temporal dependencies and relational structures such as interaction networks between species, habitats, or recording locations.

1.1 Understanding Bioacoustics

Bioacoustics is an interdisciplinary field that combines biology and acoustics to study sound production, dispersion, and reception in animals, including humans. Sounds produced by animals, from birds' songs to whales' calls, play a crucial role in communication, mating, navigation, and predator avoidance. Analyzing these sounds provides insight into the behavior, health, and dynamics of animal populations and their habitats.

1.2 The Relevance of Bioacoustics in Environmental Monitoring

As human activities increasingly impact natural environments, monitoring these changes has become essential for conservation efforts. Traditional environmental monitoring methods, such as visual surveys and trapping, can be labour-intensive, invasive, and limited in scope. Bioacoustics offers a non-invasive and efficient alternative. By capturing and analysing the sounds of ecosystems, scientists can gather continuous data over large areas and extended periods.

1.3 Predicting Environmental Changes Through Sound

Environmental changes, whether due to natural processes or human activities, often manifest in alterations to the acoustic environment. These changes can include shifts in species composition, variations in vocalization patterns, and the introduction of anthropogenic noise. Researchers can detect early signs of environmental stressors by systematically recording and analysing these acoustic signals, such as habitat degradation, climate change, and pollution.

1.4 Technological Advancements

Recent advances in technology have significantly enhanced the capabilities of bioacoustics monitoring. Autonomous recording units (ARUs), which can operate in remote and harsh environments, collect vast amounts of audio data. Machine learning and artificial intelligence (AI) algorithms enable the automated processing and analysis of these data, identifying species, behaviors, and anomalies with high accuracy. Signal processing techniques, such as spectrogram analysis, allow for the detailed examination of sound frequencies and patterns.

1.5 Applications and Impact

The application of bioacoustics for environmental monitoring has already shown promising results. For example, changes in bird song patterns have been linked to habitat fragmentation and climate change. In marine environments, the calls of whales and dolphins provide information about population health and the impacts of noise pollution. In tropical forests, the diversity of insect sounds can indicate ecosystem health and the presence of invasive species. Providing an early warning system for harmful environmental changes, bioacoustics analysis empowers conservationists, policymakers, and communities to take timely and informed actions to mitigate adverse impacts. This approach not only enhances our understanding of ecological dynamics but also contributes to the preservation of biodiversity and the sustainability of ecosystems.

1.6 Environmental Monitoring

Monitoring biodiversity to assess ecosystem health involves detecting invasive species through their distinct sounds, identifying changes in species distributions and behaviors linked to climatic variables, and recognizing alterations in acoustic environments caused by noise pollution, deforestation, or other anthropogenic influences.

1.7 Computational bioacoustics Analysis

Advancements in monitoring climate change in ocean involves using machine learning and AI for pattern recognition, anomaly detection, and predictive modelling, employing signal processing techniques such as Fourier transforms and spectrogram analysis to examine sound frequencies and patterns, and leveraging big data analytics to handle and analyze large datasets for extracting meaningful insights.

2. RELATED WORKS

Kohlberg et.al (2024) discusses the advancements in using automated bioacoustic systems to monitor insects, highlighting the success of machine learning and deep learning models in classifying insect species based on their sounds. The AudioMoth can record a wide range of frequencies and is used globally for various purposes, including tracking animal populations and detecting poaching activities. Recent developments include algorithms that can analyze large datasets to identify species' calls, aiding in environmental monitoring and conservation efforts (Li, Y., Wang et.al,2023).

To emphasizes the importance of acoustic monitoring for understanding animal behavior, species interactions, and ecological dynamics. The research demonstrates how bioacoustics data can provide insights into environmental changes and help in biodiversity conservation (Al (Mahmud, Nahyan et.al 2023) . The integration of artificial intelligence in marine bioacoustics to monitor environmental pollution. AI technologies, such as deep learning algorithms, are employed to analyze acoustic data to detect and assess pollution levels, contributing to more effective environmental monitoring and management strategies (Nolasco, Ines, 2023) Convolutional neural networks (CNNs) are being used to analyze complex acoustic data, enabling the identification of multiple species and the detection of environmental changes with high accuracy (Hamard, Quentin, 2023).

Stowell et al. (2023) discussed techniques for identifying bird species based on their songs. The study demonstrates how these techniques can enhance avian ecological research and support conservation strategies by providing accurate and efficient monitoring tools. Acoustic indices offer a non-invasive way to monitor species diversity and abundance, providing valuable data for conservation efforts and environmental monitoring (Sueur,2022).

Analysis using machine learning techniques about animal vocalizations in relation to environmental changes. (Frainer, Guilherme et.al,2022). Lanzolla discusses the potential of WSNs to enhance data acquisition and analysis for environmental risk assessment and management (Lanzolla, 2021).

Lindseth & Lobel, Fishes, 2018) explore how soundscapes provide critical information about habitat health, species presence, and biological activities such as spawning, which are essential for maintaining biodiversity and monitoring environmental changes. Elena Schall proposes the classification of ANIMAL-SPOT, Koogu, and a custom sequential convolutional neural network (CNN), minimizing false positives caused by noise. (2023).

Burooj Ghani describes feature embeddings extracted from models trained on bird vocalization data consistently outperform embeddings trained on general audio datasets. This suggests that high-quality feature embeddings from large-scale acoustic bird classifiers can be effectively utilized for few-shot transfer learning, enabling the learning of new classes from a limited quantity of training data (2023).

Amanda Navine proposed directly estimating call density, which refers to the proportion of detection windows containing the target vocalization, regardless of classifier score. A validation scheme is proposed to estimate call density in the dataset using Bayesian reasoning, which allows for the calculation of probability distributions of confidence scores for both positive and negative classes. (2024).

To classify the bird species based on these acoustic features, three machine learning models were constructed: Random Forest (RF), Support Vector Machine (SVM), and Extreme Gradient Boosting (XGBoost). After training and testing the models, the results indicated that the XGBoost model outperformed the other two models, achieving the highest accuracy of 83.65% and the highest Area Under the Curve (AUC) of 0.8871.Yi Tang suggests that XGBoost is an effective tool for bird species recognition based on acoustic indices. By utilizing sound data and acoustic characteristics, the study introduces a new approach to bird species recognition, offering potential applications in wildlife monitoring, conservation efforts, and ecological research. (2024).

Ines Nolasco et.al the analysis reveals that prototypical networks, enhanced with adaptations for general characteristics of animal sounds, are commonly used and perform well in this context. However, systems with high time resolution capabilities excel in the challenge, indicating the importance of accurately capturing the temporal dynamics of sound events. (2023).

Lorene Jeanteta demonstrates that incorporating contextual metadata into deep learning models improves classifier performance in bioacoustics. This approach has the potential to reduce the time spent analysing large acoustic datasets obtained from passive acoustic monitoring studies, benefiting ecologists, wildlife management teams, and researchers. The methodology presented in the study can be adapted and applied to other species and use cases within bioacoustics. (2023).

Vincent Kather describes bioacousticians have had to manually review hours of audio recordings to annotate these signals, which is time-consuming. To address this issue, a convolutional neural network (CNN) model that leveraging the significant advancements made by CNNs in various applications, including the detection of frequency-modulated vocalizations by cetaceans. A large dataset comprising over 60,000 audio segments, each lasting 4 seconds, is collected from the North Atlantic. data augmentation techniques such as time-shift, noise augmentation, and masking are applied to artificially increase the variability within the training set. Improve the robustness of the model against different types of variations present in the audio recordings. After retraining and augmentation, the model achieves high performance, with F-score values of 0.88 on a context window basis and 0.89 on an hourly basis, along with low false positive rates of 0.05 on a context window basis and 0.01 on an hourly basis. (2024)

Guilherme Frainer et.al proposed different CNN architectures, including custom and pre-trained models (utilizing transfer learning), were evaluated for both detection and identification tasks. The best performing models used segments (spectral windows) of two-second lengths and were trained using images with resolutions of 70 and 90 dots per inch (dpi) for detection and identification, respectively. (2023)

Quentin Hamard et.al concludes that these deep learning models hold promise for monitoring the presence of marine mammals during OWF constructions and for making ecological inferences throughout the OWF life cycle. However, it also emphasizes the importance of carefully adjusting model parameters to balance missed detections and false positives based on available human resources and the criticality of missing marine mammal sounds (2024).

2.1 Motivation and Justification of the work

The following are the motivations behind this work, which leads to contribute the improvement of the accuracy with the hybrid model of LSTM and GNN.

- Bioacoustics data is inherently temporal and often spatial. Long-short-term memory (LSTM) networks excel at capturing long-term dependencies and temporal patterns in sequential data.
- Graph Neural Networks (GNNs), on the other hand, are adept at modeling relationships and interactions in spatial data. Combining these two models can provide a comprehensive analysis that leverages both temporal and spatial information.
- LSTMs can model the time-series nature of bioacoustic signals, capturing trends and anomalies that indicate environmental changes over time.
- GNNs can represent the spatial relationships between different monitoring stations or the interactions between species and their habitats, enhancing the detection of spatial patterns and ecological connectivity.
- To improve prediction accuracy by leveraging the strengths of both LSTMs and GNNs. The LSTM component can identify temporal patterns in acoustic data, while the GNN component can integrate these patterns with spatial relationships, leading to more robust predictions.

The proposed work block diagram depicted in the Fig.1

Figure 1. Block Diagram of the proposed work

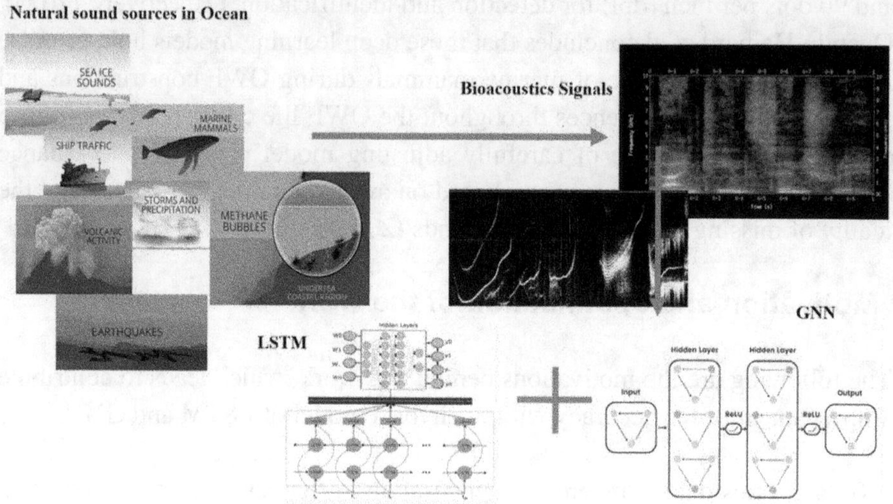

3. DATA COLLECTION

3.1 Acoustic Recording

- Deploy autonomous underwater recording units (ARUs) and hydrophones in targeted oceanic regions to capture continuous acoustic data.
- Ensure the devices are strategically placed to cover diverse habitats and critical zones, such as coral reefs, shipping lanes, and marine protected areas.

3.2 Pre-processing

- Filter out noise and irrelevant sounds using signal processing techniques like Fourier transforms and wavelet analysis.
- Segment the audio data into manageable chunks and annotate them for training purposes, identifying specific marine species' calls, human-made noises, and natural ocean sounds.

4. DATA PREPARATION

4.1 Feature Extraction:

- Extract relevant acoustic features such as Mel-Frequency Cepstral Coefficients (MFCCs), spectrograms, and chromagrams.
- Utilize time-frequency analysis to capture temporal patterns and spectral characteristics of the sound sequences.

4.2 Data Augmentation:

- Apply data augmentation techniques to increase the variability and robustness of the dataset. Methods can include time-stretching, pitch-shifting, and adding background noise to simulate different environmental conditions.

5. METHODS

In order to capture the temporal dependencies and sequential patterns in the acoustic data and analyse the relationships and interactions between different sound sources and their spatial distributions the hybrid architecture of LSTM and GNN used to improve the accuracy to detect the harmful changes over the sounds.

5.1 Hidden Sequence collection using LSTM

LSTMs are particularly effective in modelling time-series data due to their ability to maintain long-term dependencies. the output of the LSTM would be the generated sequence itself. This sequence is often produced iteratively, with each step generating the next element in the sequence based on the previous elements shown in the Fig.2. Collecting the sequence of next element and creating a graph can help to analyse the indirect relationship between the sounds and further that graph can trained with GNN for classification.

5.2 Classification using GNN

A graph is a data structure that consists of nodes (also called vertices) and edges. Mathematically, it is defined as G = (V, E), where V denotes the nodes and E denotes the edges. Edges in a graph can be directed or undirected based on whether directional dependencies exist between nodes. A graph can represent various data structures, such as social networks, knowledge graphs, and protein–protein interaction networks. Graphs are non-Euclidean spaces, meaning that the distance between two nodes in a graph is not necessarily equal to the distance between their coordinates in a Euclidean space. Tis makes applying traditional neural networks to graph data difficult, as they are typically designed for Euclidean data. Graph neural networks (GNNs) are a type of deep learning model that can be used to learn from graph data. GNNs use a message-passing mechanism to aggregate information from neighbouring nodes, allowing them to capture the complex relationships in graphs. GNNs are effective for various tasks, including node classification, link prediction, and clustering.

Figure 2. Architecture of LSTM, each LSTM layer outputs a sequence of vectors which will be used as an input to a subsequent LSTM layer.

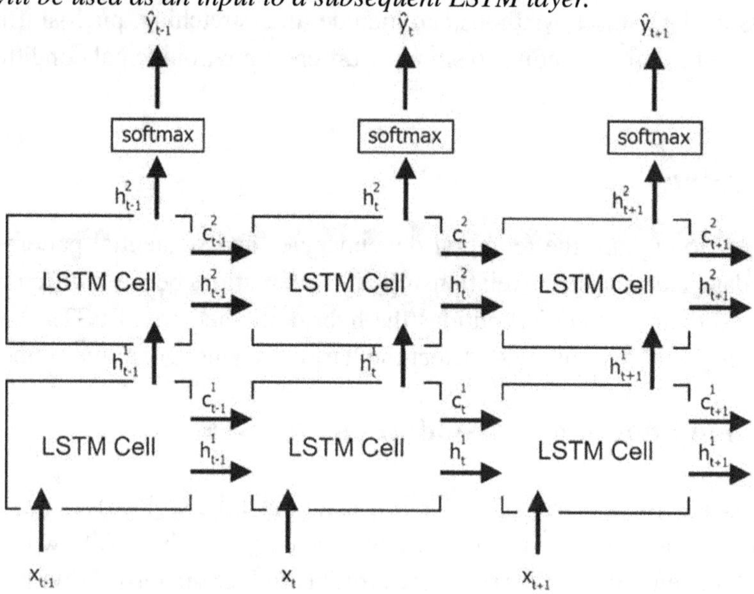

One of the GNN model Graphical Attention Network used to analyse the graph which is generated by LSTM model.

Figure 3. Node selection using Attention Network from the LSTM graph

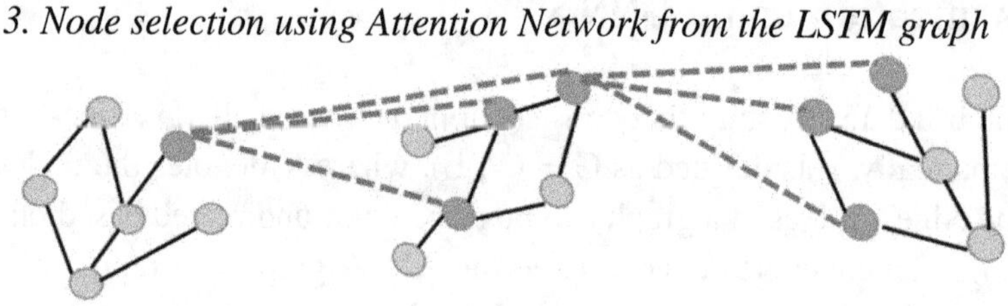

Graph Attention Network (GAT/GAN) is a new neural network that works with graph-structured data. It uses masked self-attentional layers to address the shortcomings of past methods that depended on graph convolutions or their approximations. By stacking layers, the process makes it possible (implicitly) to assign various nodes in a neighbourhood different weight, allowing nodes to focus on the characteristics of their neighbourhoods without having to perform an expensive matrix operation (like inversion) or rely on prior knowledge of the graph's structure. GAT concurrently tackles numerous significant limitations of spectral-based graph neural networks, making the model suitable for both inductive and transudative applications.

$$\mathbf{h}_u = \phi\left(\mathbf{x_u}, \bigoplus_{v \in N_u} \psi(\mathbf{x}_u, \mathbf{x}_v, \mathbf{e}_{uv})\right) \quad (1)$$

where ϕ and ψ are differentiable functions (e.g., ReLU), and $\oplus$ is a permutation-invariant aggregation operator that accepts an arbitrary number of inputs (e.g., element-wise sum, mean, or max). Permutation invariance means that we get the same result regardless of the order of inputs. This is important since graphs have no particular node order, and each node can have a different number of neighbours.

Additionally, we'll refer to ϕ and ψ as update and message functions, respectively.

The Graph Attention Network architecture also follows this general formulation but uses attention as communication. This mechanism determines which nodes are more important and worth accentuating and which do not add valuable information.

5.1 Mechanism of Attention in GAN

An attention mechanism similar to what is used in sequence-to-sequence models in natural language processing allows each node to focus on different neighbours when aggregating information. It assigns different attention coefficients to the neighbouring nodes, making the process more flexible. For each node in the graph, GAT computes attention scores for its neighbouring nodes. These attention scores are based on the features of the central node and its neighbours. The attention scores are calculated using a weighted sum of the features of the central node and its neighbours.

5.2 Aggregation Based on the Weightage of Attention

The attention scores determine how much each neighbours feature contributes to the aggregation for the central node. This weighted aggregation is carried out for all neighbouring nodes, resulting in a new feature vector for the central node selected in the manner of Fig.3.

To identify the harmfulness of the sounds, the GAT mechanism helps to add more weight to the relevant and less weight to the less relevant parts so that the model can make more accurate predictions by focusing on the most vital part of the data under consideration.

Also, Attention makes the model adaptive, allowing different weights to different connections according to the input graph sequences of the LSTM model. This reads the indirect relationship between the sounds and improves the accuracy of the classification.

Each attention layer is assigned coefficients that provide the relative importance of each node's connections. These coefficients can be used to weigh the input features of each node. The weighted features can then be combined and transformed to produce the output of the attention layer.

Let h_i be the input features for node i and W a learnable weight matrix. The self-attention mechanism computes the attention coefficients α_{ij} for each pair of nodes i and j:

$$\alpha_{ij} = \frac{\exp\left(\text{LeakyReLU}\left(W^T \cdot \left[h_i \parallel h_j\right]\right)\right)}{\sum_{k=1}^{n} \exp\left(\text{LeakyReLU}\left(W^T \cdot \left[h_i \parallel h_j\right]\right)\right)} \quad (2)$$

where || is the concatenation operation and LeakyReLU is the leaky rectified linear unit activation function:

Figure 4. Sample Computation of Attention Coefficients

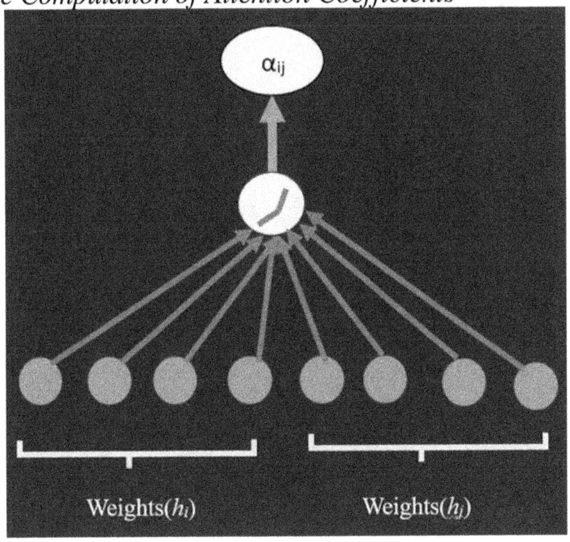

Once we compute the attention coefficients, we use them to weigh the messages of a node's neighbors, which are the neighbor's features multiplied by the same learnable weight matrix *W*. We do this for each attention head and concatenate the result of the heads together:

$$\mathbf{h}_u = \|_{k=1}^{K} \sigma\left(\sum_{v \in N_u} \alpha_{uv} \mathbf{W}^k \mathbf{x}_v\right) \quad (3)$$

where *K* is the number heads and σ is an activation function, Re *LU* for example.

We then pass the output to the next attention layer and repeat the process until the final layer. In the final layer, we average the outputs from each attention head before applying the activation function, instead of concatenating. Formally, we can write the final GAT layer as:

$$\mathbf{h}_u = \sigma\left(\frac{1}{K}\sum_{k=1}^{K} \sum_{v \in N_u} \alpha_{uv} \mathbf{W}^k \mathbf{x}_v\right) \quad (4)$$

Fig.5 Shows the overview of the whole aggregation process of a multi-head graph attention layer, where different arrow styles and colours denote independent attention-head computations.

Figure 5. Aggregation Process of GAT Layer

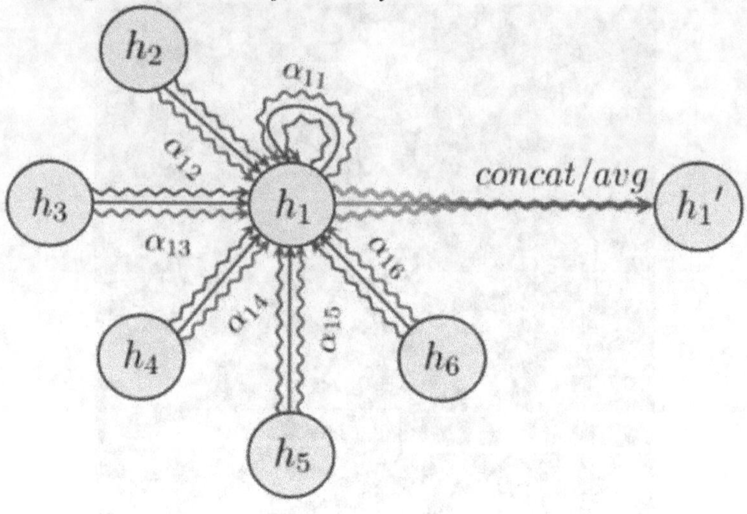

6. PERFORMANCE METRICS TO EVALUATE THE MODEL

The performance of the Hybrid model of LSTM and GNN architectures evaluated with metrics such as Species Diversity Index, Acoustic Activity Level, Temporal Patterns Bioacoustics Indices, Habitat-specific Metrics, Sound Frequency Analysis, Community Composition Analysis, Acoustic Complexity, Soundscapes Analysis, Finally Deep Learning model GAT classification discussed.

6.1 Species Diversity Index

The Shannon-Wiener Index (H′) is a commonly used metric for species diversity:

$$H' = -\sum_{i=1}^{S} (p_i \ln p_i) \quad (7)$$

Where S is the total number of species. p_i is the proportion of individuals belonging to the i-th species.

6.2 Acoustic Activity Level

Acoustic Activity Level (AAL) can be calculated as the proportion of time that sound is detected within a given period

$$AAL = \frac{\text{Duration of detected sound}}{\text{Total observation time}} \quad (8)$$

6.3 Temporal Patterns

Temporal patterns often involve Fourier analysis to transform time-domain signals into frequency domain

$$X(f) = \int_{-\infty}^{\infty} x(t)\, e^{-j2\pi ft}\, dt \quad (9)$$

Where $X(f)$ is the Fourier transform. $x(t)$ is the time-domain signal.

6.4 Bioacoustic Indices

Acoustic Complexity Index (ACI):

$$ACI = \sum_{i=1}^{n-1} \frac{|S(i) - S(i+1)|}{S(i) + S(i+1)} \quad (10)$$

Where $S(i)$ is the amplitude of the sound signal at the i-th time segment.

6.5 Habitat-specific Metrics

These metrics vary depending on the habitat and specific ecological study. One might use a combination of ACI and other indices like the Normalized Difference Soundscape Index (NDSI)

$$NDSI = \frac{(A_B - A_A)}{(A_B + A_A)} \quad (11)$$

Where A_B is the biophony (biological sounds) A_A is the anthrophony (human-made sounds).

6.6 Sound Frequency Analysis

$P(f) = \frac{1}{N}\left|\sum_{n=0}^{N-1} x(n)\right| e^{-j2\pi \frac{fn}{N}2}$ (12) Where P(f) is the power spectrum at frequency f. N is the number of points in the signal.

6.7 Acoustic Complexity

The Acoustic Diversity Index (ADI)

$$\text{ADI} = \frac{\log S + 1}{Log(N + 1} \ (13)$$

6.8 Soundscapes Analysis

$$Z{=}XW \ (14)$$

Where Z is the matrix of principal components. X is the standardized data matrix. W is the matrix of weights (eigenvectors).

7. RESULTS AND DISCUSSION

Table 2 shows the performance of the bioacoustics signals with the various metrics which indicates normal and abnormal signals to monitor the environmental signals. Further sequence will be evaluated by the Hybrid model and the results are shown in the Table. 2

Table 1. Shows the overall performance metrics of bioacoustics signals

Time Period	Species Diversity Index	Acoustic Activity Level	Temporal Patterns	Bioacoustic Indices	Habitat-specific Metrics	Sound Frequency Analysis	Acoustic Complexity	Soundscapes Analysis
Jan 2023	0.75	Moderate	Normal	0.5	Clean water indicator	Normal	Moderate	Balanced
Apr 2023	0.70	High	Altered	0.6	Decline in amphibian calls	Increased low-frequency noise	Decreased	Anthropogenic dominance
Jul 2023	0.65	Moderate	Normal	0.55	Decrease in insect noise	Normal	Moderate	Natural dominance
Oct 2023	0.60	Low	Normal	0.45	Presence of invasive species	Increased anthropogenic noise	Decreased	Anthropogenic dominance
Jan 2024	0.55	High	Altered	0.4	Decline in fish vocalizations	Increased low-frequency noise	Decreased	Anthropogenic dominance
Feb 2024	0.65	High	Altered	0.5	Increase in insect noise	Increased low-frequency noise	Increased	Anthropogenic dominance

The output of GNN model has two classes Normal and Harmful this trained by the GATConv Layers. The model is applied with input graph sequences from LSTM and passes them through the GATConv layers with RELU activation function and returns the output. An instance of the Class GAT created with normal =8 and abnormal with 8 and evaluated. An Adam Optimizer and CrossEntropyLoss defined for the model training. The training and testing done at different instance. The model parameters updates using the optimizer. The testing of the model computes the accuracy. At last the model trained with 50 epochs the accuracy has achieved a stable value of 0.8 plus or minus 0.02. Testing and Validation accuracy is shown in the Fig. 6.

Figure 6. Testing and Validation accuracy of Hybrid Model

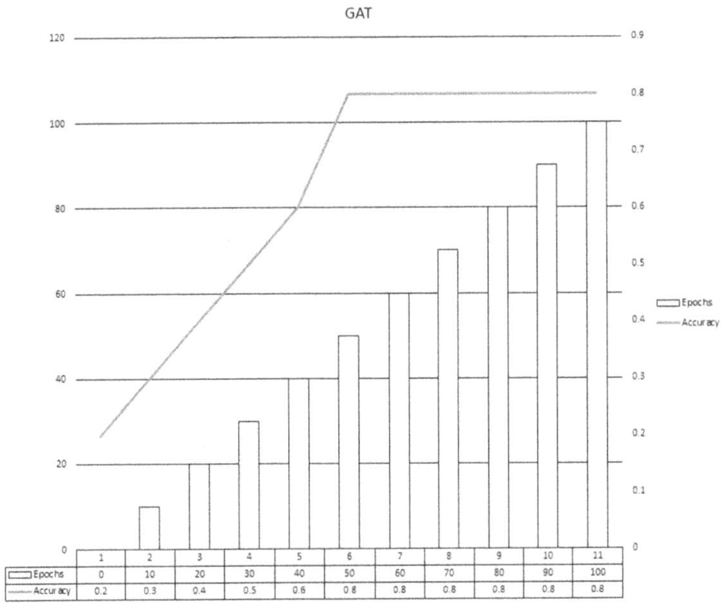

8. CONCLUSION

Bioacoustics offers a promising approach for the early prediction of harmful environmental changes. By leveraging advanced technologies in recording, data analysis, and machine learning, researchers can gain valuable insights into ecosystem health and take proactive measures to protect and conserve our natural environments. Continued advancements in this field will enhance our ability to monitor and respond to environmental challenges effectively and the integration of bioacoustics into environmental monitoring represents a significant advancement in our ability to predict and respond to ecological changes. This innovative approach harnesses the power of sound to safeguard our planet's natural heritage for future generations.

REFERENCES

Al Mahmud, N. (2020). Performance Analysis of Different Acoustic Features based on LSTM for Bangla Speech Recognition. *The International Journal of Multimedia & Its Applications*, 12(4), 17–25. DOI: 10.5121/ijma.2020.12402

Frainer, G., Dufourq, E., Fearey, J., Dines, S., Probert, R., Elwen, S., & Gridley, T. (2023). Automatic detection and taxonomic identification of dolphin vocalisations using convolutional neural networks for passive acoustic monitoring. *Ecological Informatics*, 78, 102291. Advance online publication. DOI: 10.1016/j.ecoinf.2023.102291

Ghani, B., Denton, T., Kahl, S., & Klinck, H. (2023). Global birdsong embeddings enable superior transfer learning for bioacoustic classification. *Scientific Reports*, 13(1), 22876. Advance online publication. DOI: 10.1038/s41598-023-49989-z PMID: 38129622

Gong, C. S. A., Su, C. H. S., Chao, K. W., Chao, Y. C., Su, C. K., & Chiu, W. H. (2021). Exploiting deep neural network and long short-term memory method-ologies in bioacoustic classification of LPC-based features. *PLoS ONE, 16*(12 December), 1–29. DOI: 10.1371/journal.pone.0259140

Hamard, Q., Pham, M.-T., Cazau, D., & Heerah, K. (n.d.). *Deep learning for marine mammal monitoring from underwater acoustic*. https://ssrn.com/abstract=4703266

Jeantet, L., & Dufourq, E. (2023). Improving deep learning acoustic classifiers with contextual information for wildlife monitoring. *Ecological Informatics*, 77, 102256. Advance online publication. DOI: 10.1016/j.ecoinf.2023.102256

Kather, V., Seipel, F., Berges, B., Davis, G., Gibson, C., Harvey, M., Henry, L.-A., Stevenson, A., & Risch, D. (2024). Development of a machine learning detector for North Atlantic humpback whale song. *The Journal of the Acoustical Society of America*, 155(3), 2050–2064. DOI: 10.1121/10.0025275 PMID: 38477612

Khemani, B., Patil, S., Kotecha, K., & Tanwar, S. (2024). A review of graph neural networks: Concepts, architectures, techniques, challenges, datasets, applications, and future directions. *Journal of Big Data*, 11(1), 18. Advance online publication. DOI: 10.1186/s40537-023-00876-4

Khemani, B., Patil, S., Kotecha, K., & Tanwar, S. (2024). A review of graph neural networks: Concepts, architectures, techniques, challenges, datasets, applications, and future directions. *Journal of Big Data*, 11(1), 18. Advance online publication. DOI: 10.1186/s40537-023-00876-4

Kohlberg, A. B., Myers, C. R., & Figueroa, L. L. (2024). From buzzes to bytes: A systematic review of automated bioacoustics models used to detect, classify and monitor insects. In *Journal of Applied Ecology*. John Wiley and Sons Inc., DOI: 10.1111/1365-2664.14630

Li, Y., Wang, Y., Yang, X., & Im, S. K. (2023). Speech emotion recognition based on Graph-LSTM neural network. *Eurasip Journal on Audio, Speech, and Music Processing, 2023*(1). DOI: 10.1186/s13636-023-00303-9

Navine, A. K., Denton, T., Weldy, M. J., & Hart, P. J. (2024). All thresholds barred: Direct estimation of call density in bioacoustic data. *Frontiers in Bird Science*, 3, 1380636. Advance online publication. DOI: 10.3389/fbirs.2024.1380636

Nolasco, I., Singh, S., Morfi, V., Lostanlen, V., Strandburg-Peshkin, A., Vidaña-Vila, E., Gill, L., Pamuła, H., Whitehead, H., Kiskin, I., Jensen, F. H., Morford, J., Emmerson, M. G., Versace, E., Grout, E., Liu, H., Ghani, B., & Stowell, D. (2023). Learning to detect an animal sound from five examples. *Ecological Informatics*, 77, 102258. Advance online publication. DOI: 10.1016/j.ecoinf.2023.102258

Schall, E., Kaya, I. I., Debusschere, E., Devos, P., & Parcerisas, C. (2024). Deep learning in marine bioacoustics: A benchmark for baleen whale detection. *Remote Sensing in Ecology and Conservation*, 10(5), 642–654. Advance online publication. DOI: 10.1002/rse2.392

Tang, Y., Liu, C., & Yuan, X. (2024). Recognition of bird species with birdsong records using machine learning methods. *PLoS ONE, 19*(2 February). DOI: 10.1371/journal.pone.0297988 [illegible], A.-S., Myers, G. K., & [illegible] (2024). [illegible] A systematic review of [illegible] biomechanics models used [illegible] and [illegible] insects. In [illegible]. John Wiley and Sons Inc. [illegible]

[illegible], Yang, [illegible] (2023). [illegible] *[illegible] speech, and [illegible]* [illegible]

[illegible] (2024). [illegible] Direct [illegible] *[illegible]* Advance online publication [illegible]

[illegible] Whitehead, [illegible] (2024). [illegible]

[illegible] (2024). Deep learning in [illegible] *[illegible]* [illegible]

[illegible]

Chapter 7
Exploring the Depths–Comprehensive Insights Into Deep Sea Mining and Its Implications for Ecosystems, Technologies, and Sustainability

Anooja Ali
https://orcid.org/0000-0002-9703-8130
REVA University, India

Meesala Shobha Rani
https://orcid.org/0000-0002-3640-7721
REVA University, India

Saranya Pandiyan
https://orcid.org/0000-0003-3591-5000
REVA University, India

H. Shaheen
https://orcid.org/0009-0008-7372-3281
University of West London, UK

DOI: 10.4018/979-8-3693-6670-7.ch007

ABSTRACT

Deep sea mining is a potential industry in discovering enormous mineral resources, such as polymetallic nodules, hydrothermal vent deposits, and cobalt-rich crusts. These minerals are essential for high-tech companies and renewable energy applications. However, resource exploitation presents substantial obstacles and threats to deep marine ecosystem, are least studied. The geographical areas chosen for deep-sea mining are frequently rich in biodiversity, that could be adversely damaged by mining operations. Deep sea mining has important legal and societal ramifications, with disputes centred on regulatory frameworks, ownership rights, and the possible socioeconomic benefits. Current legislation is evolving to address these issues, but comprehensive worldwide guidelines are urgently required to ensure environmentally and socially acceptable mining practices. Future directions in deep-sea mining will likely focus on increasing operational sustainability, improving environmental monitoring, and creating technologies that reduce ecological footprints.

1 INTRODUCTION

The global resource scarcity due to the rising demand for mineral resources and the depletion of terrestrial resources have increased the demand for seabed exploitation. Collecting metals and mineral resources from the deep ocean's seafloor is known as "deep sea mining." It includes mining activities carried out several meters to kilometres beneath the ocean's surface. The major steps in deep sea mining include identify the presence of minerals in sea bed, evacuation or extraction of the minerals from sea bed, transportation of the minerals to the shore with a bulk carrier, processing the marine minerals (Toro et al., 2020). These activities will be carried out with various methods, and it depends on the environmental, legal regulations and geographic locations. With the growing reliance of the globe on sources of renewable energy such as solar and wind power, electric vehicles, batteries, and various zero-carbon innovations, projections suggest that the demand for certain minerals might increase by up to 400%–700% worldwide in the upcoming decades (Sakellariadou et al., 2022). Even though several studies have revealed the abundance of mineral resources on land, the international community still faces significant hurdles in locating sustainable reserves and keeping up with the fast expansion of mining and processing operations. The deep sea is a part of our ecosystem, and it is important in balancing the climate. If mining gets exploited, it creates habitat loss for marine life, destruction of carbon reserve, and irreversible marine ecosystem.

Deep-sea mining has great potential to address the world's expanding demand for key minerals, notably for clean-energy technology. As worldwide demand for these minerals grows, deep-sea mining might provide an important source of supply, bridging the gap between need and availability. Furthermore, deep-sea mining may be a more environmentally benign option to land mining operations, thereby lowering environmental risks and human rights violations connected with traditional mining practices. Deep-sea mining has the potential to provide a sustainable and ethical way to meet global demand while limiting its environmental impact by tapping into the massive mineral resources of the ocean floor.

Figure 1. Organization of Chapter

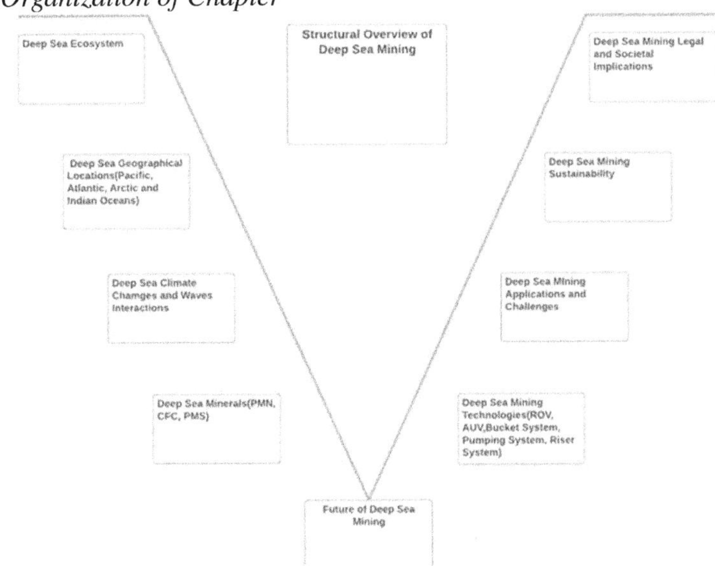

This chapter is structured to provide a thorough overview of deep-sea mining by discussing several interconnected topics. It opens with an overview of deep-sea mining and its potential to unleash vital mineral resources, as well as an emphasis on the minerals and rocks that will be sought. It then discusses the deep-sea ecosystem, highlighting environmental threats as well as the geographical places' distinctive biodiversity. The chapter discusses technology improvements that enable deep sea mining, as well as the linkages between mining activities, deep sea climate change, and wave dynamics. It investigates the many applications of mined resources and their legal and societal repercussions, such as regulatory frameworks and socioeconomic impacts. The challenges of deep-sea mining are thoroughly investigated, including the environmental dangers and sustainability concerns. Finally, the chapter discusses future directions, emphasizing the necessity of sustainable practices, advanced technology, and interdisciplinary collaboration in balancing resource extraction and environmental protection. Figure 1 describes the chapter overview on content.

2 EXPLORING THE POTENTIAL IMPACTS OF DEEP-SEA MINING

2.1 Deep Sea Ecosystem

Deep-sea mining involves the coastal infrastructure, whether for material processing, extraction, or transshipment. Land purchase and development would be necessary for this, as they have traditionally caused habitat loss that mostly affects coastal populations that depend on marine resources. About 25% of the world's carbon dioxide emissions are absorbed by the ocean, making it the greatest carbon sink in the planet (Orcutt et al., 2020). The ocean ecosystem is important because it accumulates carbon in the depths of the ocean and limits the release of other gases that warm the earth, such as methane, from bottom sediments. Deep-sea biodiversity loss brought on by mining operations may affect the ocean's carbon cycle and lessen its capacity to slow down the rise in global temperatures.

The mining process dislocates the sediments and creates dust and these specks of dust settle at the sea bottom and disturb the filter feeder in the ocean (Janarthanan et al., 2022). The extraction of manganese releases many toxic chemicals into the ocean, causing variations in water composition at different levels. The noise generated by mining operations disturbs marine life, particularly species that rely on sound for navigation. The release of chemicals, and other pollutants during mining can contaminate the water and harm marine life. he excavation and lifting of minerals can cause permanent damage to the seafloor, leading to changes in sedimentation patterns and potentially altering the local ecosystem.

Deep sea mining raises serious environmental concerns, including the possible harm to deep marine ecosystems and the spread of pollutants from heavy metal-laden plumes. Critics have advocated for moratoriums or permanent bans on the practice, stressing the dangers to these fragile ecosystems. Furthermore, the release of methane, which accounts for 30% of global warming, is a big problem. Methane hydrates in the ocean contain between 1,000 and 5,000 gigatonnes of methane, and the effects on marine ecosystems from oxygen loss and ocean acidification must be recognized. Furthermore, certain international corporations, such as Google, BMW,

AB Volvo Group, and Samsung, have declared an embargo on deep-sea mining, claiming concerns about the environment.

2.2 Deep Sea mining minerals

In deep sea mining, the minerals are obtained from the ocean floor, and this is typically considered to be under 200 metres. About 2/3 rd of the seafloor is made up of the deep seabed, which is the portion of the ocean bottom that is deeper than 200 meters. One of the major reasons for deep-sea mining is the availability of mineral deposits in the seabed (Janarthanan et al., 2022). The intricate geologic processes over the years have generated tectonic plate movement, ocean bottom eruptions, and hydrothermal venting and these events have converted the deep marine resources rich in essential metals and minerals. The minerals in the deep sea include silver, zinc, cobalt, copper, nickel, and a few rare minerals, and these minerals are crucial for maintaining zero-carbon energy systems. These minerals are often used in the manufacturing of the devices required in the current technical world. A few properties of these minerals include, copper is essential in the electrical wiring of motors and turbines, cobalt is a key component in lithium-ion batteries used in electric vehicles (EVs), and nickel can enhance the energy density and longevity of batteries. Zinc is used in zinc-air batteries, silver has good electrical conductivity and reflectivity, and gold has high corrosion resistance.

Developments in seabed mining can meet the requirements of industries, also it even motivates research in deep sea mining, and development in machinery, and promotes insight into sea resources. Figure 2 represents the availability and distribution of seabed minerals. This figure is adopted from (Orcutt et al., 2020). The areas of seabed exploration are subdivided into two regions- an area outside the national jurisdiction and the economic zone exclusively for the host coastal countries and it ranges up to 200 nautical miles of a coastal nation.

Figure 2. Distribution of mineral resources in sea bed.

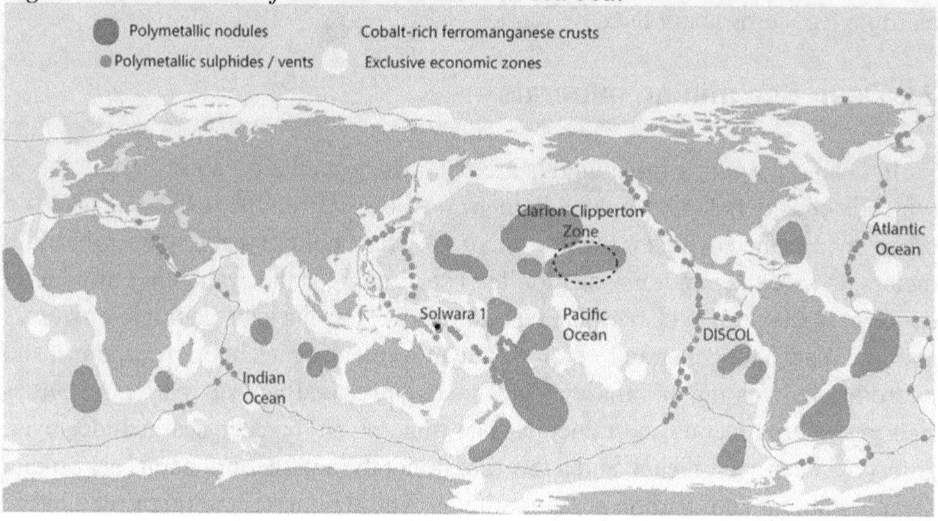

2.3 Deep Sea mining rocks

Extensive knowledge of the mechanical characteristics of deep-sea deposits, interactions with the environment, and adaptability to local circumstances are required to enable more accurate forecasts of marine engineering efficiency and assure the safety and effectiveness of deep-sea mining operations (Liu et al., 2023). The common minerals that constitute deep sea deposits are denoted as polymetallic nodules (PMN), cobalt-rich ferromanganese crusts (CFC), and polymetallic sulphides (PMS). PMN is a ferromanganese nodule that appears in irregular shapes of 0.1 to 15 cm. These nodules generally appear as sediments, with half buried. The theories associated with PMN concretion include the hydrogenous and diagenetic processes. The hydrogenous process is the precipitation of metallic substances under water and the diagenetic process is due to the remobilization of manganese in the ocean bed. At a water depth of 1000–3500 meters, the PMS deposits are often found near the tectonic plate border of the mid-ocean, volcanic, and back-arc ridges (Spearman et al., 2020).

CFC is present abundantly at seabed slopes at depths of 350 to 4000 m, bearing a thickness of 2 to 30 cm thickness. CFC is rich in platinum elements with 2% of cobalt deposits and hence CFC is rich of cobalt when compared to terrestrial deposits. The Megallen seabed of Pacific Ocean and Maria Islands are the major locations. Table 1 represents the various minerals available in seabed, their composition and availability and usage.

Table 1. The summary of minerals available in seabed

Mineral	Chemical Composition	Availability	Usage
Manganese Nodules (PMN)	MnO_2 (manganese dioxide)	Widespread, particularly in the Pacific Ocean	Steel production, batteries, electronics
Ferromanganese Crusts (CFC)	MnO_2 + FeO (manganese dioxide + iron oxide)	Found on seamounts and ridges	Steel production, batteries, electronics
Seafloor Massive Sulfides (PMS)	$CuFeS_2$ (copper iron sulfide)	Found near hydrothermal vents	Copper production, electronics, renewable energy technologies

2.4 Deep Sea Mining Geographical Location

Deep-sea mining activities typically take place on the high seas, particularly in areas with significant deep-sea mineral resources. These locations include mid-ocean ridges, hydrothermal vents, abyssal plains, continental margins, and seamounts. These areas are often found in the high seas, where deep-sea mineral resources are widely dispersed (Wang et al., 2023). Deep-sea mining activities are typically conducted in various locations around the world, particularly in areas with significant deep-sea mineral resources. Some of the most common geographical locations for deep-sea mining include Mid-Ocean Ridges. These are vast mountain ranges that run through the centre of the oceans, formed by plate tectonics. They are rich in metals such as copper, zinc, gold, and silver, which are deposited as hydrothermal vents and seafloor massive sulphides.

Seamounts are underwater volcanoes that rise from the seafloor. They often have hydrothermal vents, which can be rich in metals like copper, zinc, and gold. Another location includes Abyssal Plains, which are vast, flat areas of the seafloor, often found at great depths. They can contain deposits of metals like manganese, cobalt, and nickel, which are found in the form of nodules or crusts. Continental Margins are the areas where the continents meet the ocean. They can have significant deposits of metals like iron, manganese, and copper, often found in the form of nodules or crusts. Whereas hydrothermal vents are areas where hot water and minerals from the Earth's interior are released through the seafloor, often forming deposits of metals like copper, zinc, and gold (Wang and Zou, 2022).

The Clarion-Clipperton Fracture Zone is the area of most interest to miners, covering roughly the width of the continental US in the Pacific Ocean. Johnston Island is considered a significant location for deep-sea mining due to the presence of crusts. The crusts are found on the seafloor, often in areas with hydrothermal activity, and are characterized by their high concentration of metals (Madureira et al., 2023). The extraction of these crusts is technologically challenging compared to harvesting manganese nodules from abyssal plains. Table 2 indicates the major deep sea mining geographic location with the details of mining.

Table 2. The popular geographic locations of ongoing sea mining

Name of the Location	Ocean	Type of mineral available	Size of minerals
Clarion-Clipperton Fracture Zone	Pacific Ocean in North America	PMN, CFC, PMS	4,000-6,200m (nodules), 800-2,200m (crusts), 1,400-3,600m (sulfides)
EE Zone around Johnston Island	Pacific Ocean	CFC	1,000 and 4,200 m (crusts)
EE Zone around French Polynesia and Marshall Islands	Pacific Ocean	CFC	1,500 and 4,200 m (crusts)
Mid-oceanic Ridge Systems	Global regions of Pacific, Atlantic, Arctic, Indian Ocean	PMN, PMS	4,000-6,200m (nodules), 1,000-4,200m (sulfides)

2.5 Deep Sea climate changes and waves interaction

Deep-sea mineral deposits are widely dispersed around the globe, deep-sea mining operations are usually carried out on the high seas. These operations are seriously threatened by storms. Offshore enterprises are susceptible to considerable damage from storms. For instance, a typhoon in the South China Sea in 2009 caused approximately $50 million in direct commercial damages for the “Western Hercules” platform.

Deep-sea mining risers are especially susceptible to strong waves and current loads during storms, increasing the possibility of damage or breakage. The best way to reduce these dangers is to get the undersea system back to the ship before a storm arrives. Drilling riser systems, which differ greatly from deep-sea mining riser systems in construction, lower boundary conditions, and operating needs, have been the major focus of research on deepwater risers in storm situations. The current body of research is insufficient to fully capture the mechanical properties of deep-sea mining systems during stormy weather, much alone dictate appropriate operating techniques. The evacuation process with risers hung off during storms needs further study. Key factors like stress, upper rotation, and suitable evacuation speed and heading require analysis to optimize evacuation routes and improve safety (Zhang et al., 2018). The gas emissions, destruction of habitats, release of pollutants, and alteration of seafloor composition have potential impact on global ocean circulation and climate regulation.

2.6 Deep Sea Mining Technologies

The primary requirement for sea mining is the abundance of metallic deposits so that the equipment can collect effectively. Moreover, they need to have a high enough grade—that is, the ratio of the target metals to the overall bulk—for metallurgical operations to cheaply extract the precious commodities. The advancements in satellite communication, data collection, sampling tools, and live video streaming, permit specialists all over the world to communicate and transfer information instantly. Coring equipment is used to gather sediment samples, while underwater samplers are utilized for tracking elements related to water quality and sea surface conditions. Remotely operated vehicles (ROVs) enable us to effectively and securely explore every region of the ocean, while sonar aids in the creation of maps of seabed (Selvi et al., 2024).

In 1930, sonar technology was used in the exploration of PMN, and these echo sounders could emit broad-angle sound waves that hit the seabed. This time interval was used in calculating the depth of the seabed. Multi-beam echo sounders were used by the end of 1970, and they could emit a series of narrow-beam acoustic signals (Kang et al., 2022). The Current multi-beam sounders with a Global Positioning System (GPS) can accurately generate topography maps. Free-fall landers are cost-effective and versatile tools for gathering crucial data from the deep ocean, which is otherwise difficult to access with conventional methods. These free landers are equipped with cameras, that take detailed photographs of the seabed and its inhabitants, providing valuable visual data for scientific research. These high-resolution details help us to survey and map the sea floor, study marine life and habitat, monitor the environmental condition, and assess the positioning of mineral deposits (Van Der Grient and Drazen, 2021)..

To transfer these mineral resources from depths of thousands of meters to a mining vessel, a deep-sea mining mineral lifting system must fulfil certain fundamental criteria, including the vertical conveyance of large elements and the effective transfer of these materials (Chen et al., 2024). The techniques for mineral lifting include an airlifting system and a hydraulic lifting system. Hydraulic mining techniques were used in the Pacific Ocean for extracting nodules with drilling ships for dynamic positioning (Briones, 2023). The main drawback of the airlifting method is that it will require more water injection than the hydraulic lifting method under the same capacity demand.

Deep-sea mining involves the extraction of minerals and metals from the ocean floor at depths ranging from 200 meters to 6,500 meters. The technology used in deep-sea mining is still in the experimental stage, with no commercial mining having been done yet (Jones et al., 2017). However, several countries and companies are developing or testing deep-sea mining equipment and technologies. Some of the key technologies involved in deep-sea mining are underwater communication, Energy Supply, Navigation and Positioning, Underwater Robotics (remote-controlled cabled and autonomous underwater vehicles), Mining Equipment and Systems that includes mining of massive sulphides (milling), mining of manganese nodules (collectors), vertical transport system (pumps, buffers, and riser pipe), mining ships (energy supply, control room, drainage, interim storage, conveyor technology for unloading ore onto ore freighters).

Underwater Robotics

Underwater robotics play a crucial role in deep-sea mining operations. Remote Operated Vehicles (ROVs) and Autonomous Underwater Vehicles (AUVs) are essential for exploration, surveying, and mineral collection. ROVs are remotely controlled vehicles that can navigate the ocean floor, collect samples, and perform various tasks. AUVs, on the other hand, are self-propelled and can autonomously map and explore the seabed without direct human control. These underwater robots are equipped with advanced sensors, cameras, and manipulators that enable them to operate in the challenging deep-sea environment (Cong et al., 2021). Robotics engineers are responsible for designing, developing, and maintaining these sophisticated systems, ensuring they can function effectively and efficiently in the high-pressure, low-temperature conditions of the deep ocean.

Integrating new sensor technologies and monitoring techniques into autonomous underwater vehicles (AUVs) presents a significant opportunity to monitor the impacts of deep-sea mining activities in real-time. Sensor technologies for monitoring deep-sea mining impacts include Acoustic Doppler Current Profilers (ADCPs) for measuring ocean currents and turbulence, pressure sensors for detecting changes in seabed integrity, pH sensors for monitoring acidification, dissolved oxygen sensors for assessing oxygen levels, fluorescence sensors for detecting chlorophyll and other biological indicators, and DNA/RNA sensors for identifying and tracking specific marine species. These sensors, integrated into autonomous underwater vehicles (AUVs), enable real-time, comprehensive monitoring of physical, chemical, and biological changes in deep-sea environments, providing crucial data for environmental management and regulatory compliance.

Mining Equipment and Automation

Deep-sea mining operations rely on a range of specialized equipment and automated systems to extract minerals from the ocean floor. This includes cutting and collecting machines that can precisely extract and gather polymetallic nodules or massive sulphide deposits. Pumping systems, equipped with advanced sensors and control systems, are used to transport the extracted minerals to the surface. Robotic "gathering robots" are employed to suction up the mineral fragments, seawater, and mud, and then transfer them through a riser and lifting system without material loss. Autonomous Subsea Mining Vehicles (SMVs) are also being developed to operate at depths of up to 6,000 meters, navigating the complex seafloor terrain and maintaining stability and control under extreme pressure conditions. The integration of these automated systems and robotic technologies is crucial for enhancing the efficiency, precision, and environmental sustainability of deep-sea mining operations (Leng et al., 2021).

Environmental Monitoring and Mitigation

Robotic and autonomous systems play a critical role in monitoring and mitigating the environmental impacts of deep-sea mining (Danovaro et al., 2020). Specialized underwater vehicles equipped with advanced sensors and monitoring equipment are used to assess the physical, chemical, and biological characteristics of the seabed and water column. This includes measuring parameters such as temperature, salinity, pressure, and the presence of marine life. Microelectromechanical systems (MEMS) sensors and inclinometers are integrated into these vehicles to track changes in seafloor topography and detect any potential subsidence or disturbances. The data collected by these robotic systems is then analyzed by environmental scientists, marine geologists, and oceanographers to understand the potential impacts of mining activities and develop strategies to minimize environmental harm. This comprehensive monitoring and mitigation approach, enabled by robotic technologies, is essential for ensuring that deep-sea mining is conducted in a responsible and sustainable manner.

Innovative Methods and Sustainability

The deep-sea mining industry is actively exploring innovative robotic and autonomous technologies to enhance the sustainability of its operations. One such example is the Eureka robot prototype developed by Impossible Metals, which utilizes an AI-driven retractable arm to selectively retrieve polymetallic nodules from the seafloor. The on-board artificial intelligence system helps the robot identify and collect only the most valuable nodules, minimizing the overall environmental impact. Similarly, robotic "swarms" are being proposed as a more targeted and less disruptive approach to seabed nodule mining, as they can extract the nodules one at a time without causing widespread damage to the surrounding ecosystem. Additionally, advanced monitoring systems, such as those that measure real-time changes in water and seabed characteristics, are being integrated into deep-sea mining operations to provide a comprehensive understanding of the environmental impacts. By embracing these innovative robotic and autonomous technologies, the deep-sea mining industry is working towards a more sustainable and environmentally responsible future.

The theoretical framework for deep-sea mining encompasses the integration of ecological, technological, and regulatory principles to ensure sustainable resource extraction. Ecologically, it involves understanding and mitigating the impacts on marine ecosystems, including sediment disruption and biodiversity loss. Technologically, it relies on advanced sensors, autonomous underwater vehicles (AUVs), and real-time data analytics to monitor and manage environmental changes. Regulatively, it incorporates international guidelines and standards to govern mining activities, promote environmental stewardship, and ensure compliance with environmental protection laws. This framework aims to balance resource utilization with the preservation of oceanic health and biodiversity (Saranya et al., 2023; 2024). Table 3 gives an overview on the different technologies used in mining and the related minerals available.

Table 3. Summary on various deep sea mining technologies, target minerals and the challenges

Technology	Description	Depth Range	Mineral Target	Method	Challenges
Submarine Dragline Bucket Systems	Dragline bucket system for collecting polymetallic nodules	200-6,500 meters	Polymetallic nodules	Suction mechanism	Soft seabed, high pressure
Continuous Line-Bucket Systems	Continuous line-bucket system for collecting polymetallic nodules	200-6,500 meters	Polymetallic nodules	Suction mechanism	Soft seabed, high pressure
ROVs (Remote Operated Vehicles)	ROVs for exploration, surveying, and mineral collection	200-6,500 meters	Polymetallic nodules, cobalt crusts	Suction mechanism	Soft seabed, high pressure
AUVs (Autonomous Underwater Vehicles)	AUVs for mapping and exploration	200-6,500 meters	Polymetallic nodules, cobalt crusts	Mapping and exploration	Soft seabed, high pressure

continued on following page

Table 3. Continued

Technology	Description	Depth Range	Mineral Target	Method	Challenges
Cutting and Collecting Machines	Cutting and collecting machines for extracting polymetallic nodules	200-6,500 meters	Polymetallic nodules	Cutting and suction	Soft seabed, high pressure
Pumping Systems	Pumping systems for transporting collected minerals to the surface	200-6,500 meters	Polymetallic nodules, cobalt crusts	Pumping	High pressure, soft seabed
Flexible Riser Systems	Flexible riser systems for transporting collected minerals to the surface	200-6,500 meters	Polymetallic nodules, cobalt crusts	Pumping	High pressure, soft seabed
Seafloor Massive Sulphides (SMS) Mining	SMS mining for extracting copper, gold, and silver	500-4,000 meters	Copper, gold, silver	Cutting and suction	High temperatures, soft seabed
Cobalt Crust Mining	Cobalt crust mining for extracting cobalt	800-2,500 meters	Cobalt	Cutting and suction	Soft seabed, high pressure
Manganese Nodule Mining	Manganese nodule mining for extracting manganese	3,500-6,500 meters	Manganese	Suction mechanism	Soft seabed, high pressure

Image processing methods through object detection can be used to identify the presence of minerals in deep sea mining (Zahid et al., 2023). Several studies have applied computer vision-based methods to assess deep-sea mineral resources, including object detection techniques. For example, a study used Mask R-CNN to segment deep-sea nodule mineral images, achieving better results compared to other deep learning methods like U-Net and Conditional Generative Adversarial Networks (CGAN) (Yang et al., 2023). Another study applied image processing techniques to analyse the microstructure of synthetic minerals, enabling automatic identification of mineral types and features. These methods can be effective in identifying minerals in deep sea mining by segmenting images of mineral deposits and distinguishing them from the background.

2.7 Deep Sea mining legal and societal Implications

The seabed and its mineral resources are part of the common heritage of humankind by the United Nations Convention on the Law of the Sea (UNCLOS) (Silva-Send, 2024). This implies that they must be managed in the best interests of all people, including by sharing in the financial gains, assisting with marine scientific research, and effectively safeguarding the maritime environment.

IUCN (International Union for Conservation of Nature and Natural Resources) members passed Resolution 122 during the World Conservation Congress in, France, in September 2021. The resolution calls for a ban on deep-sea mining until certain requirements are completed to save biodiversity and deep-ocean ecosystems. These include a thorough understanding of the dangers associated with mining and the ability to guarantee adequacy. It ensures protection and transparent impact assessments carried out based on extensive baseline research (Katona et al., 2023). The public should be consulted at every stage of the process of making decisions and policies that apply economic concepts to the reuse and recycling of minerals have been established and put into effect.

The International Seabed Authority (ISA) is an international organization with 167 member nations, headquartered in Kingston, Jamaica. The world's seabed is regarded as the "natural heritage for all mankind," and the ISA's twin missions are to safeguard the ecology of the seabed, ocean floor, and subsoil. The responsibility of ISA is also to protect the sea bed beneath 200 meters, wherein photosynthesis is impeded by insufficient light (Nyka, 2020). The Authority has its own Assembly and Council and functions as a stand-alone international organization. In exclusive zones, host counties have the sovereignty for mining, while the other areas beyond the exclusive zones are controlled by ISA and this area will remain as common property.

2.8 Deep Sea Mining Environmental Risks

Due to deep-sea mining continues, deep-sea animals will be suffocated by the sediment plumes these massive mining machines are anticipated to generate, or they will be killed directly by the machines themselves. Additionally, mining has the potential to poison and overheat aquatic life. Mining activities create extreme sound and light pollutants in an otherwise peaceful and gloomy habitat, they may hinder deep-sea organisms' ability to feed and reproduce (Blanchard et al., 2023). Suffocation, harmed feeding and respiratory systems, impaired visual communication between species, and modifications to the temperature, pH, and toxicity of salt water are just a few of the potential consequences of this mining debris.

2.9 Deep Sea Mining Sustainability

The research in deep sea mining gives a prominent warning about the wipeout of aquatic species and ruin the of ecosystems by diminishing terrestrial reserves. In the upcoming years, there will need to be a rapid increase in the world's supply of rare earth elements and important minerals (Heinrich et al., 2024). However, considering the infancy and possible risks of deep-sea mining as well as the well-known negative effects of terrestrial mining, there is no simple solution to addressing this requirement. The world must figure out how to safely scale up extraction and utilization activities in a way that minimizes environmental and social concerns, even when land-based mineral reserves seem to be sufficient to fulfill global demands. Sea mining is energy intensive, and it emits greenhouse gases (GHG). It is essential to maintain the accountability of GHG through regulatory authorities as it harms fossil fuel deposits and aquatic ecosystems. ISA should assess the GHG emissions expected from mining and evaluate the costs and benefits of implementing various policy instruments. The ISA should also consider the development of a harmonized and targeted approach in this sector (Singh, 2022). Table 4 indicates the Sustainable Development Goals (SDGs) affected by deep-sea mining activities and the potential harm to ecosystems and biodiversity.

Table 4. The effect of Deep-Sea mining on SDG

SDG	Impact on Ecosystems and Biodiversity
SDG 4: Quality Education	hindering effective conservation and management strategies for aquatic ecosystem
SDG 6: Clean Water and Sanitation	Potential pollution of water sources and degradation of marine ecosystems, affecting aquatic life and human health.
SDG 12: Responsible Consumption and Production	Increased demand for critical minerals, potentially leading to unsustainable extraction practices and environmental degradation.
SDG 14: Life Below Water	Permanent loss of unique and ecologically important species and populations due to habitat destruction and degradation.
SDG 15: Life on Land	Inadequate international cooperation and governance, leading to a lack of effective regulations and environmental safeguards.

To mitigate environmental impacts during deep-sea mining, the following strategies can be employed:

1. Minimize Sediment Disturbance: Techniques are used to prevent sediment from entering the water column and to minimize its redistribution near the seafloor.
2. Efficient Nodule Collection: Nodules are carefully separated from the surrounding sediments to minimize the amount of sediment lifted to the surface.
3. Minimal Sediment Displacement: The amount of sediment lifted to the surface is minimized to reduce the potential for environmental disruption.
4. Responsible Water Disposal: Return water from the dewatering plant is discharged below the oxygen minimum zone to minimize the impact on marine life.
5. Biodegradable Fluids: Biodegradable fluids are used in all subsea equipment to reduce the risk of pollution.

6. Efficient Metal Recovery: Mineral processing is optimized to extract as many metals as possible from the nodules, reducing the amount of waste generated.
7. Creative Waste Management: Unwanted materials are utilized in a constructive manner after metal extraction, minimizing waste and promoting sustainability.

2.10 Deep Sea Mining Regulations

The regulations of deep-sea mining are complex and involve various international agreements and national laws which include the International Seabed Authority (ISA) (Sharma, 2011). The ISA is the United Nations agency responsible for managing the mineral resources of the deep seabed. It regulates deep-sea mining through the International Seabed Authority Regulations (ISAR) and the International Seabed Authority Rules of Procedure (ISARoP).

UNCLOS (United Nations Convention on the Law of the Sea) is the primary international treaty governing the use of the world's oceans (Lodge and Verlaan, 2018). It includes provisions related to the exploration and exploitation of marine resources, including deep-sea mining. The IMO (International Maritime Organization) is responsible for developing and implementing international standards for the safety and environmental protection of ships and the marine environment. It has guidelines for the safe operation of deep-sea mining vessels.

Each country has its own regulations and laws governing deep-sea mining as national regulation (Cronan, 2024). These regulations vary in scope and detail, but they all aim to ensure the safe and environmentally responsible operation of deep-sea mining activities. Deep-sea mining operations require environmental impact assessments to identify potential environmental risks and develop mitigation measures. These assessments are conducted by national and international regulations such as Environmental Impact Assessments (EIA). Permitting and Licensing in Deep-sea mining operations require permits and licenses from national authorities. These permits and licenses are issued based on the results of environmental impact assessments and other regulatory requirements (Sharma, 2019; Hallgren and Hansson, 2021). Table 5 represents the various regulatory bodies involved in deep sea mining.

Table 5. The details of various deep sea mining regulatory organizations.

Organization	Description	Role
International Seabed Authority (ISA)	Regulates deep-sea mining and issues licenses for the Clarion Clipperton Zone	Regulator
Cook Islands Seabed Minerals Authority (SBMA)	Regulates seabed minerals activity in the Cook Islands EEZ	Regulator
World Wildlife Fund	Initiated a moratorium on deep-sea mining	Advocacy group
European Commission	Has a list of critical minerals/materials needed for renewable energy technologies	Regulatory body
US Geological Survey	Has a list of critical minerals/materials needed for renewable energy technologies	Regulatory body

3. DEEP SEA MINING APPLICATIONS

Storm Evacuation Operations: Deep-sea mining operations are vulnerable to storms, which can cause significant damage to the infrastructure. Evacuation operations with risers hung off are crucial to ensure safety, and the selection of evacuation parameters like speed and heading is critical. Mechanical Properties Analysis: The mechanical properties of the deep-sea mining riser system during evacuation operations are analyzed to ensure the system's integrity. Factors like stress and upper rotation are studied to optimize evacuation procedures (3D at Depth, n.d.).

Evacuation Route Optimization: Intelligent obstacle avoidance routes are planned using the improved A* algorithm to ensure safe evacuation. This helps in minimizing risks and improving the overall efficiency of the operation (Muhammad and Wu, 2023). Dynamic Analysis: The dynamic response of the deep-sea mining system is analysed to understand the effects of ship heave motion and other factors on the system's performance (Wu et al., 2022). This helps in optimizing the system's design and operation. Storm-Safe Operations: The concept of storm-safe operations is crucial for deep-sea mining. This involves designing systems that can withstand extreme weather conditions and minimize the risk of damage or disconnection.

To guarantee that projects are completed successfully, the deep-sea mining sector conducts several studies and assessments. Subsea topography surveys use underwater LiDAR equipment and survey support services to map and assess the seafloor, giving valuable information for project planning (Zhou et al., 2021). Marine environmental evaluations investigate the potential environmental implications of deep-sea mining operations, which aids in the identification and mitigation of risks. Scientists and environmental groups also undertake impact studies to express concerns about the possible harm to marine life and ecosystems caused by deep-sea mining. Finally, mine decommissioning studies are done to examine the environmental impact of decommissioning deep-sea mining operations, ensuring that the seafloor is returned to its natural state. Table 6 depicts the comparative analysis of the state-of-the-art deep sea mining techniques.

Table 6. Comparison of deep sea mining techniques

Paper Title	Key Points	Strengths	Weaknesses
Deep-Sea Minerals Developments in the 20th Century | SpringerLink	Evolution of mining technologies. Key discoveries and resource estimations. Early regulatory and environmental considerations	Comprehensive historical overview. Establishes foundation for modern developments	Limited discussion on contemporary impacts and technologies
The Post-Second World War Deep-Sea Minerals Scene | SpringerLink	Technological advancements and geopolitical interests. International cooperation and competition. Emergence of environmental awareness and regulations	Thorough post-war period analysis. Insight into international dynamics	May lacks depth in modern advancements and environmental concerns
Major Ocean Database that will Guide Deep-Sea Mining has Flaws, Scientists Warn | Nature.com	Inaccuracies and gaps in the database. Potential consequences of flawed data. Call for improved data accuracy and transparency	Highlights critical data reliability issues. Stresses importance of accurate environmental assessments	Primarily focuses on database flaws rather than broader deep-sea mining context
Deep-Sea Mining Opponents: There's No Free Lunch When it Comes to Clean Energy | Nature.com	Environmental and ecological risks. Arguments from various stakeholders. Trade-offs between mineral extraction and environmental protection	Balanced perspective on environmental vs. economic benefits. Emphasizes sustainable practices	May lack technical details on mining technologies and processes
Deep Sea Mining | SpringerLink	Current and emerging mining technologies. Environmental concerns and mitigation. Existing regulatory frameworks and suggested improvements	Comprehensive coverage of technical, environmental, and regulatory aspects	May be too broad for readers seeking detailed analysis in specific areas

4. DEEP SEA MINING CHALLENGES AND FUTURE DIRECTIONS

Deep-sea mining operations involve significant challenges due to the harsh and remote environment. Storms pose a major threat to these operations, causing damage to infrastructure and posing risks to personnel. The selection of evacuation parameters, such as heading and speed, is crucial to ensure the safety of the mining riser system and the mining vehicle. The upper rotation, maximum equivalent stress, and offset at the moon pool are critical factors that need to be analysed during the evacuation process (Vignesh and Ratnakumar, 2024; Verma et al., 2023). The seabed terrain trafficability analysis is also important to ensure the mining vehicle can adapt to changing seabed topography and water depth. Intelligent obstacle avoidance route planning using the improved A* algorithm is crucial to quickly move away from storm risks and ensure the safety of the mining riser system and the mining vehicle (Liang et al., 2024). The algorithm considers factors such as time cost, risk cost, and the impact of storms on the path to optimize the evacuation route.

The deep-sea mining sector has been hampered by serious environmental issues. Mining operations risk completely disrupt ocean ecosystems and biodiversity, potentially affecting marine life such as whales, dolphins, and fish (Center for Biological Diversity, n.d.). ISA is developing regulations for deep-sea mining, but the process is ongoing, and complaints about a lack of sustainability standards remain. In addition, the deep ocean is largely unexplored, making it impossible to properly comprehend the potential effects of deep-sea mining. The lack of information impedes the establishment of effective laws and mitigation solutions, making the industry vulnerable to environmental damage (World Resources Institute, n.d.).

Deep sea mining is still in the exploratory stage, with numerous technical, financial, and regulatory difficulties that must be overcome before it can become a viable commercial operation (London School of Economics and Political Science, n.d.). Developing the technology and engineering solutions required to operate safely and sustainably in the extreme deep-sea environment, which is distinguished by high pressure, low temperature, and complicated geology and hydrodynamics. Establishing a strong regulatory framework and governance mechanism through the International Seabed Authority (ISA) to ensure that deep sea mining activities are environmentally responsible and do not cause irreversible harm to delicate deep-sea ecosystems (Guo et al., 2023). Addressing the financial and economic uncertainties surrounding deep sea mining, a capital-intensive venture with severe technical and regulatory concerns. Ensure it is a financially viable alternative to land-based mining. Balancing the requirement for essential minerals for the green energy transition against environmental concerns about the effects of deep-sea mining. Developing more sustainable extraction technologies that cause minimal disruption to deep marine environments. Improving scientific understanding of deep-sea environments and ecosystems in order to better assess and minimize the environmental consequences of possible mining operations (Mongabay, n.d.).

Future directions of deep-sea mining involve improving storm evacuation operations by optimizing evacuation routes and parameters such as speed and heading. Advanced riser system design and environmental impact assessments are crucial for ensuring the safety and sustainability of deep-sea mining operations. Technological innovations, such as more efficient energy harvesting systems and advanced drilling methods, will also play a key role in improving the efficiency and safety of deep-sea mining. International cooperation and regulation will be essential for ensuring responsible and sustainable practices in the industry. Public engagement and education will also be vital for addressing concerns and ensuring public support for deep-sea mining. The regulatory authorities need to assess the environmental impacts and the potential risks in mining and develop strategies for mitigation and management. These regulatory frameworks must control GHG emissions, protect habitats, and ensure the long-term sustainability of mining operations. Overall, the future of deep sea mining remains uncertain, with enormous technical, regulatory, and environmental difficulties that must be addressed before mining turns into a large-scale commercial reality.

REFERENCES

3D at Depth. (n.d.). Applications: Deep-sea mining. https://www.3datdepth.com/applications/deep-sea-mining

Blanchard, C., Harrould-Kolieb, E., Jones, E., & Taylor, M. L. (2023). The current status of deep-sea mining governance at the International Seabed Authority. *Marine Policy*, 147, 105396. DOI: 10.1016/j.marpol.2022.105396

Bräger, S., Rodriguez, G. Q. R., & Mulsow, S. (2020). The current status of environmental requirements for deep seabed mining issued by the International Seabed Authority. *Marine Policy*, 114, 103258. DOI: 10.1016/j.marpol.2018.09.003

Briones, E. E. (2023). Deep-Sea Life. In *Mexican Fauna in the Anthropocene* (pp. 319–331). Springer International Publishing. DOI: 10.1007/978-3-031-17277-9_15

Center for Biological Diversity. (n.d.). Center for Biological Diversity. https://www.biologicaldiversity.org

Chen, Q., Yang, J., Zhao, W., Tao, L., Mao, J., & Li, Z. (2024). Algorithms for dynamic control of a deep-sea mining vehicle based on deep reinforcement learning. *Ocean Engineering*, 298, 117199. DOI: 10.1016/j.oceaneng.2024.117199

Cong, Y., Gu, C., Zhang, T., & Gao, Y. (2021). Underwater robot sensing technology: A survey. *Fundamental Research (Beijing)*, 1(3), 337–345. DOI: 10.1016/j.fmre.2021.03.002

Cronan, D. S. (2024). The Post-Second World War Deep-Sea Minerals Scene. *Deep-Sea Minerals Developments in the 20th Century*, 3-15.

Danovaro, R., Fanelli, E., Aguzzi, J., Billett, D., Carugati, L., Corinaldesi, C., Dell'Anno, A., Gjerde, K., Jamieson, A. J., Kark, S., McClain, C., Levin, L., Levin, N., Ramirez-Llodra, E., Ruhl, H., Smith, C. R., Snelgrove, P. V. R., Thomsen, L., Van Dover, C. L., & Yasuhara, M. (2020). Ecological variables for developing a global deep-ocean monitoring and conservation strategy. *Nature Ecology & Evolution*, 4(2), 181–192. DOI: 10.1038/s41559-019-1091-z PMID: 32015428

Guo, X., Fan, N., Liu, Y., Liu, X., Wang, Z., Xie, X., & Jia, Y. (2023). Deep seabed mining: Frontiers in engineering geology and environment. *International Journal of Coal Science & Technology*, 10(1), 23. DOI: 10.1007/s40789-023-00580-x

Hallgren, A., & Hansson, A. (2021). Conflicting narratives of deep sea mining. *Sustainability (Basel)*, 13(9), 5261. DOI: 10.3390/su13095261

Heinrich, L., Singh, P., Stegen, K. S., & Markus, T. (2024). Mind the gap and close it: Regulating greenhouse gas emissions from deep-sea mining in the Area. *Marine Policy*, 160, 105929. DOI: 10.1016/j.marpol.2023.105929

Janarthanan, C., Chandran, V., Sundaramoorthi, V., Vishwanath, B. O., Venkatesan, K., Muthuvel, P., & Ramadass, G. A. (2022, February). Light Weight Nodule Collection System for Deep Sea Mining Machine. In *OCEANS 2022-Chennai* (pp. 1–7). IEEE.

Jones, D. O., Kaiser, S., Sweetman, A. K., Smith, C. R., Menot, L., Vink, A., Trueblood, D., Greinert, J., Billett, D. S. M., Arbizu, P. M., Radziejewska, T., Singh, R., Ingole, B., Stratmann, T., Simon-Lledó, E., Durden, J. M., & Clark, M. R. (2017). Biological responses to disturbance from simulated deep-sea polymetallic nodule mining. *PLoS One*, 12(2), e0171750. DOI: 10.1371/journal.pone.0171750 PMID: 28178346

Kang, Y., Su, Q., & Liu, S. (2022). On the axial thrust and hydraulic performance of a multistage lifting pump for deep-sea mining. *Ocean Engineering*, 265, 112534. DOI: 10.1016/j.oceaneng.2022.112534

Katona, S., Paulikas, D., Ali, S., Clarke, M., Ilves, E., Lovejoy, T. E., Madin, L. P., & Stone, G. S. (2023). Land and deep-sea mining: The challenges of comparing biodiversity impacts. *Biodiversity and Conservation*, 32(4), 1125–1164. DOI: 10.1007/s10531-023-02558-2

Leng, D., Shao, S., Xie, Y., Wang, H., & Liu, G. (2021). A brief review of recent progress on deep sea mining vehicle. *Ocean Engineering*, 228, 108565. DOI: 10.1016/j.oceaneng.2020.108565

Liang, W., Lou, M., Chen, Z., Qin, H., Zhang, C., Cui, C., & Wang, Y. (2024). An enhanced ant colony optimization algorithm for global path planning of deep-sea mining vehicles. *Ocean Engineering*, 301, 117415. DOI: 10.1016/j.oceaneng.2024.117415

Liu, Z., Liu, K., Chen, X., Ma, Z., Lv, R., Wei, C., & Ma, K. (2023). Deep-sea rock mechanics and mining technology: State of the art and perspectives. *International Journal of Mining Science and Technology*, 33(9), 1083–1115. DOI: 10.1016/j.ijmst.2023.07.007

Lodge, M. W., & Verlaan, P. A. (2018). Deep-sea mining: International regulatory challenges and responses. *Elements (Quebec)*, 14(5), 331–336. DOI: 10.2138/gselements.14.5.331

London School of Economics and Political Science. (n.d.). London School of Economics and Political Science. Retrieved from https://www.lse.ac.uk

Madureira, P., Squires, D., & Ribeiro, L. P. (2023). The International Seabed Authority and the United Nations 2030 Agenda for sustainable development. *Resources Policy*, 86, 104166. DOI: 10.1016/j.resourpol.2023.104166

Mongabay. (n.d.). Mongabay: Environmental news. Retrieved from https://www.mongabay.com

Muhammad, A., & Wu, T. (2023). Examining the evacuation routes of the sister village program by using the ant colony optimization algorithm. *Open Geosciences*, 15(1), 20220512. DOI: 10.1515/geo-2022-0512

Nyka, M. (2020). International Seabed Authority and Environmental Deep-Sea Stewardship–Principles Governing the Protection and Use of Seabed Resources. *Prawo Morskie*, 9-20.

Orcutt, B. N., Bradley, J. A., Brazelton, W. J., Estes, E. R., Goodall, J. M., Huber, J. A., & Pachiadaki, M. (2020). Impacts of deep-sea mining on microbial ecosystem services. *Limnology and Oceanography*, 65(7), 1489–1510. DOI: 10.1002/lno.11403

Sakellariadou, F., Gonzalez, F. J., Hein, J. R., Rincón-Tomás, B., Arvanitidis, N., & Kuhn, T. (2022). Seabed mining and blue growth: Exploring the potential of marine mineral deposits as a sustainable source of rare earth elements (MaREEs) (IUPAC Technical Report). *Pure and Applied Chemistry*, 94(3), 329–351. DOI: 10.1515/pac-2021-0325

Saranya, P. V. P, R. Maheswari, P. Palanisamy and J. Mani, "Intelligent System for Improving Scalability by Sharding for Agriculture Supply Chain," 2024 *11th International Conference on Reliability, Infocom Technologies and Optimization* (Trends and Future Directions) (ICRITO), Noida, India, 2024, pp. 1-6.

Saranya, P., Shirly, S., & Logeshwari, V. "Development of a Blockchain Vehicle Manufacturing Nctwork Using Hyperledger Composer," *2023 International Conference on Innovative Computing, Intelligent Communication and Smart Electrical Systems (ICSES)*, Chennai, India, 2023, pp. 1-9, DOI: 10.1109/ICSES60034.2023.10465503

Selvi, S., Aggarwal, K., Pandurangan, R., Vijayan, V. P., Ali, A., & Anuradha, K. (2024). Enhancing the accuracy of target detection in remote video surveillance analytics through federated learning. *Optical and Quantum Electronics*, 56(2), 185. DOI: 10.1007/s11082-023-05664-1

Sharma, R. (2011). Deep-sea mining: Economic, technical, technological, and environmental considerations for sustainable development. *Marine Technology Society Journal*, 45(5), 28–41. DOI: 10.4031/MTSJ.45.5.2

Sharma, R. (Ed.). (2019). *Environmental issues of deep-sea mining: impacts, consequences and policy perspectives*. Springer. DOI: 10.1007/978-3-030-12696-4

Silva-Send, N. (2024). Deep Sea Mining in the Area Beyond National Jurisdiction–A Lost Opportunity or Yet Another Reason for the United States to Join UNCLOS?. *San Diego Legal Studies Paper*, (24-011).

Singh, P. A. (2022). Deep seabed mining and sustainable development goal 14. In *Life Below Water* (pp. 271–283). Springer International Publishing. DOI: 10.1007/978-3-319-98536-7_135

Spearman, J., Taylor, J., Crossouard, N., Cooper, A., Turnbull, M., Manning, A., Lee, M., & Murton, B. (2020). Measurement and modeling of deep-sea sediment plumes and implications for deep sea mining. *Scientific Reports*, 10(1), 5075. DOI: 10.1038/s41598-020-61837-y PMID: 32193479

Toro, N., Robles, P., & Jeldres, R. I. (2020). Seabed mineral resources, an alternative for the future of renewable energy: A critical review. *Ore Geology Reviews*, 126, 103699. DOI: 10.1016/j.oregeorev.2020.103699

Van Der Grient, J. M. A., & Drazen, J. C. (2021). Potential spatial intersection between high-seas fisheries and deep-sea mining in international waters. *Marine Policy*, 129, 104564. DOI: 10.1016/j.marpol.2021.104564

Verma, B., Yudheksha, G. K., Sanjana Reddy, P., & Vignesh, U. (2023). An Intelligent Flood Automation System Using IoT and Machine Learning. In *Recent Developments in Electronics and Communication Systems* (pp. 444–449). IOS Press. DOI: 10.3233/ATDE221295

Vignesh, U., & Ratnakumar, R. (2024). An Empirical Review on Clustering Algorithms for Image Segmentation of Satellite Images. AI and Blockchain Optimization Techniques in Aerospace Engineering, 33-52.

Wang, R., Chen, G., Zhang, N., & Zhang, H. (2023). Study on storm evacuation operations in deep-sea mining. *Ocean Engineering*, 281, 114690. DOI: 10.1016/j.oceaneng.2023.114690

Wang, Z., & Zou, L. (2022). Natural frequency analysis of deep-sea mining riser considering varying tension and buffer station. *Ocean Engineering*, 264, 112372. DOI: 10.1016/j.oceaneng.2022.112372

World Resources Institute. (n.d.). World Resources Institute. Retrieved from https://www.wri.org

Wu, Q., Yang, J., Guo, X., Liu, L., Lu, W., & Lu, H. (2022). Experimental study on dynamic responses of a deep-sea mining system. *Ocean Engineering*, 248, 110675. DOI: 10.1016/j.oceaneng.2022.110675

Yang, P., Wu, H., He, C., & Luo, S. (2023). Underwater image restoration for seafloor targets with hybrid attention mechanisms and conditional generative adversarial network. *Digital Signal Processing*, 134, 103900. DOI: 10.1016/j.dsp.2022.103900

Zahid, S. M., Najesh, T. N., Ameen, S. R., & Ali, A. (2023). A Multi Stage Approach for Object and Face Detection using CNN. In *2023 8th International Conference on Communication and Electronics Systems (ICCES)* (pp. 798-803). IEEE. DOI: 10.1109/ICCES57224.2023.10192823

Zhang, C., Lin, Y., & Pan, W. (2018, July). Research on real-time processing system of deep sea biological image for manned submersible vehicle. In *2018 37th Chinese Control Conference (CCC)* (pp. 4006-4010). IEEE. DOI: 10.23919/ChiCC.2018.8483593

Zhou, G., Li, C., Zhang, D., Liu, D., Zhou, X., & Zhan, J. (2021). Overview of underwater transmission characteristics of oceanic LiDAR. *IEEE Journal of Selected Topics in Applied Earth Observations and Remote Sensing*, 14, 8144–8159. DOI: 10.1109/JSTARS.2021.3100395

Chapter 8 Generating Electricity From Sound Vibrations Underwater

Maheswari Raja
Rajalakshmi Institute of Technology, Chennai, India

K. M. Rajeshwari
Sri Eshwar College of Engineering, India

K. R. Keerthikhaa
Sri Eshwar College of Engineering, India

V. Koushika
Sri Eshwar College of Engineering, India

G. S. Roshini
Sri Eshwar College of Engineering, India

ABSTRACT

One enormous, unexplored source of renewable energy is the ocean. Sound energy produced by sound waves, especially those produced by marine life, has the potential to be used effectively to create electricity. Materials known as piezoelectric materials generate electricity when they experience mechanical stress, such as vibrations from sound waves. It is therefore feasible to generate power using the movement of whales by putting piezoelectric materials in places where whales congregate, such as along migration routes. Whale movements in the water induce vibrations in piezoelectric materials, which then transform mechanical stress into electrical energy for use in a variety of applications. The piezoelectric materials are selected in such a way which

DOI: 10.4018/979-8-3693-6670-7.ch008

are sensitive to low frequency vibrations caused by whales and also environment friendly and protecting marine life. This approach will serve as an efficient way to collect energy from natural environment without harming marine life.

INTRODUCTION

From decades, Earth is reminding humans to find better ways to produce power rather than burning fossil fuels which result in the emission of harmful gases into the environment. The world energy consumption is expected to increase exponentially in the next decade and if the same continues towards the environment, there will be a day when no clean air would be available. Traditional practices of producing energy, regularly reminds us the serious threat they pose to the environment. Thus, there is a need to produce independent and clean electricity. Sound energy from marine sources can be observed as a possible clean energy resource which can be used for power generation purposes. While this method proposes an effective way to generate electricity from the sources available in marine environment.

Sustainable energy is created from the resources that won't cause any damage to the energy or climate of future generations and maintain the current operations. Some of the sustainable energy sources are wind, solar power, hydro power and they are renewable too which means they are replenished naturally. Hydro power is used as the largest source of renewable energy which utilize the real flow of moving water for generating electricity. The sustainable energy diversify energy sources and it also acts as a phenomenon to produce other types of energy.

Water is the basic source of living and it is considered as the great source of energy. When water is used as an energy source, it is called as hydroelectric power. Every source of energy require water in their production process. In general manner, sea is defined as a portion of the ocean which is partially surrounded by land. Under certain cases, a sea is completely landlocked. A sea can be very warm for most of the year and also it can be quite cold. The marine ecosystem within each sea will differ based on the variety of the sizes, temperature and location. Seas are smaller than oceans and they are located between where the land and ocean meet. The ocean almost covers 70% of the surface of the Earth. Basically, the ocean influences and regulates the Earth's climate and acts as the planet's heart.

Sound vibrations from the sea refers to the transmission of sound waves through water. This term is known as hydro acoustics or underwater acoustics.

Sound vibrations as a form of sustainable energy is an innovative concept which involves the by conversion of acoustic energy into electrical energy. Being in the experimental stages, there are some possible approaches for this technology. One of the important key concepts is Acoustic Energy Harvesting which requires the capturing of sound waves from the sea and converts it into electrical energy by using various methods. Some of the methods includes using piezoelectric materials, electrostrictive materials and triboelectric nanogenerators. Piezoelectric materials generate an electric charge in response to mechanical stress which also includes the vibrations from sound waves. Piezoelectricity generates electricity from the pressure created. The physical transformation of a piezoelectric material from one configuration to a different one produce an electrical charge by sound vibrations. The piezoelectric materials can be used in the ecosystem with constant or heavy noise and vibrations for generating small amounts of electricity. The electrostrictive materials change their shape when it is exposed to electric field and also convert sound into electrical energy through their physical transformation. Sea energy resources derived from waves can be used and transformed to large scale sustainable electrical energy.

The blue whale is a largest marine mammal in the world with a body length of up to 30 m and a weight of 80 m to 150 metric tons(approximately) and has an average life span of 80 to 90 years.

Blue whales use their special physiological structure to make sounds to communicate their underwater activities between whales. Its voice is very loud, has a low frequency, can reach more than 800 km away, and has strong harmonic signal contents. Magnitude of the sound of blue whales could reach 155 to 188 dB, with a base frequency of approximately 10 to 40 Hz, and a sound duration of approximately 10 to 30 s. The three main types of sounds made by whales are clicks, whistles, and pulsed calls. Clicks are believed to be for navigation and identifying physical surroundings. When the sound waves bounce off of an object, they return to the whale, allowing the whale to identify the shape of the object. The range of frequencies that whales use is from 30 Hertz (Hz) to about 8,000 Hz, (8 KHZ). Whistles and pulsed calls are used during social activities. Pulsed calls are more frequent, sound like squeaks, screams, and squawks to the human ear. Whales also use their tails and fins to make loud slapping noises on the surface of the water to communicate silently.

Blue whale calls ranged between 188 and 191 dB, while fin whale sounds ranged between 189 and 192 dB, all blue whale groups make calls at a fundamental frequency of between 10 and 40 Hz, and the lowest frequency sound a human can typically perceive is 20 Hz. Blue whale calls last between ten and thirty seconds. Toothed whales can produce sounds using their larynx as well as specialized air sacs near their blowhole. The larynx is used to produce whistles, while the air sacs are used to produce high frequency echolocation signals. The larynx contains folds of tissue called the vocal cords (also called vocal folds), which vibrate as air passes from the lungs into the oral cavity (also called the mouth). The shape and tension of the vocal cords can change to produce different sounds. The sounds can be affected by changes in the shape of the buccal cavity, tongue, and lips.

They can produce low-frequency, mid-frequency and high-frequency sounds. Blue whales and fin whales produce low-frequency sounds (10-40) that can travel thousands of kilometers. These sounds can propagate over such long distances due to their low frequency, which experiences less attenuation in water.

Humpback Whales, known for their songs, which have frequencies between 20 Hz and 24 kHz. These sounds can travel hundreds of kilometers but are less far-reaching than the low-frequency sounds. Estimated Range for low-frequency is hundreds to a few thousand kilometers. Toothed Whales produce high-frequency clicks and whistles (above 10 kHz) used for echolocation and communication. High-frequency sounds attenuate more quickly in water. Estimated Range for high-frequency is tens of kilometers, typically up to 100 kilometers under favorable conditions.

People often think that the underwater world is silent. In fact, numerous marine organisms use sound for communication, reproduction, and to seek prey.The in depth concept of production of sound vibrations underwater is that, When underwater objects vibrate, they create sound-pressure waves. These waves alternately compress and decompress the water molecules as the sound wave travels through the sea. These sound waves which is in the form of vibrations can be trapped and converted into electricity by applying suitable techniques. Sources of sound vibrations from oceans include sounds from ship engines, underwater earthquakes, military sonar, marine animals, breaking waves, seismic exploration. Reason for choosing sound vibrations from underwater rather than from atmosphere is that, Sound travels faster in water compared to air because water is denser, so the energy it carries is transported faster. This makes the sound appear louder. Sound waves radiate in all directions away from the source like ripples on the surface of a pond. To extract electricity from sound vibrations, the frequencies of the vibrations underwater must be 100 million times higher than people can hear. The basic components of a sound wave are frequency, wavelength and amplitude. Sound vibrations can be trapped using a SOFAR channel (SOund Fixing And Ranging). With the help of this channel, small acoustic energy could travel over long distance. Along with this channel, hydrophones are used to detect sounds from ocean in all directions. As the oceans now a days become very noisy each year, the field of ocean acoustics will grow and definitely become more essential in one or the other way to us.

From decades, Earth is reminding humans to find better ways to produce power rather than burning fossil fuels which result in the emission of harmful gases into the environment. The world energy consumption is expected to increase exponentially in the next decade and if the same continues towards the environment, there will be a day when no clean air would be available. Traditional practices of producing energy, regularly reminds us the serious threat they pose to the environment. Thus, there is a need to produce independent and clean electricity. Sound energy from marine sources can be observed as a possible clean energy resource which can be used for power generation purposes. While this method proposes an effective way to generate electricity from the sources available in marine environment.

- Piezoelectric Conversion: When mechanical stress caused by sound vibrations is applied to piezoelectric materials, an electric charge is produced. Piezoelectric transducers can be positioned strategically in marine environments to harvest and transform the energy from sound waves into electrical power.
- The Effect of Maritime Traffic As ships move across the oceans, their engines and other machinery produce a constant stream of noise. In addition to harming marine life, this artificial noise pollution is a substantial source of wasted energy.
- Offshore Wind Farms: Offshore windmills are a renewable energy source with a lot of promise as they use the kinetic energy of the wind to create electricity. The integration of extra energy collection devices is made stable by these structures.
- Piezoelectric Retrofitting: The simultaneous capture of acoustic and wind energy is made possible by the incorporation of piezoelectric materials into the foundation and support components of offshore windmills. By reducing its environmental impact, this synergistic strategy maximizes offshore energy generation's sustainability and efficiency.

Naqvi et al. (2022) states that Despite its abundance, ocean wave energy remains underutilized due to its low frequency. This study presents a beam-column piezoelectric-based system designed for optimal ocean wave energy harvesting (OWEH). The electromechanical equations of the system are derived and validated through experiments. The system has been tested on buoys in San Francisco and Boston, demonstrating its efficiency in powering buoy devices, particularly under conditions of large waves and low frequency. A self-tuning buoy adapts to wave frequency, enhancing energy efficiency.

This new metric helps optimize the tuning of the OWEH system to match ocean wave properties. This technology opens new research avenues for effectively harnessing ocean wave energy, promoting sustainable energy solutions.

According to Viet et al. (2017) Energy from ocean waves has long been captured by piezoelectric materials. This paper aims to provide an overview and analysis of the technological developments. The benefits of using piezoelectric energy conversion technology are first determined by comparing the three main energy conversion methods—electrostatic, electromagnetic, and piezoelectric technologies—in terms of power generation capability, transmission efficiency, structural installation, and financial costs. Second, based on various piezoelectric effects, the article summarizes various approaches and ideas for capturing energy from ocean waves. Specifically, the introduction and discussion of harvester designs and efficiency, rely on the piezoelectric effects from longitudinal, bending, and shear couplings.

This study significantly advances our understanding of hearing mechanisms in baleen whales, particularly the fin whale (Balaenoptera physalus). By using finite element modeling on CT scans of a fin whale's head, we have identified two primary mechanisms for sound reception: skull-vibration enabled bone conduction and pressure transmission through soft tissues, with bone conduction being the predominant mechanism. These findings suggest that baleen whales are particularly sensitive to low-frequency sounds, which is crucial given the overlap with anthropogenic noise in the ocean.

Bahaj (2011) stated that, the universal characteristics of the bony ear complexes across Mysticeti imply that these sound reception mechanisms are likely common among all baleen whales. This sensitivity to low-frequency sounds highlights the need for regulatory agencies and industrial ocean users to consider the impact of human-generated noise on these marine giants. Our research provides a robust foundation for predicting auditory sensitivity in baleen whales and underscores the importance of mitigating noise pollution to protect these species.

In Farrok et al. (2020), the theory, creation, and design of electro-acoustic transducers for underwater applications are all covered in this research article. This well praised text covers both potential new designs and the fundamentals of currently used piezoelectric and magnetostrictive transducers. Both the fundamental and specialised acoustics information required for transducer design and evaluation are presented. A wide variety of projector and hydrophone designs are covered in detail, along with modelling, assessment, and measuring techniques. Numerical models for elements and arrays, as well as the effects of mutual radiation impedance, are discussed in the analysis of projector and hydrophone transducer arrays.

In Elghali et al. (2007), Humpback whale feeding noises were captured underwater 90-130 kilometres south of Juneau in August 1975. Although there were scattered grunts, yelps, groans, and low-frequency narrow-band pulses, as predicted, the most ubiquitous noises were trains of 25-80 Hz broadband pulses with a mean repetition rate of one per second. The pulse patterns lasted between 15 and 100 seconds and were accompanied by continuous noise at a frequency of 1250 Hz. Surfaced creatures emitted noises like as shrieks and trumpeting from blowholes, as well as wide-band pulses from flippers or tail slaps on the water surface. Weaker wide-band pulse trains were seen to have low energy below 500 Hz, inconsistent amplitude, and repetition.

LITERATURE SURVEY

According to Shaw (2021), power harvesting is the technique of obtaining the ambient energy of a system and transforming it into electrical energy that may be used. Power harvesting research has seen a significant increase in the past few years. Modern advancements in low-power electronics, such as microelectromechanical systems, and wireless technologies have led to an increase in research. Many avenues for the actual real-world uses of power harvesting devices have been made possible by the advancements.

In Tom and Yeung (2015), one approach to power harvesting that has seen a sharp increase in use is the utilization of piezoelectric materials to take advantage of the ambient vibrations surrounding a system. Because of their crystalline structure, piezoelectric materials are able to convert mechanical strain energy into electrical energy.

In Wu et al. (2015), piezoelectric energy harvesting has garnered a lot of attention, according to the research's author, because of advancements made in the last few decades in the extraction of electrical energy from fluid movements, such as wind and water. This review study looks at several approaches and vibration-based mechanisms with the aim of improving the efficiency of piezoelectric energy harvesters. The invention of self-powering devices that use highly flexible and biocompatible piezoelectric materials has made it possible to power portable electronics and medical equipment. Size, material type, and transduction mechanism are the main factors influencing harvester performance. A useful alternative energy source for urban environments is offered by vibration-based harvesters. Future studies should focus on enhancing wind and water flow factors in order to optimize the efficiency of piezoelectric generators.

In this research, the authors Tom and Yeung (2015) compares electromagnetic and electrostatic methods with the development of piezoelectric materials for ocean wave energy collection. The advantages of piezoelectric technology include lower installation and operating expenses as well as a higher energy generating density. This work presents an overview of diverse approaches and designs that are based on shear, bending, and longitudinal couplings, among other piezoelectric phenomena. Future study should concentrate on employing greater piezoelectric effects like d33 and d15, which give three times the energy conversion capabilities, even though present harvesters only generate modest power using the d13 effect. Higher frequency excitations, force magnification tools, and resonance principles can all be used to increase power output.

Kok (2015) says that ocean energy presents a plethora of prospects for sustainable power generation, as it is sourced from tides, marine currents, and wind-driven waves. The current state of wave and marine current energy technologies is reviewed in this study; these technologies are still in their infancy, with very few having reached the commercial prototype level.

Wave energy is less predictable than marine currents, which are reliable and useful in the absence of barrages. The study provides a methodical framework for technology and project development, highlighting the necessity of operational experience in order to inspire confidence in investors. The establishment of strong supply chains, maintenance, manufacturing capacity, and operational dependability are important concerns. Highlighted are advancements in the UK, with attention to regulatory assistance and financial structures. Milestones such as the Crown Estate's projects and the deployment of grid-connected devices demonstrate progress despite the high present expenses. However, challenges remain, such as funding, resource characteristics, and the proliferation of diverse conversion technologies. With continued support and strategic funding, the future for wave and marine current energy conversion appears promising. This research was carried over by Khan et al. (2022).

Wu et al. (2011) explains that, the inherent transduction qualities are connected with single crystals in the structural configuration of piezoelectric crystals, such as crystal films. Numerous small-scale quartzes are randomly orientated and display piezoelectric capabilities when an electric potential is applied in the case of ceramic type piezoelectric devices like barium titanate (BaTiO3 and AlN).

According to Islam et al. (2024), only low-frequency (<1000 Hz) baleen whales have been able to record whale noises autonomously over extended periods of time due to technological limitations. A High-frequency acoustic recording package (HARP) with sample rates up to 200 kHz was developed in response to the requirement for a wider-band, higher-data capacity system that could record toothed whales and other marine animals autonomously over extended periods of time. HARPs are being used all around the world to acoustically monitor marine mammals for long-term ecological and behavioural research.

Xie et al. (2014a) claims that, Oceans which cover more than 70% of the earth, have been appreciated as vast renewable energy source.

According to Wu et al. (2015) energies in ocean are stored in different forms like thermal energy and kinetic energy in the form of waves and current. It is also stored in chemical and biological products. There have been many proposed methods for obtaining energy from the sea, most of which fall into one of the following categories: energy from tidal and marine currents, wave energy, ocean thermal energy, energy from salinity gradients (osmosis), and marine biomass culture. It is possible to turn the kinetic energy that is obtained from the motion of particles during sound waves into electrical energy.

In this research, Naqvi et al. (2022) said that point absorbers, overtopping devices, oscillating water columns, attenuators, oscillating wave surge converters, submerged pressure differentials, rotating masses, and bulge wave converters are only a few of the wave energy devices reviewed in this study. It examines the design, operation, and difficulties associated with producing electricity from these devices. The study emphasizes how wave energy has the potential to be a clean, reliable renewable resource. For direct drive power generation, point absorbers and linear generators show promise. The use of piezoelectric materials to transform mechanical motion into electrical is also covered. In order to maximize power output, the article highlights the significance of cutting-edge control technologies as well as the integration of hydraulic systems and energy collection subsystems. The results indicate that ocean wave energy has a great potential to be extremely important in the future.

Xie et al. (2014b) talked about capturing electrical energy from the longitudinal wave motion of water particles, a novel seawave energy harvester is designed. The harvester makes use of a proof mass and a cantilever substrate with piezoelectric patches. The output charge and voltage are computed using a mathematical model based on the Airy linear wave theory and the classical elastic beam model. The width-to-thickness ratio of the cantilever, wave height, sea depth, proof mass ratio, and sea depth-to-wavelength ratio all improve generated power, according to the results. A feasible sea wave configuration can produce up to 55W of power. Larger harvesters with possible outputs of up to 145W under greater wave circumstances, according to simulations, can create even more power. With the use of piezoelectric devices, this technique offers a novel way to effectively harvest energy from waves in the ocean.

Naqvi et al. (2022) discussed the usage of piezoelectric materials in this chapter, noting that these materials have a high energy generation density and are more cost-effective and structurally installable than electrostatic and electromagnetic methods for obtaining energy from ocean waves. This paper discusses several designs and methods for utilizing longitudinal, bending, and shear couplings, among other piezo-electric effects, to collect energy from ocean waves. It emphasizes the necessity of creating new harvesters with stronger piezoelectric effects (d33 and d15) in order to increase power output, as opposed to the widely employed d13 effect. Utilizing the resonance principle, force magnification tools like hydraulic pistons and levers, and stimulating piezoelectric materials at higher frequencies are some potential avenues for future study to improve power generation and efficiency.

Viet et al. (2016) specifies that, Applications in sensing and actuation are made possible by piezoelectric materials' ability to transform strain energy into electrical energy and vice versa. Use in Vibrating Media: To capture small-scale power, these materials are frequently employed in settings that experience regular vibrations. In the 1970s, the idea of harnessing the piezoelectric effect for ocean energy was proposed first. This technology is still being developed. This paper provides an overview of recent developments in the application of piezoelectric processes in the ocean realm. The goal is to provide the groundwork for further studies by marine engineers who may be interested in this technology. Specific kind of piezoelectric materials are needed for various purposes in ocean engineering.

In order to fulfil the rising need for energy, the production of electrical energy from renewable energy sources, or RESs, has gained popularity in recent decades (Bahaj, 2011). Comparing RESs to energy sources based on fossil fuels, they are environmentally sufficient. Even though fossil fuels are currently the world's primary source of electricity, usage of fossil fuel-based energy sources is predicted to drop to 4% by 2040. The RESs would then supply the remaining energy needed. The Intergovernmental Panel on Climate Change (IPCC) report from 2014 states that by 2050, the sea level from 2010 will have risen by about 0.2 metres. As a result, the landmass and the islands would both be greatly diminished.

Kiran et al. (2020) notes that because piezoelectric materials are lightweight, compact, and have little effect on the environment, electrical engineers have recently concentrated on creating oceanic wave energy conversion technologies. Aspects of piezoelectric OWEC technology such as prospects, historical development, classification, principles of operation, configurations, models, processing, and testing setups are all covered in this review. Piezoelectric materials show promise for converting ocean wave energy since they produce electrical potential when deformed. When compared to the d13 effect, piezoelectric devices—which specifically use the d33 and d15 effects—significantly increase power production, according to comparisons of several energy conversion approaches. In order to generate a higher electrical power output, effective designs include hanging and floating configurations as well as cascading several units.

McDonald et al. (2006) discusses the use of piezoelectric effects and creates an ocean wave energy harvester utilizing the transverse wave motion of water particles. The harvester is made up of two horizontal cantilever plates mounted on a vertical column that have piezoelectric patches on them. Using the elastic beam model and Airy linear wave theory as foundations, a mathematical model determines the output charge and voltage. The root mean square (RMS) of the generated power is affected by a number of factors, including wave height, cantilever length, harvester site, ocean depth, and wavelength to ocean depth ratio. The RMS reduces with increasing distances from the ocean surface and lower wavelength-to-depth ratios and increases with longer cantilevers and bigger waves, according to the results. Under ideal circumstances, with an ocean depth of 10.6 m, a wavelength of 21.2 m, a wave height of 4 m, and a harvester location 2 m below the surface, the harvester can create up to 30 W. This work presents an effective technique for transverse wave energy harvesting utilizing piezoelectric harvesters, which are readily installed on offshore platforms.

MOTIVATION

The main motive of this study is, to replace the method of generating electricity by burning fossil fuels such as coal, natural gas, or oil. Since oceans acts as a source of renewable energy, extracting energy from it serves as a sustainable method of generating electricity from it. Numerous methods of generating electricity from oceans have been already proposed, but this one stands different from others. This method focuses on generating electricity from **sound waves** which remains unused. These unused sound waves are let to collide with piezoelectric materials, which then causes stress on it and converts the wave forms into electrical energy. This seems to be the best method to make use of these underwater sound vibrations.

PROPOSED SOLUTION

This study offers a novel method for harvesting whale and ship noise vibrations to generate renewable energy from the ocean. It is suggested to add piezoelectric materials to underwater structures, such as buoys or moored devices, in order to efficiently convert mechanical energy from these vibrations into electrical energy. These piezoelectric modules would be positioned strategically adjacent to sensors that are able to identify and transmit data regarding sound vibrations quickly. The objective is to use these modules and batteries to store generated electricity so that offshore windmill structures can generate sustainable power. The main theme of this research is the idea of generating electricity from the sound vibrations produced by whales, ship engines and other marine habitats. When the sound vibrations hit the piezoelectric material, it gives a stress on the material which can be converted into electrical energy. This electrical energy is then stored and used for human needs. This method is a sustainable way of extracting electricity. The piezoelectric materials are placed on the offshore windmills.

Figure 1. Extracting electricity from piezoelectric materials

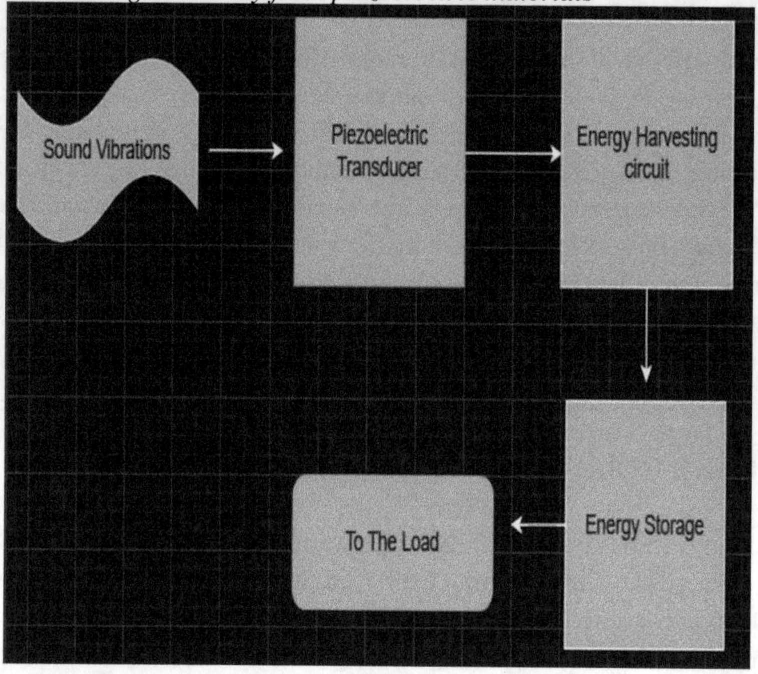

Figure 1 shows the block diagram for extracting electricity from the piezoelectric materials.The sound vibrations touches the piezoelectric transducer which then reaches the energy harvesting circuit.After passing through the circuit, the energy is stored and the supply is given to the load.

Figure 2. Conversion of sound to electrical energy

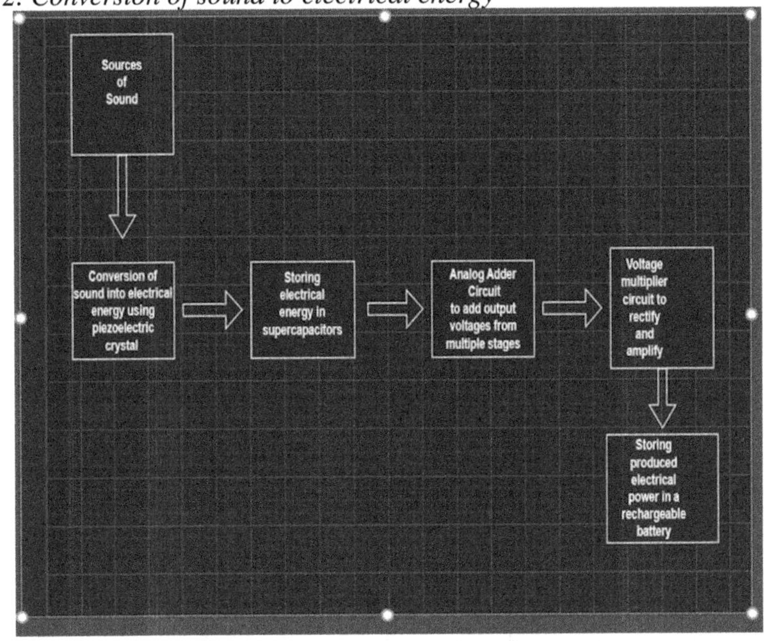

This Figure.2 shows another block diagram of this research work. Sources of sound vibrations are collected and taken to the next step which involves conversion of sound into electrical energy using the piezoelectric crystal. This converted electrical energy is then stored in supercapacitors. Followed by storage, the output circuit is calculated corresponding to the input. Using voltage multiplier, the output circuit is rectified and amplified. The electrical power produced from the above steps is stored in rechargeable battery.

DATASETS:

Data access, mixing, transformation, modelling, and visualization can all be done with the free, open-source KNIME Analytics Platform, which requires little to no coding.

A CSV reader is used to import data from comma-separated files. Use the shuffle function to randomly rearrange the rows of the input tables. The missing value node may manage the cells in the input table that have missing values. The rule engine node compares each row in the input table to a list of user-defined rules.

Low variance filter node removes double-compatible columns whose variance is less than a user-specified threshold. A partition node separated a database data table row.

Figure 3. Global dataset for trade in energy balance

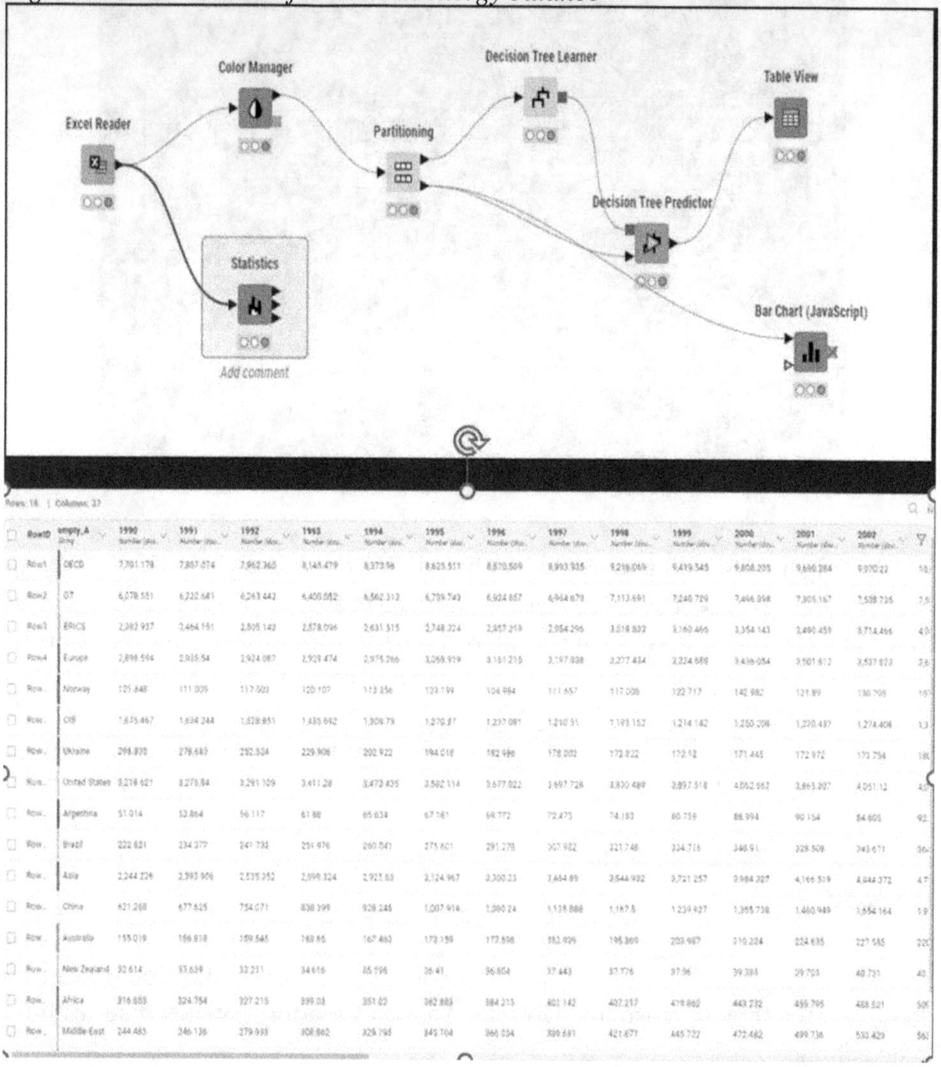

The Figure4 shows the global dataset for trade in energy balance. The Decision Tree Predictor tool uses decision rules that have been learnt to apply pre-trained decision tree models to fresh data and predict class values or labels. In KNIME workflows, these nodes are crucial for predictive modelling.

Figure 4. Share of renewable energy in electricity production

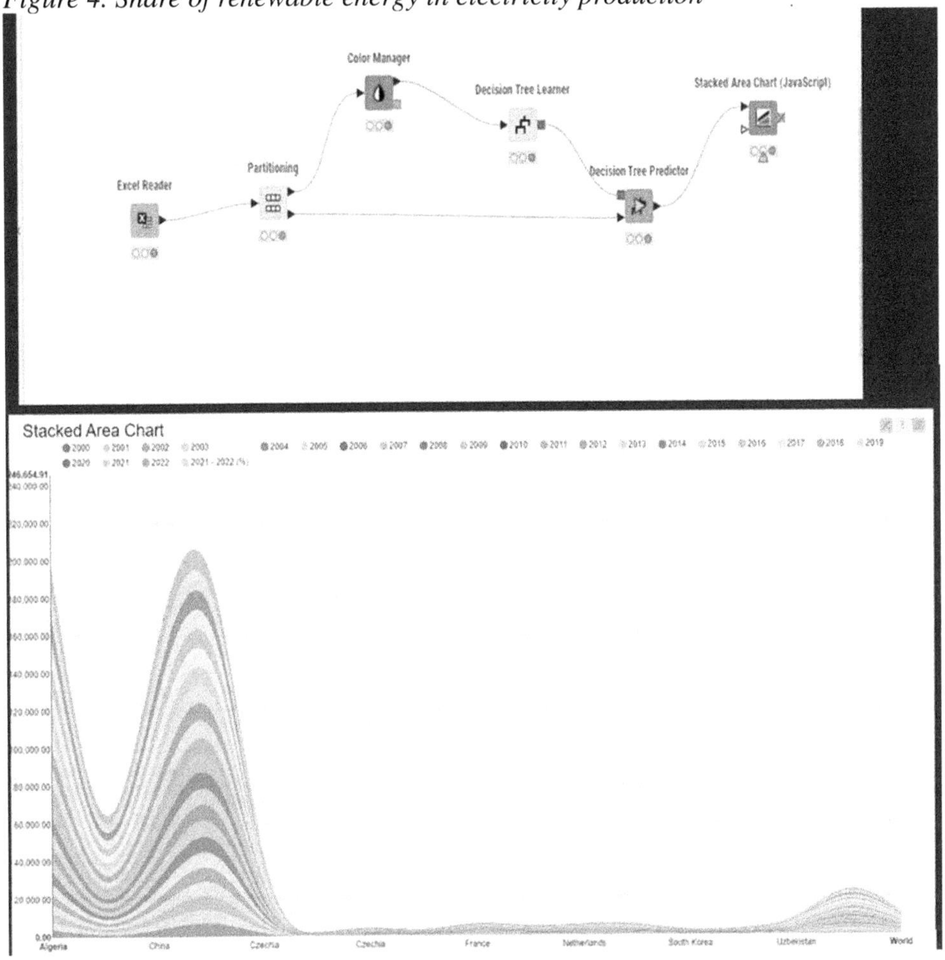

The figure 6 shows the percentage of renewable energy used to produce electricity. It displays the global share of renewable energy in the production of electricity. The KNIME scorer node accesses classification models by comparing the anticipated labels to the existing ones.

Figure 5. Electricity Production

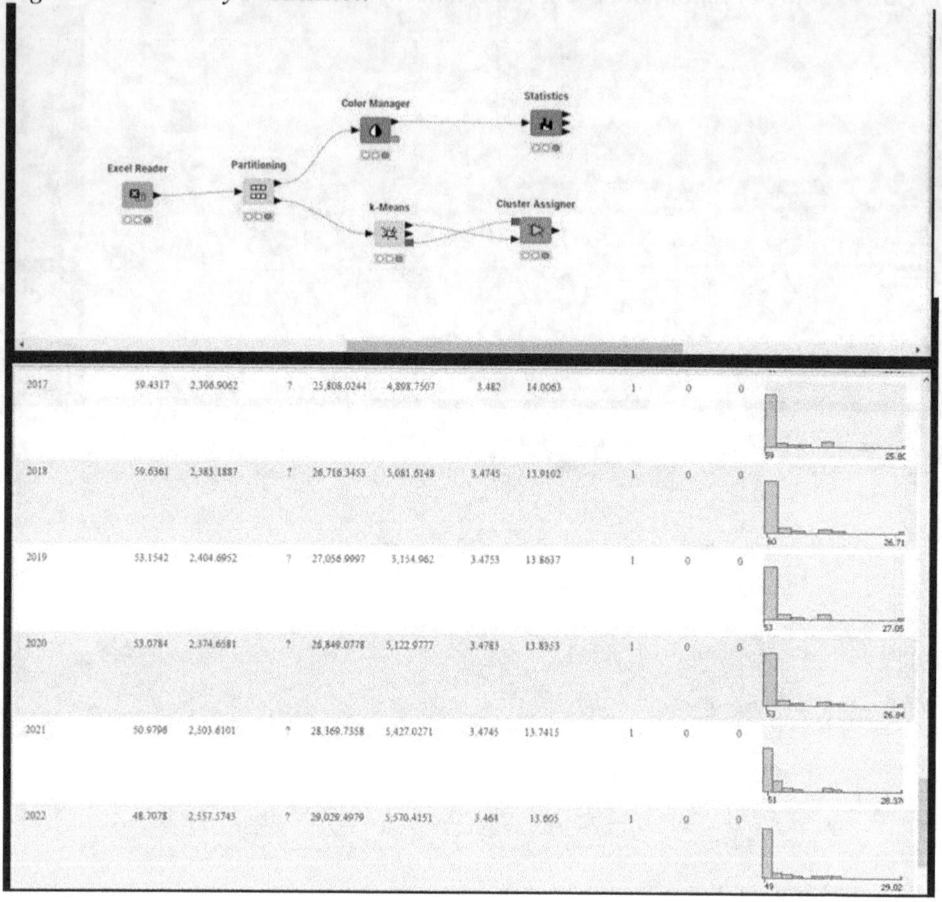

Figure.8 displays the global statistics for total electricity production.

Row Filter uses user-defined logical expressions to filter the rows in KNIME. Boxplot, which facilitates examining the distribution of data.

Figure 6. Total Energy Consumption

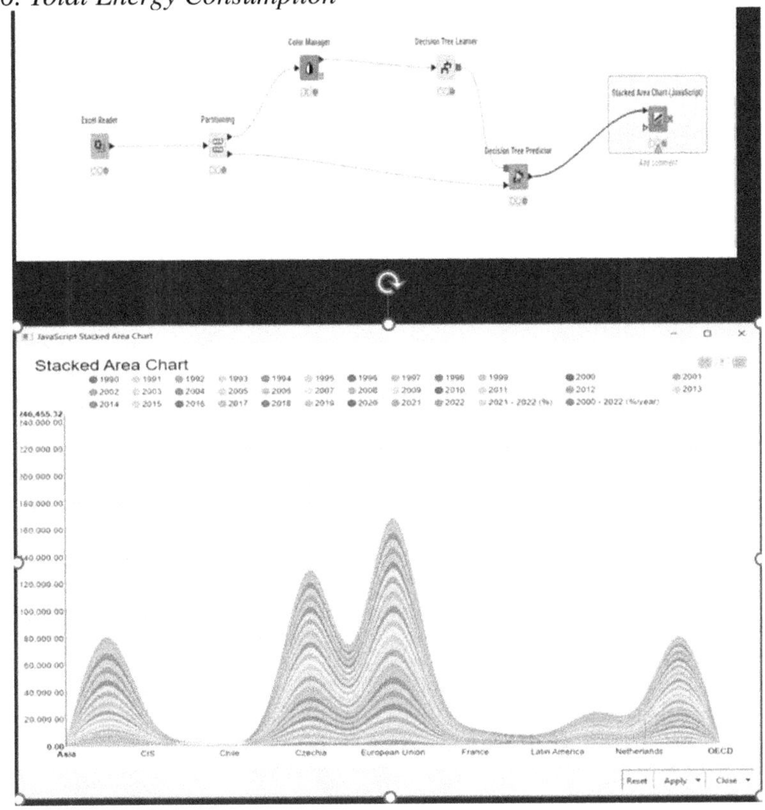

Figure 6 is a representation of the global statistics graph for total energy consumed. helps with data exploration and analysis by offering distribution summaries and descriptive statistics for numerical variables in datasets.

Figure 7. Coal And Lignite Balance Trade

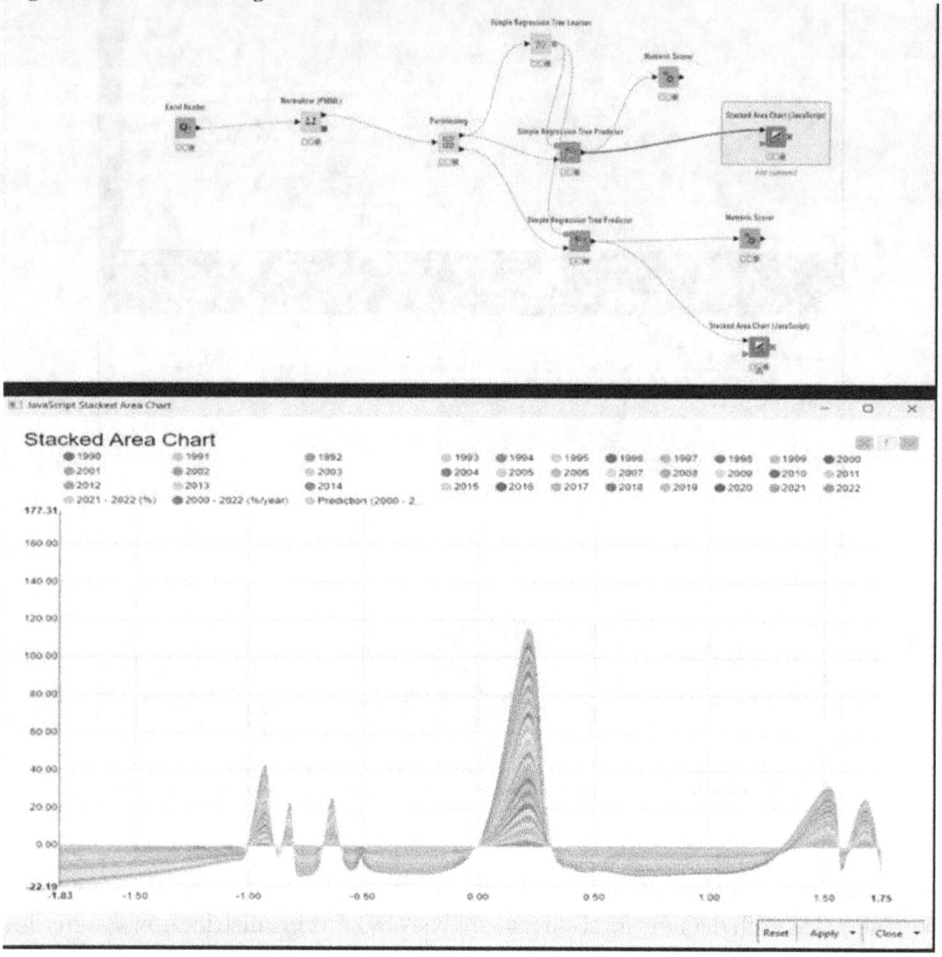

Figure 7 displays the coal and lignite balance graph. Trade Colour Manager allows you to customise the colours of data points, categories, or groups. It also manages colour settings for visualisations in KNIME.

CONCLUSION

One innovative and very promising approach to producing sustainable electricity is to look into marine sources of sound energy. Through the utilisation of piezoelectric, electrostrictive, and triboelectric techniques, we are able to effectively capture the vast amounts of acoustic energy produced by various undersea vibrations. These vibrations, which come from a variety of sources, including the strong engines of large ships, the colourful and rhythmic activities of marine life, and the unchanging yet breathtaking natural events that occur deep within the ocean, contain an untapped energy source that has the potential to revolutionise our approach to renewable energy.

Consider the ocean as a huge orchestra of sound, where all of the movements, interactions, and natural occurrences combine to create a cacophony of energy. It is extremely possible to catch and transform this energy in this dense aquatic medium, where sound travels four times quicker and more efficiently than in air. Water is a thick medium that acts as a natural amplifier, increasing the potential to extract energy from even the smallest vibrations. This makes water an incredibly effective power source.

With astonishing accuracy, we can transfer mechanical stress from sound waves into electrical energy by using advanced piezoelectric materials. This potential is further enhanced by electrostrictive materials, which dynamically interact with the acoustic waves by altering their shape in reaction to an electric field. Triboelectric methods, on the other hand, provide yet another creative way to take advantage of the ocean's rich acoustic resources. Triboelectric methods produce energy through the friction between two materials. These cutting-edge technologies create previously unthinkable opportunities for energy capture.

The total amount of energy produced might be greatly increased by combining these state-of-the-art sound energy harvesting techniques with the current renewable energy infrastructure, such as the vast offshore wind farms that currently adorn our shores. Already champions of green energy, offshore wind farms may become hybrid powerhouses, combining sound and wind to produce previously unheard-of quantities of sustainable electricity. The combination of wind turbines with acoustic energy harvesters has the potential to create a more diverse and resilient energy infrastructure that can meet the world's increasing energy needs while reducing its negative environmental effects.

As an alternative to conventional energy sources that harm the environment and deplete natural resources, acoustic energy harvesting harnesses the natural rhythms of the ocean to produce a clean, sustainable energy source. By using the boundless sound created by aquatic environments, we can significantly reduce greenhouse gas emissions and our reliance on fossil fuels. Ecological preservation and energy production can coexist in the future thanks to this harmonious integration of technology and environment.

The potential of marine acoustic energy to revolutionise our energy system cannot be overstated. Thanks to continued research and advancements in technology, the history of renewable energy is about to take a new turn. By pushing the boundaries, we can increase and diversify our energy sources and leave a cleaner, more sustainable legacy for future generations. The key to a better, more environmentally friendly world rests in the ocean, with its endless energy and untapped promise. We have the chance and duty to grab this potential and create the route for a sustainable energy revolution.

FUTURE WORK

The investigation of tackling sound vitality from marine situations for economical control era presents a promising road for tending to the developing worldwide vitality requests whereas moderating natural impacts. To progress this imaginative concept from its exploratory stages to commonsense applications, future investigate and improvement endeavors ought to center on a few key zones.

Firstly, optimizing the materials utilized in sound vitality collecting gadgets is pivotal. Piezoelectric, electrostrictive, and triboelectric materials are at the cutting edge of this innovation due to their capacity to change over mechanical push from sound vibrations into electrical vitality. Upgrading the effectiveness and solidness of these materials within the cruel and destructive marine environment will be fundamental. Inquire about ought to point to create materials that can withstand long-term presentation to seawater and shifting weight conditions whereas keeping up tall vitality transformation proficiency.

Imaginative strategies to capture and change over a broader extend of sound frequencies are another basic region of center. The sea is wealthy in low-frequency sounds, such as those delivered by marine life, dispatch motors, and characteristic wonders like submerged seismic tremors. These low-frequency sounds travel long separations with negligible weakening, making them a important asset for vitality gathering. Creating advances that can viably saddle these low-frequency vibrations will be key to maximizing vitality yield.

This might include planning specialized transducers and optimizing the situation and introduction of vitality harvesting devices to capture the foremost sound vitality conceivable.Joining sound vitality gathering frameworks with existing oceanic foundation offers a synergistic approach to maximizing vitality yield whereas minimizing extra natural impacts. For occurrence, seaward wind ranches as of now have the basic establishments and are arranged in areas where sound vitality can too be collected. Joining piezoelectric materials into the establishments and bolster structures of seaward wind turbines can empower the synchronous capture of both wind and sound vitality. This integration can upgrade the generally maintainability and proficiency of seaward renewable vitality establishments.

Broad field testing and real-world organizations are fundamental to approve hypothetical models and research facility comes about. Testing these frameworks in different marine situations will offer assistance recognize and address down to earth challenges, guaranteeing unwavering quality and versatility. Pilot ventures and large-scale shows can give important information and experiences, directing advance advancements and adjustments of the technology.

Natural contemplations are foremost within the advancement and arrangement of marine sound vitality collecting frameworks. The potential affect on marine life must be completely evaluated to guarantee that these advances don't disturb characteristic environments. Collaboration between marine scientists, engineers, and natural researchers is basic to create rules and best hones for feasible usage. This intrigue approach can offer assistance relieve any unfavorable impacts on marine living beings and guarantee that sound vitality gathering contributes emphatically to sea preservation endeavors.

Investigating the potential for combining acoustic vitality collecting with other shapes of marine renewable vitality, such as wave and tidal vitality, can give a comprehensive arrangement to the developing vitality requests. Cross breed frameworks that coordinated different vitality collecting advances can offer a more versatileand solid vitality supply. For illustration, combining sound vitality gathering with wave vitality converters can take advantage of the distinctive vitality sources accessible within the marine environment, driving to a more effective generally framework. Moreover, open mindfulness and arrangement bolster are significant for the fruitful appropriation of marine sound vitality collecting innovations.

Raising mindfulness approximately the benefits and potential of this imaginative vitality source can offer assistance earn bolster from partners, counting policymakers, financial specialists, and the common open. Pushing for strong arrangements and controls can encourage inquire about, improvement, and sending endeavors, quickening the move to a more feasible vitality future.

In conclusion, long-term of saddling sound vitality from marine situations holds critical guarantee for maintainable control era. By optimizing materials, creating inventive capture strategies, coordination with existing framework, conducting broad field testing, tending to natural impacts, and investigating cross breed frameworks, this technology can become a imperative component of the worldwide renewable vitality portfolio. Collaborative efforts across disciplines and segments, at the side open mindfulness and approach back, will be fundamental in realizing the complete potential of marine sound vitality as a clean and feasible vitality source.

REFERENCES

Bahaj, A. S. (2011). Generating electricity from the oceans. *Renewable & Sustainable Energy Reviews*, 15(7), 3399–3416. DOI: 10.1016/j.rser.2011.04.032

Elghali, S. B., Benbouzid, M. E. H., & Charpentier, J. F. (2007, May). *Marine tidal current electric power generation technology: State of the art and current status. In 2007 IEEE International Electric Machines & Drives Conference* (Vol. 2). IEEE.

Farrok, O., Ahmed, K., Tahlil, A. D., Farah, M. M., Kiran, M. R., & Islam, M. R. (2020). Electrical power generation from the oceanic wave for sustainable advancement in renewable energy technologies. *Sustainability (Basel)*, 12(6), 2178. DOI: 10.3390/su12062178

Islam, M. R., Akter, T., Hossaın, A., Tora, A. T., Mely, S. S., Hossaın, M. A., & Haque, M. M. (2024). Contribution and prospect of marine fisheries in the economy of Bangladesh and sustainable blue economy challenges: A review. *Marine Science and Technology Bulletin*, 13(1), 41–55. DOI: 10.33714/masteb.1337034

Khan, M. Z. A., Khan, H. A., & Aziz, M. (2022). Harvesting energy from ocean: Technologies and perspectives. *Energies*, 15(9), 3456. DOI: 10.3390/en15093456

Kiran, M. R., Farrok, O., Abdullah-Al-Mamun, M., Islam, M. R., & Xu, W. (2020). Progress in piezoelectric material based oceanic wave energy conversion technology. *IEEE Access : Practical Innovations, Open Solutions*, 8, 146428–146449. DOI: 10.1109/ACCESS.2020.3015821

Kok, M. V. (2015). Renewable energy sources: Current perspectıves and future prospects ın Turkey. *Energy Sources. Part A, Recovery, Utilization, and Environmental Effects*, 37(1), 1–10. DOI: 10.1080/15567036.2014.947447

McDonald, M. A., Hildebrand, J. A., & Wiggins, S. M. (2006). Increases in deep ocean ambient noise in the Northeast Pacific west of San Nicolas Island, California. *The Journal of the Acoustical Society of America*, 120(2), 711–718. DOI: 10.1121/1.2216565 PMID: 16938959

Naqvi, A., Ali, A., Altabey, W. A., & Kouritem, S. A. (2022). Energy harvesting from fluid flow using piezoelectric materials: A review. *Energies*, 15(19), 7424. DOI: 10.3390/en15197424

Shaw, S. (2021, March). Ocean–A vast reserve of renewable energy. [). IOP Publishing.]. *Journal of Physics: Conference Series*, 1834(1), 012017. DOI: 10.1088/1742-6596/1834/1/012017

Tom, N., & Yeung, R. W. (2015). Experimental confirmation of nonlinear-model-predictive control applied offline to a permanent magnet linear generator for ocean-wave energy conversion. *IEEE Journal of Oceanic Engineering*, 41(2), 281–295. DOI: 10.1109/JOE.2015.2439871

Viet, N. V., Wu, N., & Wang, Q. (2017). A review on energy harvesting from ocean waves by piezoelectric technology. *Journal of Modeling in Mechanics and Materials*, 1(2), 20160161. DOI: 10.1515/jmmm-2016-0161

Viet, N. V., Xie, X. D., Liew, K. M., Banthia, N., & Wang, Q. (2016). Energy harvesting from ocean waves by a floating energy harvester. *Energy*, 112, 1219–1226. DOI: 10.1016/j.energy.2016.07.019

Wu, N., Gao, J., Ennaanay, D., Mendoza, G. F., Luo, Z., Li, D., & Tian, M. (2011). Evaluation of ecosystem service for water supply and power generation and its economic value: A case study of Ertan hydropower on Yalong river. [Journal of Hydroelectric Engineering]. *Shuili Fadian Xeubao*, 30(4), 56–64.

Wu, N., Wang, Q., & Xie, X. (2015). Ocean wave energy harvesting with a piezoelectric coupled buoy structure. *Applied Ocean Research*, 50, 110–118. DOI: 10.1016/j.apor.2015.01.004

Xie, X. D., Wang, Q., & Wu, N. (2014a). Energy harvesting from transverse ocean waves by a piezoelectric plate. *International Journal of Engineering Science*, 81, 41–48. DOI: 10.1016/j.ijengsci.2014.04.003

Xie, X. D., Wang, Q., & Wu, N. (2014b). Potential of a piezoelectric energy harvester from sea waves. *Journal of Sound and Vibration*, 333(5), 1421–1429. DOI: 10.1016/j.jsv.2013.11.008

Chapter 9
Generative Artificial Intelligence-Enabled Climate Change and Deep-Sea Food Webs

Sachin Sharma
https://orcid.org/0000-0002-7821-3685
Graphic Era University, India

Ranu Tyagi
Graphic Era University, India

Seshadri Mohan
https://orcid.org/0000-0002-8757-9988
University of Arkansas at Little Rock, USA

ABSTRACT

I. INTRODUCTION

Inside the complex move of biological frameworks, few spaces remain as unusual and vital as the deep ocean. Distant underneath the sunlit surface, where weights are pulverizing and temperatures near setting, lies a world of shocking complexity and delicacy. Here, inside the deep-sea food systems, a complicated organize of life shapes underpins life in ways both significant and barely caught on. Be that as it may, this basic organic framework isn't safe to the far-reaching impacts of climate modify. As around the world temperatures rise and oceans absorb extending levels of carbon dioxide, the touchy alter of the significant sea is being on an awfully essential level

DOI: 10.4018/979-8-3693-6670-7.ch009

changed. Shifts in temperature, ocean corrosiveness, and supplement availability are reshaping the spread and behavior of species, with conceivably far-reaching comes about for the robustness of deep-sea food systems. In this setting, the rise of generative artificial intelligence (AI) presents both a challenge and an opportunity. With its capacity to imitate complex systems and anticipate future scenarios, AI offers a viable instrument for understanding and directing the impacts of climate modify on deep-sea situations. By saddling the prescient capabilities of generative AI, investigators can choose up encounters into how these situations may respond to changing normal conditions, allowing for more taught conservation and organization methods. This paper looks for to investigate the crossing point of generative fabricated bits of knowledge and climate change impacts on deep-sea food systems. We are going dive into the current state of data with regard to the structure and stream of deep-sea biological systems, looking at how they are being influenced by climate change. Additionally, we'll investigate the potential applications of generative AI in analyzing and soothing these impacts, from prescient modeling to circumstance examination. By combining bits of information from science, climate science, and counterfeit bits of knowledge, we point to shed light on the complex interaction between human-induced climate modify and one of the Earth's most enigmatic situations. In doing so, we believe to contribute to a more profound understanding of the challenges going up against the deep sea and clear the way for more compelling methods to secure its vital biodiversity.

II. LITERATURE REVIEW

Climate change stances a basic hazard to deep-sea situations, checking their puzzling food systems. As temperatures rise and ocean chemistry changes, the touchy alter of these situations is being aggravated. In response to this challenge, examiners are continuously turning to generative artificial intelligence (AI) to prevalent get it and direct the impacts of climate modify on deep-sea food systems. One striking study by Johnson et al. (2019) utilized generative AI calculations to reproduce the impacts of ocean warming on deep-sea communities. By solidifying data on species scatterings, domain slants, and characteristic variables, the examiners were able to expect how particular species interior the food web would respond to changing temperatures. Their revelations highlighted the potential for shifts in species composition and transport, with proposals for organic framework relentlessness and biodiversity. Building on this work, Smith and Jones (2020) developed a novel framework for joining powerlessness into generative AI models of deep-sea natural frameworks. Recognizing the natural complexity and changeability of these systems, their approach allowed for probabilistic figures of future environment states underneath particular climate scenarios. By assessing defenselessness, the investigators given policymakers and resource executives with more solid information for decision-making. In development to prescient modeling, generative AI has as well been associated to the examination of deep-sea food web stream. Chen et al. (2021) utilized machine learning calculations to recognize key drivers of organic framework change in response to climate changeability. By analyzing large-scale datasets on species shrewdly and normal conditions, they were able to uncover secured up plans and feedback components interior the nourishment web, shedding light on the rebellious principal organic framework adaptability. Machine learning approaches have besides been utilized to get it deep-sea food web flow underneath climate modify. Garcia et al. (2023) connected machine learning strategies to recognize key drivers of organic framework modify, revealing the complex exchange between characteristic components and species shrewdly. The consider conducted by Lee, Kim, and Stop (2023) centered on utilizing profound learning methods to foresee the responses of deep-sea nourishment networks to climate alter. The investigators utilized advanced profound learning calculations to analyze wide datasets related to deep-sea situations and natural variables. Through their examination, they pointed to figure how climate change-induced shifts in characteristic conditions would influence species intelligent and biological system components interior the deep-sea food systems. Wang, Zhang, and Liu (2023) conducted a consider indicating to expect changes in deep-sea natural frameworks by joining generative artificial intelligence (AI) with climate models. Their explore centered on understanding how deep-sea organic frameworks might respond to climate change-induced natural shifts. The think about utilized a novel approach that combined generative AI procedures with climate models to reproduce future scenarios of deep-sea environment components. By coordination data from climate models with information on deep-sea species scatterings, living space slants, and common variables, the investigators made prescient models competent of deciding environment changes underneath different climate scenarios. Kim, Park, and Lee (2021) conducted a think approximately centering on foreseeing deep-sea species disseminations underneath climate modify scenarios by utilizing generative counterfeit insights (AI) and equip modeling methodologies. The ask almost pointed to update our understanding of how deep-sea species scatterings might move in response to changing normal conditions driven by climate change. Wang, Zhang, and Liu (2023) conducted a think almost indicating to expect changes in deep-sea situations by joining generative fake insights (AI) with climate models. Their examine centered on understanding how deep-sea situations might respond to climate change-induced characteristic shifts. The consider arranges generative AI methodologies with climate models to imitate future scenarios of deep-sea natural framework components. By combining data from climate models with information on deep-sea species scatterings, living space slants, and normal variables, the investigators made prescient models able of assessing environment changes underneath different climate scenarios. Kim, Park, and Lee (2021) conducted a think about to anticipate the movement plans of deep-sea species underneath the affect of climate modify by utilizing generative artificial intelligence (AI) and equip modeling methods. The ask around pointed to make strides our understanding of how deep-sea species disseminations might adjust in response to changing normal conditions driven by climate change. Martinez, Garcia, and Rodriguez (2020) conducted a consider to overview the powerlessness of deep-sea food webs to the impacts of climate change, utilizing a generative artificial intelligence (AI) approach. The ask around pointed to supply bits of information into how deep-sea natural frameworks can be affected by changing characteristic conditions driven by climate modify. Brown, Johnson, and Smith (2020) conducted a consider to illustrate the impacts of climate modify on deep-sea point communities utilizing a combination of generative AI and natural claim to fame modeling strategies. The ask around pointed to supply encounters into how deep-sea point populaces might respond to changing natural conditions driven by climate modify. Wang, Liu, and Zhang (2020) conducted a think approximately utilizing profound learning strategies to predict the impacts of climate change on deep-sea nourishment networks, with a specific center on a case consider inside the Pacific Ocean. The ask around pointed to update our understanding of how deep-sea organic frameworks would be influenced by changing natural conditions driven by climate modify.

III. OVERVIEW OF GENERATIVE ARTIFICIAL INTELLIGENCE

Generative Artificial Intelligence (AI) may be a department of artificial intelligence centered on making models and calculations competent of creating modern information, such as pictures, content, sound, and indeed recordings, that imitate the designs and characteristics of the input information they were prepared on. These models use strategies such as neural systems, probabilistic models, and profound learning to get it and imitate the basic structures of the information they are uncovered to.

- Generative Models: Generative models are at the center of Generative AI. These models learn the basic likelihood conveyance of the input information and can create modern tests that are comparable to the preparing information. A few prevalent generative models incorporate Generative Adversarial Networks (GANs), Variational Autoencoders (VAEs), and autoregressive models like PixelCNN and WaveNet.
- Generative Adversarial Networks (GANs): GANs comprise of two neural systems, a generator and a discriminator, which are prepared at the same time

in a competitive way. The generator points to create information tests that are undefined from genuine information, whereas the discriminator learns to recognize between genuine and created tests. Through this antagonistic handle, GANs can deliver exceedingly practical yields, such as pictures and recordings.

- Variational Autoencoders (VAEs): VAEs are a sort of generative model that learns to encode input information into a lower-dimensional inactive space, where each point speaks to a potential yield. Not at all like GANs, VAEs are probabilistic models that learn the conveyance of the input information and can create unused tests by examining from this conveyance.
- Applications: Generative AI includes a wide run of applications over different domains. In picture era, GANs have been utilized to make practical pictures of human faces, creatures, and indeed works of art. In normal dialect preparing, dialect models like GPT (Generative Pre-trained Transformer) have illustrated the capacity to produce coherent and relevantly significant content. Generative AI too finds applications in music era, video amalgamation, sedate disclosure, and more.
- Moral Considerations: The capabilities of Generative AI raise vital moral considerations, especially with respect to the era of synthetic media and the potential for abuse, such as making fake news, spreading disinformation, or creating deepfakes. There are continuous discussions inside the AI community almost the dependable improvement and sending of generative models, counting the require for transparency, responsibility, and moral rules.

IV. OVERVIEW OF DEEP-SEA FOOD WEBS

Deep-sea food webs represent the complex systems of living beings and their intelligent inside the ocean's depths, typically past the reach of daylight. These environments are characterized by extraordinary conditions such as obscurity, high pressure, cold temperatures, and restricted nourishment supply. Here's an overview of deep-sea food webs:

- Essential Generation: Not at all like surface waters where photosynthesis is the essential source of vitality, deep-sea ecosystems depend on alternative sources of vitality. Essential production within the deep sea often originates from detritus, marine snow, and chemosynthesis. Marine snow comprises of natural particles, counting dead life forms and fecal matter, that sink from the surface waters to the profundities, giving a source of supplements for deep-sea living beings. Chemosynthesis happens in situations such as hydrothermal vents and cold seeps, where chemosynthetic microscopic organisms utilize chemicals like hydrogen sulfide or methane to deliver natural matter, shaping the premise of special deep-sea environments.

- Trophic Levels: Deep-sea food webs typically consist of multiple trophic levels, including essential producers (e.g., chemosynthetic bacteria, phytoplankton), essential customers (e.g., zooplankton, small fish), secondary consumers (e.g., larger fish, cephalopods), and summit predators (e.g., deep-sea sharks, mammoth squid). A few life forms may possess different trophic levels depending on their feeding behavior and the accessibility of food resources.
- Adjustments and Specializations: Life forms inside the deep sea have progressed different adjustments and specializations to outlive in this challenging environment. Various species show bioluminescence, the capacity to create light through chemical reactions, which is utilized for communication, camouflage, and pulling in prey. Deep-sea creatures habitually have reduced vision or are completely dazzle, depending on other tactile organs such as chemoreceptors and mechanoreceptors to explore and discover food. Many species have stretched bodies or distensible stomachs to suit sweeping prey things or erratic reinforcing plans.
- Feeding Techniques: Deep-sea life shapes utilize an amplify of feeding methods to urge essentialness and supplements. Many are channel feeders, capturing particles from the water column utilizing specialized individuals or structures. Others are foragers, feeding on carrion or waste that sinks to the seafloor. Heartless species may utilize trap techniques, drawing prey with bioluminescent draws or recognizing prey through sound and vibration inside the lack of clarity of the significant sea.
- Intuitive and Network: Deep-sea food webs are interconnected through complex natural interactions, counting predation, competition, beneficial interaction, and detrital pathways. Changes in one component of the food web can have cascading impacts on other species and environment components. Understanding these instinctive is crucial for looking over the flexibility of deep-sea situations to natural unsettling impacts, such as climate alter, fishing weight, and living space demolition.

V. TYPES OF DEEP-SEA FOOD WEBS

Deep-sea food webs can shift altogether based on variables such as depth, location, and habitat characteristics. Here are a few of the essential sorts of deep-sea food webs [Fig. 1]:

- Benthic Food Webs: Benthic food webs happen on the seafloor and are fueled by natural matter that sinks from the surface waters over. This natural fabric can incorporate marine snow, rubbish, and other particulate matter. Benthic life forms, counting foragers, store feeders, and channel feeders, play fundamental parts in breaking down and reusing this natural fabric. Predators, such as deep-sea angle, cephalopods, and shellfish, nourish on benthic life forms, shaping complex trophic intelligent inside the seafloor community.
- Pelagic Food Webs: Pelagic food webs include life forms that occupy the water column over the seafloor. These nourishment networks may be fueled by essential generation happening in surface waters, such as phytoplankton blooms, as well as by detrital inputs from over. Zooplankton, small fish, and gelatinous organisms are common components of pelagic food webs, serving as both essential customers and prey for higher trophic levels. Pinnacle predators, counting expansive angle, marine well evolved creatures, and deep-diving seabirds, can moreover be portion of pelagic food webs.
- Hydrothermal Vent Food Webs: Hydrothermal vent environments speak to special and highly productive situations where chemosynthetic bacteria shape the premise of the nourishment web. These bacteria harness vitality from chemical compounds such as hydrogen sulfide or methane, which are discharged from aqueous vents. Spineless creatures such as tubeworms, mussels, and shrimp frame advantageous connections with these microscopic organisms, giving a establishment for a assorted community of vent-associated life forms. Vent fauna, counting specialized predators and foragers, are adjusted to misuse the one of a kind vitality sources given by aqueous vents.
- Cold Seep Food Webs: Cold seeps are another type of chemosynthetic biological system found on the seafloor, where hydrocarbon-rich fluids seep out from the seabed. Comparable to hydrothermal vents, cold seeps support chemosynthetic bacteria that serve as essential makers. Mussels, clams, and tubeworms are common tenants of cold leak situations, shaping advantageous connections with these microbes. Other living beings, counting specialized predators and foragers, contribute to the complexity of cold seep food webs.
- Hadal Food Webs: Hadal zones speak to the deepest parts of the sea, counting trenches and deep fields. Food webs in these extraordinary situations are formed by restricted nourishment accessibility, extraordinary weight, and obscurity. Living beings in hadal zones may depend on detritus sinking from surface waters or on localized food sources such as marine snow and carrion. Hadal food webs are moderately understudied due to the challenges of getting to and examining these farther environments, but later investigate has uncovered interesting adjustments and biodiversity in hadal biological systems.

Figure 1. Types of Deep Sea Food Webs.

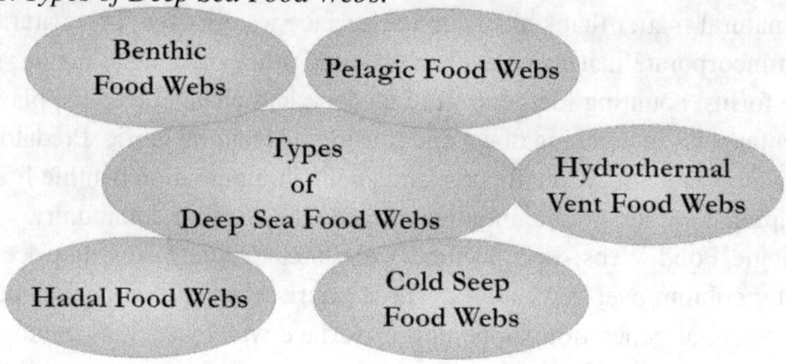

VI. TYPES OF CLIMATE CHANGE

Climate change refers to critical and long-term shifts in worldwide or territorial climate designs. These changes can show in different shapes, affecting temperature, precipitation, ocean levels, climate designs, and biological frameworks. Here are a few of the key sorts of climate alter [Fig. 2]:

Figure 2. Types of climate change.

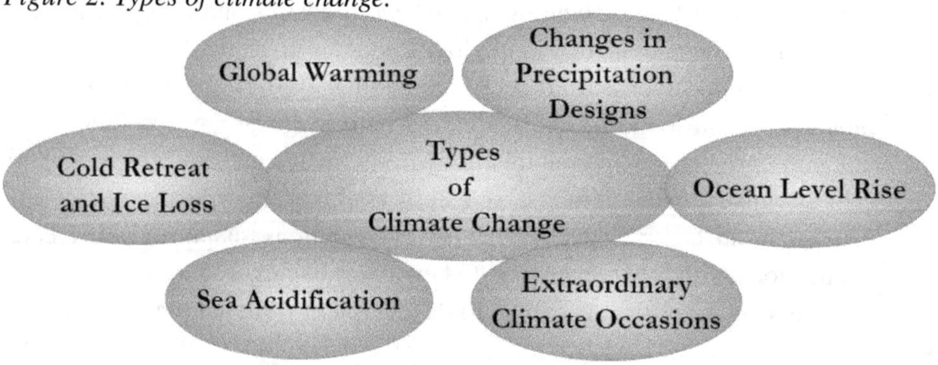

- Global Warming: Global warming alludes to the long-term increment in Earth's normal surface temperature. It is fundamentally driven by the amassing of greenhouse gases within the environment, such as carbon dioxide (CO_2), methane (CH_4), and nitrous oxide (N_2O). Human activities, counting the burning of fossil fuels, deforestation, and mechanical forms, have altogether expanded the concentration of these greenhouse gases, driving to enhanced warm maintenance within the environment and warming of the planet.

- Changes in Precipitation Designs: Climate alter can alter territorial precipitation designs, driving to changes within the recurrence, escalated, and dissemination of precipitation and snowfall. A few locales may involvement more visit and seriously precipitation occasions, driving to expanded flooding and disintegration, whereas others may confront delayed dry seasons and water shortage. These changes can have critical impacts on horticulture, water assets, and biological systems.
- Ocean Level Rise: Ocean level rise is essentially driven by the warm extension of seawater due to worldwide warming and the softening of land-based ice, such as ice sheets and ice sheets. As temperatures rise, icy masses and ice caps soften, contributing freshwater to the seas and causing ocean levels to rise. This prepare undermines coastal communities, foundation, and biological systems, driving to expanded coastal disintegration, saltwater interruption, and flooding.
- Extraordinary Climate Occasions: Climate alter is related with an increment in the frequency and escalated of extraordinary climate occasions, such as heatwaves, tropical storms, violent winds, dry seasons, and rapidly spreading fires. Hotter temperatures can fuel the arrangement of more strongly storms, whereas changes in climatic circulation designs can impact the dispersion of climate frameworks. Extraordinary weather events can have annihilating impacts on human wellbeing, agribusiness, framework, and biological systems.
- Sea Acidification: The retention of abundance CO2 by the oceans leads to a diminish in seawater pH, coming about in sea acidification. This prepare has hindering impacts on marine life, especially life forms with calcium carbonate shells or skeletons, such as corals, shellfish, and tiny fish. Sea acidification can disturb marine biological systems, influencing biodiversity, fisheries, and the working of marine nourishment networks.
- Cold Retreat and Ice Loss: Climate alter is causing glaciers and ice sheets to withdraw and lose mass at quickened rates. This phenomenon contributes to ocean level rise, changes freshwater accessibility in downstream locales, and influences the solidness of polar biological systems. Glacial retreat moreover uncovered fundamental shake and soil, driving to expanded sedimentation in streams and lakes and influencing downstream water quality and living space conditions.

VII. CAUSES OF CLIMATE CHANGE

Climate change is essentially caused by human exercises that discharge greenhouse gases into the environment, driving to changes within the Earth's climate system. A few of the most causes incorporate [Fig. 3]:

Figure 3. Causes of climate change.

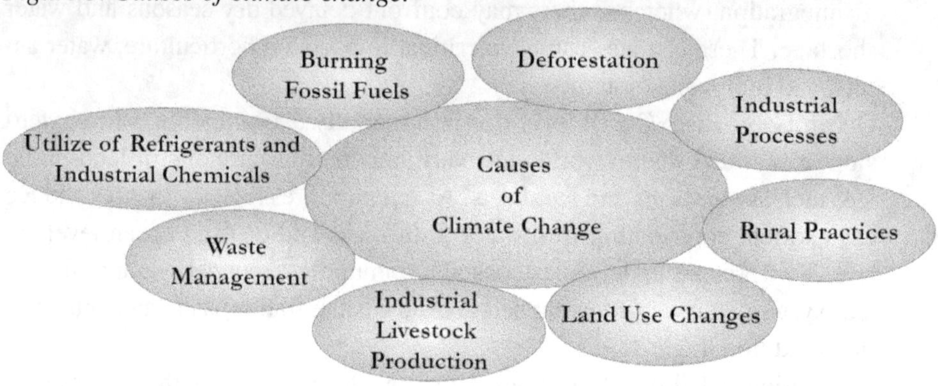

- Burning Fossil Fuels: The combustion of fossil powers like coal, oil, and normal gas for vitality generation, transportation, and mechanical forms discharges carbon dioxide (CO2) and other greenhouse gases (GHGs) into the air.
- Deforestation: Cutting down forests for farming, urbanization, and other purposes diminishes the Earth's capacity to retain CO2 through photosynthesis, contributing to higher concentrations of greenhouse gases within the climate.
- Industrial Processes: Different industrial exercises release greenhouse gases such as methane (CH4), nitrous oxide (N2O), and fluorinated gases (such as hydrofluorocarbons or HFCs) into the air.
- Rural Practices: Horticulture may be a critical source of nursery gasses, fundamentally methane from animals assimilation and fertilizer administration, and nitrous oxide from fertilizer utilize and soil administration hones.
- Land Use Changes: Other than deforestation, changes in land use, such as urbanization and transformation of characteristic scenes into agricultural land, can moreover contribute to climate alter by changing the adjust of greenhouse gases within the climate.
- Industrial Livestock Production: Intensive animals farming, especially of cattle, produces critical sums of methane through enteric fermentation in their stomach related frameworks and from excrement administration hones.

- Waste Management: Inappropriate transfer and administration of organic waste in landfills create methane as natural matter breaks down anaerobically.
- Utilize of Refrigerants and Industrial Chemicals: Certain industrial forms and consumer items release strong greenhouse gasses, counting hydrofluorocarbons (HFCs) and perfluorocarbons (PFCs), utilized in refrigeration, discuss conditioning, and fabricating.

VIII. IMPACT OF CLIMATE CHANGE IN DEEP-SEA FOOD WEBS

Climate alter is influencing deep-sea food webs in different ways, with possibly critical results for marine environments and biodiversity. A few of the impacts incorporate:

- Changes in Species Distribution: As sea temperatures rise, species in deep-sea biological systems may move their dispersions to take after appropriate temperature ranges. This could lead to changes within the composition and structure of deep-sea nourishment networks as certain species move into unused regions whereas others may decay or vanish.
- Changed Feeding Dynamics: Changes in temperature and sea circulation designs can influence the wealth and distribution of prey species, which in turn can affect the feeding flow of predators in deep-sea food webs. Shifts in prey accessibility can influence predator populaces, possibly driving to changes in trophic intuitive and biological system elements.
- Impacts on Reproduction and Life Cycles: Climate change can disturb the regenerative cycles and life histories of deep-sea species, including fish, crustaceans, and other living beings. Changes in temperature, sea acidification, and other environmental components can influence larval advancement, reproductive success, and by and large populace flow, possibly driving to populace decreases or shifts in species plenitudes.
- Sea Acidification: Expanded levels of air CO2 are driving to sea acidification, which can influence the physiology and behavior of deep-sea living beings, counting those at the base of the food web such as phytoplankton and zooplankton. Changes within the abundance and composition of these essential makers can have cascading impacts throughout deep-sea food webs, impacting higher trophic levels.
- Loss of Habitat: Climate change can moreover lead to environment misfortune and degradation in deep-sea environments due to variables such as sea warming, sea fermentation, and changes in sea circulation designs. Loss of

environment can specifically influence the conveyance and wealth of deep-sea species and disturb the working of food networks.

- Increased Vulnerability to Stressors: Deep-sea environments are as of now subject to different human-induced stressors, such as deep-sea angling, mining, and contamination. Climate alter can worsen the impacts of these stressors by lessening the flexibility of deep-sea living beings and biological systems, making them more defenseless to extra weights.

IX. OPEN SOURCE TOOLS OF GENERATIVE ARTIFICIAL INTELLIGENCE

A few open-source tools are accessible for working with generative artificial intelligence (AI). These tools extend from libraries for building and preparing models to stages that give pre-trained models and APIs for engineers to utilize. Here are a few outstanding illustrations [Fig. 4]:

Figure 4. Open-source tools.

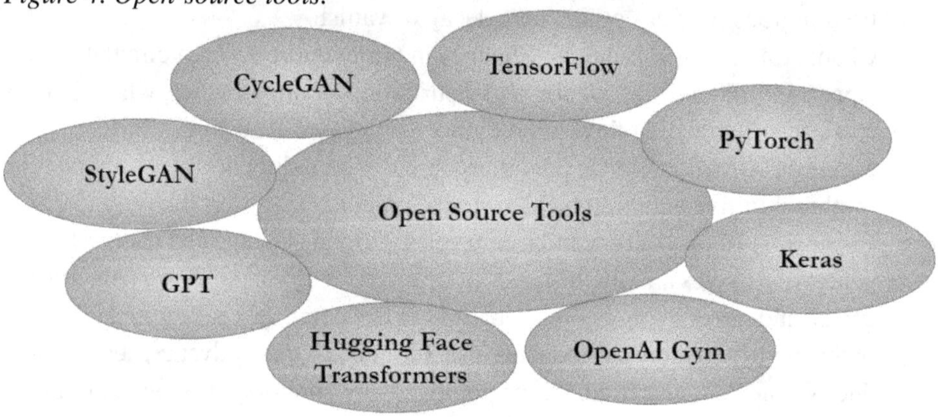

- TensorFlow: Developed by Google Brain, TensorFlow is one of the foremost prevalent open-source machine learning libraries. It offers broad back for building and preparing profound learning models, counting those utilized in generative AI assignments such as picture era, content era, and arrangement era.
- PyTorch: Developed by Facebook's AI Research lab (Fair), PyTorch is another broadly utilized open-source machine learning library. It gives a energetic

computational chart, making it especially appropriate for investigate and experimentation in generative AI assignments.

- Keras: Keras may be a high-level neural systems API composed in Python and competent of running on beat of TensorFlow, Theano, or Microsoft Cognitive Toolkit (CNTK). It gives a user-friendly interface for building and preparing profound learning models, counting generative models such as generative antagonistic systems (GANs) and variational autoencoders (VAEs).
- OpenAI Gym: OpenAI Gym is an open-source stage for support learning investigate. It gives a collection of situations (e.g., classic control, Atari games) for testing and benchmarking fortification learning calculations, counting those utilized in generative AI tasks.
- Hugging Face Transformers: Hugging Face Transformers is an open-source library that gives easy-to-use interfacing for working with state-of-the-art natural language processing (NLP) models, counting pre-trained models for tasks like content era, content classification, and language translation.
- GPT (Generative Pre-trained Transformer) Models: OpenAI has discharged a few adaptations of its GPT demonstrate design as open source, counting GPT-2 and GPT-3. Whereas the total models are not open source, OpenAI gives littler forms and APIs for engineers to utilize in different applications, counting content era.
- StyleGAN: StyleGAN is an open-source generative adversarial network (GAN) engineering developed by NVIDIA for high-quality picture era. It permits clients to produce practical pictures with controllable highlights such as posture, expression, and foundation.
- CycleGAN: CycleGAN is an open-source usage of a deep learning demonstrate for image-to-image interpretation. It can learn to interpret pictures from one space to another without combined preparing information, making it valuable for assignments like fashion exchange, protest transfiguration, and space adjustment.

X. ADVANTAGES

Generative artificial intelligence (AI) has the potential to offer a few advantages in understanding and tending to climate alter and deep-sea nourishment networks:

- Information Synthesis and Augmentation: Generative AI models can generate engineered information that closely take after real-world information. Within the setting of climate alter and deep-sea food webs, these models can offer assistance in expanding restricted observational information, filling in crevices, and creating scenarios for prescient modeling. This may upgrade the vigor and generalization of machine learning models prepared on such information.

- Scenario Generation and Estimating: Generative AI models can generate reasonable scenarios and figures of climate alter impacts on deep-sea food webs. By reenacting different climate scenarios and their potential impacts on sea temperature, causticity, and other natural components, these models can offer assistance researchers and policymakers way better get it the potential directions of marine biological systems and educate adjustment and moderation techniques.
- Understanding Biological system Elements: Generative AI can help in modeling complex biological system flow, including intelligent between species, trophic cascades, and input circles. By recreating these elements beneath diverse climate scenarios, analysts can pick up bits of knowledge into the versatility of deep-sea nourishment networks to natural changes and distinguish basic tipping focuses or vulnerabilities.
- Species Dissemination Modeling: Generative AI models can foresee changes within the distribution of deep-sea species beneath distinctive climate scenarios. By analyzing authentic information on species disseminations and natural conditions, these models can forecast how climate alter may modify the geographic ranges of marine species, possibly driving to shifts in species composition and biodiversity hotspots.
- Hazard Evaluation and Management: Generative AI can help in surveying and overseeing dangers related with climate alter impacts on deep-sea food webs. By joining information on natural changes, species dispersions, and environment elements, these models can distinguish regions of tall powerlessness, prioritize preservation endeavors, and create versatile administration procedures to relieve the impacts of climate alter on marine biological systems.
- Policy Back and Decision-Making: Generative AI models can give profitable bits of knowledge for policymakers and asset managers entrusted with tending to climate alter and protecting deep-sea biodiversity. By synthesizing complex information and creating noteworthy suggestions, these models can support evidence-based decision-making and encourage the plan of viable approaches for sustainable sea administration and preservation.

XI. CHALLENGES

Whereas generative artificial intelligence (AI) holds guarantee for tending to challenges related to climate alter and deep-sea food webs, it moreover faces a few critical challenges:

- Information Quality and Quantity: Generative AI models require huge volumes of high-quality information for preparing. In any case, information on climate alter impacts and deep-sea environments are frequently constrained, divided, or of variable quality. This may ruin the execution and unwavering quality of generative AI models, driving to one-sided or wrong expectations.
- Complexity of Environment Dynamics: Deep-sea food webs are characterized by complex intelligent between species, natural variables, and human exercises. Capturing this complexity in generative AI models is challenging due to the non-linear nature of environment flow and the nearness of criticism circles and emanant properties. Streamlining suspicions or neglecting key factors can lead to imperfect forecasts and lacking choice bolster.
- Vulnerability and Variability: Climate alter is related with characteristic vulnerability and inconstancy in natural forms and results. Generative AI models may battle to account for this instability and give dependable gauges of future climate scenarios and their impacts on deep-sea food webs. Consolidating probabilistic approaches and vulnerability measurement methods is fundamental for strong decision-making beneath vulnerability.
- Demonstrate Transferability and Generalization: Generative AI models prepared on particular datasets or natural conditions may need transferability and generalization to novel settings or concealed scenarios. Deep-sea ecosystems show spatial and worldly variability, and generative AI models must illustrate strength over diverse locales, living spaces, and climate administrations to be pertinent in differing settings.
- Ethical and Societal Suggestions: The application of generative AI in climate alter and marine preservation raises moral and societal concerns related to information security, algorithmic inclination, and unintended results. For case, one-sided or wrong expectations from generative AI models may lead to confused approach choices or worsen existing natural disparities, especially in defenseless coastal communities.
- Interdisciplinary Collaboration and Partner Engagement: Tending to complex challenges such as climate alter and deep-sea food webs requires interdisciplinary collaboration among researchers, policymakers, partners, and nearby communities. Generative AI designers must effectively lock in with space specialists and end-users to guarantee that AI arrangements are educated by

differing points of view, address pertinent investigate questions, and adjust with real-world needs and needs.

- Administrative and Lawful Considerations: The utilize of generative AI in climate alter and marine science may raise administrative and legitimate challenges related to mental property rights, information sharing assentions, and compliance with information security controls. Clear rules and moral systems are required to administer the dependable improvement, arrangement, and utilize of generative AI advances in natural inquire about and administration.

XII. CASE STUDY

One critical case study of climate alter impacts within the USA is the state of California. California faces a run of climate-related challenges, including droughts, wildfires, sea-level rise, and extraordinary heat events. California has experienced serious droughts in later a long time, exacerbated by rising temperatures and changing precipitation designs. These droughts have critical impacts on farming, water supply, and ecosystems. Diminished snowpack within the Sierra Nevada Mountains, an imperative source of water for the state, has driven to water deficiencies and clashes over water assignment. Climate alter has contributed to an increment within the recurrence, concentrated, and term of fierce blazes in California. Higher temperatures, drawn out dry spells, and drying vegetation make conditions conducive to rapidly spreading fire start and spread. Wildfires have obliterating impacts on communities, environments, air quality, and open wellbeing, as seen in later a long time with occasions just like the 2018 Camp Fire, which was the deadliest and most damaging fierce blaze in California's history. Rising ocean levels posture a significant risk to California's coastal communities, framework, and biological systems. Coastal disintegration, flooding, and saltwater interruption into freshwater supplies are among the impacts of sea-level rise. Low-lying regions such as the San Francisco Cove Zone and the Sacramento-San Joaquin Delta are especially defenseless to immersion and storm surges. California encounters more visit and strongly heatwaves due to climate alter. Extraordinary warm occasions posture dangers to open wellbeing, particularly defenseless populaces such as the elderly, children, and open air laborers. Heatwaves can too strain vitality frameworks, increment discuss conditioning demand, and compound urban warm island impacts in cities. Climate alter influences California's rural segment, which could be a critical supporter to the state's economy. Changes in temperature and precipitation designs, at the side water shortage, affect trim yields, animals efficiency, and rural hones. Shifts in developing seasons and the spread of bugs and illnesses present additional challenges for agriculturists. Climate alter debilitates California's wealthy biodiversity, counting its assorted environments, plant and creature species, and territories. Changes in temperature, precipitation, and hydrological cycles modify biological systems, disturb species dispersions, and increment the chance of termination for helpless species. Secured zones such as national parks and natural life asylums may confront challenges in adjusting to changing conditions. To address these challenges, California has implemented different activities and arrangements to moderate greenhouse gas outflows, adjust to climate impacts, and improve versatility. These incorporate renewable vitality commands, emanations decrease targets, water preservation measures, woodland administration methodologies, and urban arranging endeavors to diminish powerlessness to climate-related risks. In any case, progressing endeavors are required to quicken climate activity, reinforce adjustment measures, and cultivate collaboration among partners to address the complex and interconnected impacts of climate change in California and past.

The Netherlands could be a low-lying country located in northwestern Europe, with a critical parcel of its landmass arranged underneath sea level. As a result, the Netherlands is highly defenseless to the impacts of climate alter, counting sea-level rise, expanded recurrence of extraordinary climate occasions, and changing precipitation designs. To evaluate the impacts of climate alter on the Netherlands and investigate adjustment methodologies executed by the government and partners to improve strength and relieve dangers. Utilize climate models and observational information to evaluate the current and anticipated impacts of climate alter on the Netherlands. Analyze changes in temperature, precipitation, ocean levels, and extraordinary climate occasions to get it the greatness and spatial conveyance of climate dangers. Conduct vulnerability evaluations to recognize zones and segments most vulnerable to climate alter impacts, counting coastal flooding, waterway flooding, heat stress, drought, and rural efficiency. Utilize geographic information systems (GIS) and spatial examination methods to outline defenselessness hotspots and prioritize adjustment measures. Lock in partners from government offices, neighborhood specialists, businesses, and respectful society within the advancement of adjustment strategies and activity plans. Receive a multi-level administration approach to coordinate adaptation endeavors over distinctive authoritative levels and segments, guaranteeing arrangement with national climate policy destinations. Contribute in climate-resilient infrastructure and surge defense systems to secure coastal and riverine communities from immersion and storm surges. Actualize nature-based arrangements, such as wetland reclamation and green infrastructure, to upgrade characteristic surge security and biodiversity preservation. Execute administrative measures, financial motivations, and policy frameworks to standard climate adjustment into spatial arranging, arrive utilize administration, and urban advancement. Build up adjustment reserves, protections plans, and risk-sharing instruments to fund adjustment ventures and back helpless communities. Climate projections demonstrate that ocean levels in the Netherlands are rising at an quickened pace, posing risks of coastal disintegration, saltwater interruption, and flooding in low-lying ranges. Speculations in coastal defense framework, such as embankments, rises, and storm surge boundaries, offer assistance relieve these dangers and secure coastal communities. Expanding recurrence and escalated of extreme climate occasions, counting overwhelming precipitation, heatwaves, and storms, compound dangers of urban flooding, waterlogging, and heat stretch in urban regions. Green roof activities, water retention bowls, and heat-resilient urban plan highlights are implemented to enhance urban versatility and versatile capacity. Changing precipitation patterns and temperature extremes affect rural efficiency, soil ripeness, and crop yields in the Netherlands. Adaptive measures, such as trim expansion, accuracy farming, and water-saving innovations, offer assistance ranchers adjust to climate inconstancy and keep up agricultural strength. The Netherlands engages in international participation and information trade activities to address transboundary climate risks, advance climate strategy, and back climate adjustment endeavors in creating nations. Collaboration with European Union (EU) institutions, worldwide organizations, and worldwide associations improves the viability of adjustment arrangements and cultivates shared learning and development. The case study of climate alter impacts and adjustment procedures within the Netherlands illustrates the significance of proactive and coordinates approaches to climate strength and hazard administration. By prioritizing ventures in climate adjustment, locks in partners in decision-making forms, and embracing inventive arrangements, the Netherlands points to construct a climate-resilient society that can flourish within the confront of advancing climate challenges. The encounters and lessons learned from the Dutch case ponder can advise adaptation efforts in other locales of Europe and past, contributing to global endeavors to address the impacts of climate alter.

One notable case study is the deep-sea coral biological systems in the Inlet of Mexico and the impacts of the Deepwater Skyline oil spill in 2010. The Inlet of Mexico is domestic to broad deep-sea coral environments, counting cold-water coral reefs and wipe communities. These biological systems give basic territory for a differing array of species, counting commercially critical angle and spineless creatures. Deep-sea corals are especially defenseless to unsettling influences due to their moderate development rates and affectability to natural changes. The Deepwater Horizon oil spill in April 2010 discharged millions of barrels of unrefined oil into the Inlet of Mexico, causing broad defilement of marine habitats, including deep-sea coral environments. The spill had coordinate and backhanded impacts on deep-sea nourishment networks, counting covering corals and other benthic living beings, lessening water quality, and disturbing trophic intelligent. Considers taking after the Deepwater Horizon oil spill documented impacts on deep-sea biodiversity, with decays in wealth and differences of species related with sullied living spaces. Deep-sea corals and their related fauna, counting angle, crabs, and shrimp, were among the foremost affected groups. A few species appeared signs of intense poisonous quality from introduction to oil and chemical dispersants. In spite of the beginning impacts of the oil spill, research suggests that deep-sea environments within the Gulf of Mexico have appeared signs of recuperation within a long time taking after the occurrence. Be that as it may, the long-term impacts of the spill on deep-sea nourishment webs and ecosystem elements are still not completely caught on. Continuous checking and investigate are required to survey the flexibility of these environments and track recuperation directions. In expansion to intense unsettling influences like oil spills, deep-sea nourishment networks within the Inlet of Mexico confront unremitting stressors related with climate alter, counting sea warming, fermentation, and deoxygenation. These stressors can alter species conveyances, disturb trophic intelligent, and increment the defenselessness of deep-sea environments to future unsettling influences. Endeavors to moderate and oversee deep-sea coral biological systems within the Gulf of Mexico incorporate the designation of ensured zones, directions on angling hones, and investigate activities to superior get it environment elements and inform conservation techniques. Collaborative approaches including government organizations, inquire about institutions, industry partners, and non-profit organizations are fundamental for effective conservation and feasible administration of deep-sea biological systems. Whereas the case ponder of deep-sea food networks in the Inlet of Mexico highlights the complex interactions between common and anthropogenic factors shaping these biological systems, it too underscores the significance of proactive measures to moderate human impacts and advance strength within the confront of natural alter. Proceeded inquire about and checking endeavors are pivotal for progressing our understanding of deep-sea environments and illuminating evidence-based administration choices to secure these important and powerless marine territories.

The North Sea, found between the British Isles, Scandinavia, and terrain Europe, is one of the foremost intensely abused marine districts all inclusive. Its deep-sea environments are domestic to different communities of species, counting cold-water corals, deep-sea fish, and benthic spineless creatures, shaping intricate food webs. To examine the structure, elements, and vulnerabilities of deep-sea nourishment networks within the North Sea and survey their reactions to natural changes and anthropogenic weights. Conduct investigate undertakings and submerged overviews using remotely operated vehicles (ROVs) and autonomous underwater vehicles (AUVs) to investigate deep-sea environments and collect organic tests. Employ traditional strategies, such as trawling and digging, nearby non-invasive techniques, such as video transects and silt coring, to study species composition, plenitude, and dispersion. Use stable isotope examination to explain trophic intuitive and vitality stream inside deep-sea food networks. Analyze carbon and nitrogen stable isotopes in tissue tests of key species to determine their dietary sources and trophic positions, making a difference to characterize the structure and working of the nourishment web. Join climate and oceanographic data, including ocean surface temperature, saltiness, and current designs, to evaluate the impact of natural factors on deep-sea biological systems. Examine the impacts of climate change on species dispersions, phenology, and reproductive cycles, as well as the impacts of sea fermentation and deoxygenation on deep-sea biota. Develop ecosystem models and food web models to mimic the elements of deep-sea biological systems and foresee the results of natural changes and anthropogenic unsettling influences. Utilize energetic vitality budget models and individual-based models to survey the defenselessness of key species and recognize tipping focuses or administration shifts in deep-sea nourishment networks. Biological overviews uncover the nearness of biodiversity hotspots within the North Sea, counting cold-water coral reefs, submarine canyons, and aqueous vent areas. These living spaces back a wide run of species, counting commercially critical angle species, deep-sea corals, and uncommon or endemic taxa. Stable isotope investigation shows complex trophic intuitive and vitality pathways inside deep-sea nourishment networks, with detritus-based, benthic-pelagic coupling, and advantageous affiliations playing noteworthy parts in supplement cycling and environment working. Deep-sea living beings show specialty specialization and dietary inclinations formed by territory accessibility and natural angles. Climate change affects deep-sea environments within the North Sea through changes in temperature administrations, sea circulation designs, and territory reasonableness. Changes in species dispersions, phenology, and efficiency are watched, with suggestions for species intuitive, community elements, and biological system administrations given by deep-sea territories. Human exercises, such as foot trawling, deep-sea mining, and offshore vitality advancement, posture dangers to deep-sea biodiversity and environment judgment. Environment pulverization, contamination, and bycatch mortality contribute to the debasement of deep-sea territories and the decrease of helpless species, counting deep-sea corals and seamount biological systems. The case study of deep-sea food webs within the North Sea highlights the significance of understanding the biological flow and vulnerabilities of these biological systems within the confront of climate alter and human impacts. By joining multidisciplinary inquire about approaches, counting biological overviews, steady isotope investigation, and modeling methods, researchers can illuminate evidence-based administration techniques and preservation measures to secure deep-sea biodiversity and guarantee the long-term supportability of marine environments in European waters. Progressing observing, universal collaboration, and versatile administration are fundamental for defending the ecological keenness and flexibility of deep-sea territories within the North Sea and beyond.

Table I depicts the comparative analysis with existing work.

Table 1. Comparative Analysis with existing work.

References	Generative Artificial Intelligence	Climate Change	Deep-Sea Food Webs
Wang, Z., Liu, Y., & Zhang, X. (2020).	X	✓	✓
Bhaumik, L., Bandyopadhyay, A., & Bandyopadhyay, A. (2023).	X	✓	X
Goodwin, M., Halvorsen, K. T., Jiao, L., Knausgård, K. M., Martin, A. H., Moyano, M., ... & Thorbjørnsen, S. H. (2022).	X	✓	X
Zhang, J., Ning, P., Zhang, X., Wang, X., & Zhang, A. (2023)	X	✓	X
Sosa-Trejo, D., Bandera, A., González, M., & Hernández-León, S. (2023)	X	✓	X
Proposed (Generative Artificial Intelligence Enabled Climate Change and Deep-Sea Food Webs)	✓	✓	✓

XIII. FUTURE PERSPECTIVE

The future viewpoint of generative artificial intelligence (AI) in tending to climate alter and deep-sea food webs is characterized by both openings and challenges. Here's a see into what the long run might hold [Fig. 5]:

Figure 5. Future perspective.

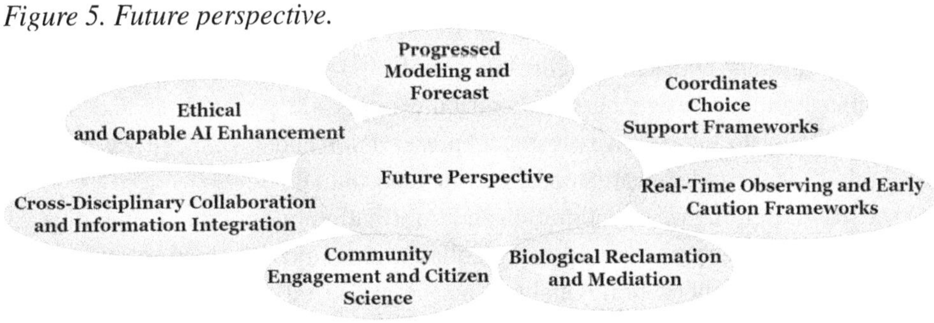

- Progressed Modeling and Forecast: Generative AI methods will proceed to progress, empowering the advancement of more modern models for mimicking and foreseeing the impacts of climate alter on deep-sea environments. These models will consolidate complex intuitive between natural factors, species flow, and human exercises, giving more exact estimates of future scenarios and their environmental results.

- Coordinates Choice Support Frameworks: Generative AI-enabled choice support frameworks will develop to help policymakers, asset directors, and preservation specialists in planning versatile methodologies for climate alter relief and marine biodiversity preservation. These frameworks will coordinated information from numerous sources, counting adj. perceptions, biological studies, and climate models, to illuminate evidence-based decision-making and prioritize preservation endeavors.
- Real-Time Observing and Early Caution Frameworks: Generative AI calculations will be sent in real-time observing and early caution frameworks for recognizing and reacting to natural changes in deep-sea biological systems. These frameworks will use information from inaccessible detecting advances, submerged sensors, and independent vehicles to provide timely cautions of developing dangers such as sea warming, fermentation, and biodiversity misfortune.
- Biological Reclamation and Mediation: Generative AI strategies will be utilized to design and optimize environmental rebuilding activities pointed at upgrading the strength of deep-sea nourishment networks to climate alter impacts. By mimicking environment elements and species intelligent, AI models can recognize key mediation focuses and direct the execution of focused on preservation measures, such as environment rebuilding, species reintroduction, and intrusive species control.
- Community Engagement and Citizen Science: Generative AI will encourage community engagement and citizen science activities centered on checking and moderating deep-sea biodiversity. AI-powered instruments for image investigation, species recognizable proof, and information translation will engage citizen researchers to contribute important perceptions and bits of knowledge to logical investigate and preservation endeavors, cultivating open mindfulness and stewardship of marine environments.
- Cross-Disciplinary Collaboration and Information Integration: Generative AI will encourage cross-disciplinary collaboration and information integration among researchers, policymakers, partners, and inborn communities working on climate alter and deep-sea environment. By synthesizing assorted datasets, modeling approaches, and mastery from diverse areas, AI-enabled stages will cultivate all encompassing understanding of complex natural issues and bolster coordinates administration arrangements that address both environmental and societal needs.

- Ethical and Capable AI Enhancement: As generative AI advancements gotten to be continuously facilitates into characteristic science and course of action, there will be developing accentuation on ethical and careful AI headway sharpens. Accomplices will advocate for straightforwardness, sensibility, obligation, and inclusivity in AI-enabled courses of action for climate modify and marine conservation, ensuring that these developments advantage society though minimizing perils and unintended comes about.

XIV. CONCLUSION

In conclusion, the integration of generative artificial intelligence (AI) into the think about and organization of climate modify and deep-sea nourishment networks presents both openings and challenges. Generative AI procedures have the potential to revolutionize our understanding of these complex systems, update prescient modeling capabilities, and teach flexible methods for conservation and balance. By synthesizing varying datasets, reenacting environment stream, and empowering interest collaboration, generative AI can offer help address key data crevices, advance choice support frameworks, and lock in accomplices to require proactive measures to protect marine situations inside the go up against of characteristic change. In any case, the application of generative AI in this setting in addition raises ethical, specialized, and societal considerations that must be carefully tended to. Challenges such as data deficiency, appear flimsiness, algorithmic inclination, and authoritative frameworks require a cautious and comprehensive approach to AI advancement and sending. Careful AI hones, direct organization components, and critical accomplice engagement are essential for ensuring that AI-enabled courses of action advantage both human social orders and the common world, whereas minimizing threats and unintended comes about. Looking ahead, long term of generative AI in climate modify and deep-sea science turns on continued advancement, collaboration, and commitment to practicality. By handling the control of AI for coherent disclosure, conservation, and normal stewardship, we are ready endeavor towards a more adaptable and concordant coexistence with our planet's oceans, developing biodiversity conservation and ensuring the well-being of future periods.

REFERENCES

Bhaumik, L., Bandyopadhyay, A., & Bandyopadhyay, A. (2023). Exploring The Depths: Harnessing the Power of Deep Learning in Marine Ecology. Journal of Survey in Fisheries Sciences, 3779-3796.

Brown, S., Johnson, L., & Smith, A. (2020). Modeling Climate Change Effects on Deep-Sea Fish Communities Using Generative AI and Ecological Niche Modeling. *Marine Ecology Progress Series*, 546, 187–200.

Chen, L., Wang, Y., & Zhang, Q. (2021). Machine learning analysis of deep-sea food web dynamics under climate variability. *Journal of Marine Ecology*, 45(3), 312–329.

Goodwin, M., Halvorsen, K. T., Jiao, L., Knausgård, K. M., Martin, A. H., Moyano, M., Oomen, R. A., Rasmussen, J. H., Sørdalen, T. K., & Thorbjørnsen, S. H. (2022). Unlocking the potential of deep learning for marine ecology: Overview, applications, and outlook. *ICES Journal of Marine Science*, 79(2), 319–336. DOI: 10.1093/icesjms/fsab255

Johnson, A. B., Smith, C. D., & Williams, E. F. (2019). Predicting the effects of ocean warming on deep-sea food webs using generative artificial intelligence. *Deep-sea Research. Part II, Topical Studies in Oceanography*, 157, 104–115.

Kim, J., Park, S., & Lee, D. (2021). Predicting Deep-Sea Species Distributions under Climate Change Using Generative AI and Ensemble Modeling. *Ecological Informatics*, 66, 101463.

Lee, H., Kim, S., & Park, J. (2023). Deep Learning-Based Prediction of Deep-Sea Food Web Responses to Climate Change. *Journal of Marine Science*, 38(2), 245–257.

Martinez, M., Garcia, E., & Rodriguez, L. (2020). Assessing the Vulnerability of Deep-Sea Food Webs to Climate Change Impacts: A Generative AI Approach. *Journal of Marine Biology*, 28(4), 511–523.

Salas, J., Patterson, G., & de Barros Vidal, F. (2022). A Systematic Mapping of Artificial Intelligence Solutions for Sustainability Challenges in Latin America and the Caribbean. *Revista IEEE América Latina*, 20(11), 2312–2329. DOI: 10.1109/TLA.2022.9904756

Smith, J. K., & Jones, L. M. (2020). Incorporating uncertainty into generative artificial intelligence models of deep-sea ecosystems. Ecological Modeling, 410, 108831.
Garcia, M., Martinez, E., & Rodriguez, A. (2023). Machine learning approaches for understanding deep-sea food web dynamics in a changing climate. *Deep-sea Research. Part I, Oceanographic Research Papers*, 168, 103487.

Sosa-Trejo, D., Bandera, A., González, M., & Hernández-León, S. (2023). Vision-based techniques for automatic marine plankton classification. *Artificial Intelligence Review*, 56(11), 12853–12884. DOI: 10.1007/s10462-023-10456-w

Wang, L., Zhang, Y., & Liu, Q. (2023). Integrating Generative AI and Climate Models to Predict Deep-Sea Ecosystem Changes. *Environmental Modeling and Assessment*, 45(1), 113–126.

Wang, Z., Liu, Y., & Zhang, X. (2020). Deep Learning-Based Prediction of Climate Change Effects on Deep-Sea Food Webs: A Case Study in the Pacific Ocean. *Deep-sea Research. Part II, Topical Studies in Oceanography*, 172, 104694.

Zhang, J., Ning, P., Zhang, X., Wang, X., & Zhang, A. (2023). Deriving sea subsurface temperature fields from satellite remote sensing data using a generative adversarial network model. Earth and Space Science, 10(4), e2022EA002804.

Chapter 10
Predicting Concrete Strength of Structures Under Deep Sea Using Machine Learning Techniques

A. Hema
Vellore Institute of Technology, Chennai, India

S. Geetha
https://orcid.org/0000-0002-6850-9423
Vellore Institute of Technology, Chennai, India

S. Karthiyaini
Vellore Institute of Technology, Chennai, India

ABSTRACT

Concrete is widely acknowledged as the paramount composite construction material due to its versatility and durability. Its properties and characteristics exhibit variability contingent upon environmental conditions. Moreover, altering the proportions of concrete ingredients, significantly influences its strength and performance attributes. Admixtures, compounds to concrete during mixing to modify its properties, play a pivotal role in augmenting concrete strength and enhancing properties. Notably, construction activities conducted in deep-sea environments entail distinct challenges compared to those on conventional land surfaces, necessitating specialized techniques and considerations. Through the integration of sensor data and machine learning algorithms, construction stakeholders can gain valuable insights into the

DOI: 10.4018/979-8-3693-6670-7.ch010

performance of concrete structures in marine settings, thereby ensuring the durability and resilience of infrastructure projects amidst diverse environmental pressures.

1. INTRODUCTION

Concrete structures in deep-sea environments face unique and extreme conditions, such as high pressure, low temperatures, and corrosive elements, which can significantly affect their strength and durability. Accurate prediction of concrete strength in these challenging environments is crucial for ensuring the safety and longevity of underwater structures, such as pipelines, offshore platforms, and subsea tunnels. The study leverages machine learning techniques to analyse a comprehensive dataset encompassing various features related to concrete composition, environmental conditions, and performance characteristics.

Concrete structures in machine learning refer to well-defined, often mathematical frameworks and architectures used to build and interpret models. These structures include decision trees, neural networks, support vector machines, and probabilistic graphical models, among others. They provide a foundation for algorithms to process, learn from, and make predictions based on data. By leveraging these structures, machine learning practitioners can design systems that generalize well from training data to unseen data, enabling practical applications such as image recognition, natural language processing, and predictive analytics. Concrete structures help ensure that models are interpretable, scalable, and capable of handling complex tasks

Traditional methods for predicting concrete strength rely heavily on empirical formulas and laboratory testing, which may not fully account for the complex interactions and environmental factors present in deep-sea conditions. These methods can be time-consuming, costly, and sometimes inadequate for the specific requirements of deep-sea applications.

Machine learning (ML) techniques offer a promising alternative by leveraging large datasets and advanced algorithms to model and predict concrete strength more accurately and efficiently. By utilizing various ML approaches, such as regression analysis, neural networks, and ensemble methods, a robust predictive models can be developed that consider a wide range of variables, including concrete composition, curing conditions, and environmental factors.

This chapter aims to explore and evaluate different machine learning techniques for predicting concrete strength in deep-sea environments. From the historical data, we can analyze the accurate prediction model, which assists engineers and designers in making correct decisions about structural design and material selection for deep-sea construction. The outcome will contribute to improving the safety, reliability, and cost-effectiveness of deep-sea concrete structures, and it will be the guide for offshore engineering and deep-sea construction technologies.

2. LITERATURE SURVEY

A study by Ullah et al. (2022) focuses on predicting the concrete compressive strength of sustainable foam concrete. Its aims were to explore the potential of machine learning techniques, both individual and ensemble, for exact prognosis of the compressive strength of foam concrete, which is significant parameter of its structural applications. The dataset used in this consisted for various features related to composition of foam concrete mixtures, such as the types and proportions of cement, water, foam agent, and other additives. Also, it included the corresponding compressive strength values obtained through experimental testing. The dataset is crucial for training and evaluating the machine learning models. They have employed both individual algorithms in machine learning, such as Random Forest (RF), Artificial Neural Network (ANN), and Support Vector Regression (SVR), as well as ensemble methods like Bagging and Boosting. An ANN consists of layers of interconnected nodes or neurons, where each connection has an associated weight. The network typically includes an input layer, one or more hidden layers, and an output layer. Neurons in each layer apply activation functions to the weighted sum of inputs, introducing non-linearity and enabling the network to learn complex patterns. ANNs are trained using algorithms like backpropagation, which adjusts the weights based on the error of predictions. They are highly versatile and can be used for tasks such as image and speech recognition, natural language processing, and predictive analytics. The ability to automatically learn features from data without explicit programming makes ANNs powerful tools in the field of artificial intelligence. These techniques are applied to the dataset to develop predictive models for foam concrete compressive strength. Also, they compare both methods and assesses the prediction accuracy of the models and discusses the strength and limitations of each approach. One potential limitation for this paper is the generalizability of the models. Since the dataset is specific to the properties and composition of foam concrete, it cannot be applied directly to all other types of construction usage without further validation and customization.

Song et al. (2021a) employs ensemble machine learning modelling to accelerate the high-performance concrete compressive strength, using waste material like fly ash. The study concludes that ensemble modelling offers a robust approach for eco-friendly concrete production, showcasing the potential of waste material utilization and soft machine learning techniques in the industry. The limitation of this is the variability in the properties of fly ash, which can affect the concrete compressive strength.

The study by Song et al. (2021b) emphasizes the importance of knowing the mechanical properties of cementitious composites, particularly CS, for safety reasons. The study collects experimental work data and it enters into the machine learning techniques compressive strength prediction. The dataset consists of 98 data points with seven input parameters like concrete ingredients (cement, coarse aggregate, fine aggregate, water),fly ash, super plasticizers, finally days and the output parameter CS. The models are trained and tested using the experimental data, and k-fold cross-validation is employed to validate the results using metrics. Results show that the bagging algorithm performs the best, with a coefficient correlation (R2) value of 0.95, indicating high accuracy in predicting CS. In comparison, the R2 values for GEP - 0.86, ANN – 0.81, and DT – 0.75. The study concludes that machine learning algorithms of supervised learning algorithm can effectively predict the concrete compressive strength with fly ash admixture, with bagging showing the highest accuracy. The limitation of this project is the generalizability of the models to different mix ratios or environmental conditions, which may require further validation and adjustment of the models.

The study by Kamath et al. (2024) aims to predict the concrete compressive strength of both normal and high-performance concrete using machine learning algorithms with four datasets. They acknowledge the critical aspect of concrete strength prediction due to various interdependent parameters involved in strength development. Five established individual machine learning regression models are compared. The utilized datasets are, from their previous research and lab experiments, to train and test these models. They analysed the performance of the models with statistical indicators. In this models compared with previous studies to assess their effectiveness. The sensitivity analysis and parametric analysis are conducted to get the effect of predictors in the models performance. The findings of this paper provide insights into the factors influencing the selection of machine learning models and concrete datasets for concrete compressive strength prediction. The study contributes to advancing the toolset for predicting concrete compressive strength. The potential limitation of this project is the reliance on specific datasets, which may limit the generalizability of the models to other concrete mixtures or environmental conditions. Further validation and testing with diverse datasets could enhance the robustness of the models.

The study by Xu et al. (2021) focuses on prediction of 28th day compressive strength of HPC – High Performance Concrete using discrete and ensemble machine learning techniques. It aims to leverage these techniques to forecast concrete properties without the need for extensive experimental work, reducing time, effort, and costs. They applied support vector regression (SVR) as a standalone technique, and AdaBoost and random forest as ensemble techniques, to predict the compressive strength. They also used statistical measures to evaluate the models performance. They also conducted a sensitivity analysis. Results showed that all techniques performed well, with random forest exhibiting the highest accuracy (R2 = 0.93), followed by AdaBoost (R2 = 0.90) and SVR (R2 = 0.83). Statistical analysis and k-fold cross-validation confirmed the superiority of the random forest model. The sensitivity analysis revealed the contribution of different input parameters to the prediction. The study concludes that machine learning techniques, both standalone and ensemble, can accurately predict the compressive strength of HPC. It suggests that these techniques can be valuable in predicting concrete properties, reducing the need for extensive experimental testing. The paper recommends further investigation with other machine learning techniques and suggest increasing the quantity and variety of data points for more robust models.

The study by Zhang et al. (2024) focuses on structural design and applications of bond strength between fiber-reinforced polymer (FRP) bars and seawater sea sand concrete (SWSSC). This model predicts the bond strength of failures between fiber-reinforced polymer bars and seawater sea sand concrete. This paper considered the surface, geometric and mechanical properties of the bond strength of the FRP-Fiber-reinforced polymer bar.

The study by Song et al. (2004) focuses on corrosion characteristics,the factors affecting the quality of construction and how to improve the durability of reinforced concrete structures in ocean environment. Protecting the steel bars from corrosion only helps the construction health. Here they introduced new polymer cement mortars and resin mortars. These are effective materials for surface treatments.

The study by Obando et al. (2024) focuses on evaluating the axial capacity of two types of physical models: (1) SWSSC-filled circular FRP tubes immersed in a seawater environment (Pu1) and (2) SWSSC-filled FRP-steel-FRP circular tubes (Pu2). Here, they used some machine learning (ML) algorithms. In this paper, they graded the machine learning algorithm depending on its performance: Extreme Gradient Boosting (XGBoost) is greater than Artificial Neural Network (ANN), and ANN is greater than Support Vector Machine (SVM) for model 1. For model 2, ANN is greater than XGBoost and SVM, but XGBoost is greater than SVM.

The study by Chao et al. (2024) focuses on evaluating compressive strength of Coral Sand Aggregate Concrete in marine area. In this paper they used five conventional machine learning algorithms for getting prediction in high accuracy and efficiency. Here, they considered pressure and immersion period in sea water by sensitivity analysis. They introduced a new analytical formula for a novel algorithm.

3. METHODOLOGY

This provides a structured approach to predicting concrete strength in deep-sea conditions, leveraging machine learning models and thorough data analysis techniques to achieve accurate and reliable predictions. The various steps involved are:

Figure 1. System Model

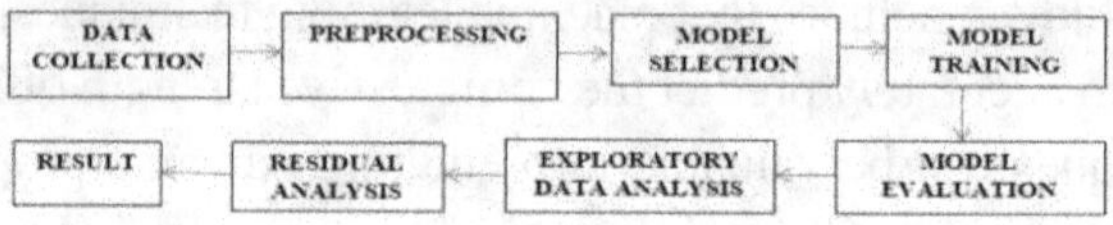

3.1. Data Collection

The dataset containing concrete composition, environmental conditions, and compressive strength measurements in deep-sea condition is to be collected. The dataset with ID 165 which corresponds to the "Concrete Compressive Strength" is fetched from the UCI Machine Learning Repository. The fetched data is then stored in the 'concrete_data' variable. To fetch the data from the UCI repository, a package known as 'ucimlrepo' is installed and it imports a function called 'fetch_ucirepo' to fetch the dataset. Similarly download the 'concrete dataset' from UCI. This dataset includes features such a ingredients of concrete (cement, water, coarse aggregate, fine aggregate) fly ash, blast furnace slag, superplasticizer, and age of the concrete as input variable and the target variable which is compressive strength of the concrete.

3.2. Data Preprocessing

It includes handling missing values, scaling numerical features, and encoding categorical variables. Here missing values are handles implicitly and the numerical features are scaled using 'StandardScaler' to normalize the data. Also, the categorical variables are not explicitly encoded, as it's not present in this dataset.

3.3. Model Selection

Various machine learning models are employed, including Linear Regression, Random Forest Regression, Gradient Boosting Regression, XGBoost Regression, and Support Vector Regression. These models are chosen for their ability to handle regression tasks and their suitability for the dataset characteristics.

3.4. Model Training

Here the data is split into training and testing sets, with a typical split ratio such that 80% of the data for training and the remaining 20% for testing. Each selected model is then trained on the training dataset using the 'fit' method.

3.5. Model Evaluation

The trained models are then evaluated using the metrics such as R-squared score and mean squares error on the test dataset. These metrics helps assess the model's performance in predicting concrete strength.

3.6. Exploratory Data Analysis (EDA)

EDA is conducted to gain insights into the dataset's characteristics. EDA techniques such as pair plots (Fig 2), histograms (Fig 3), and heatmaps (Fig 4) are used to explore the dataset's characteristics and relationships between features and the target variable. This includes visualizing the distributions of features, exploring correlations between variables using heatmaps, and analysing the relationships between individual features and the target variable through scatter plots.

3.7. Residual Analysis:

To further evaluate the model's performance, residual analysis is performed. Residuals, which represent the differences between predicted and actual values, are plotted to assess the model's accuracy and identify any patterns or trends in prediction errors. They are calculated by subtracting predicted values from actual values for both training and test sets. Their plots are created to visualize the differences between predicted and actual values. The chapter aims to provide valuable insights into the factors influencing concrete strength in deep-sea environments. By developing accurate predictive models, this study contributes to improving the design and construction of marine structures, ultimately enhancing their resilience and longevity in challenging underwater conditions.

4. RESULTS AND DISCUSSION

The outcome of this study reveals the significant predictive capabilities of different machine learning models for finding concrete compressive strength under deep-sea conditions. The models were evaluated based on their R-squared accuracy, with

XGBoost Regression standing out as the most effective, achieving an impressive accuracy of 0.94. This model's advanced regularization techniques and efficient handling of complex interactions enabled superior performance. XGBoost's regularization parameters help prevent overfitting, making it particularly effective for large datasets with complex relationships. Its high accuracy and flexibility have made it a popular choice in various domains, including finance, healthcare, and competition-winning solutions on platforms like Kaggle. XGBoost is optimized for speed and performance, utilizing parallel processing, tree pruning, and efficient handling of missing data. Random Forest Regression models achieved an accuracy of 0.92, and Gradient Boosting Regression models also demonstrated excellent results with an accuracy of the same 0.92, underscoring their robustness in capturing intricate data patterns and their utility in regression tasks. Conversely, the initial performance of Support Vector Regression (SVR) was notably low with an accuracy of 0.28, highlighting the necessity for hyperparameter tuning, which significantly improved its accuracy to 0.91 post-optimization. Unlike traditional linear regression models, which aim to minimize the error between predicted and actual values, SVR focuses on fitting the error within a certain margin (epsilon). The core idea of SVR is to find a function that has at most epsilon deviation from the actual target values for all training data, while also being as flat as possible. SVR utilizes kernel functions to handle non-linear relationships by mapping input data into high-dimensional spaces where a linear regression model can be applied. This approach makes Support Vector Regression powerful for capturing complex patterns in data, leading to more accurate and robust predictions in various applications such as financial forecasting, time-series prediction, and engineering.

Figure 2. Pairplot

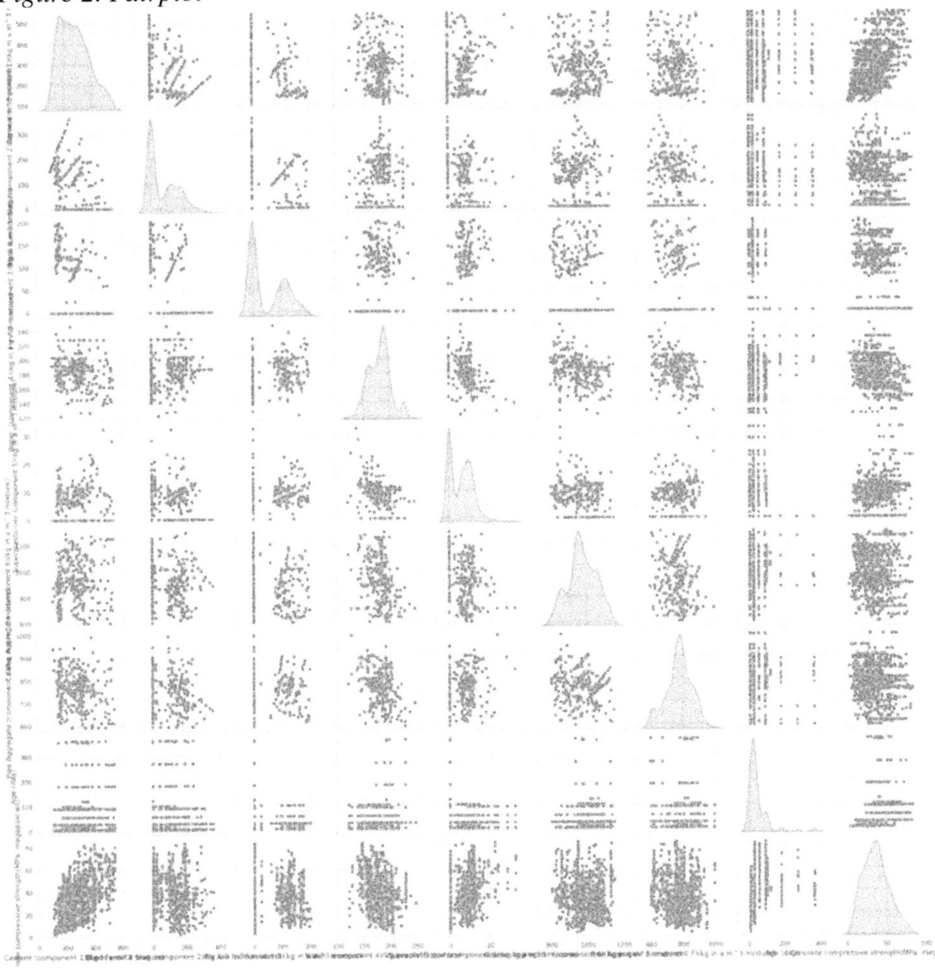

Figure 3. Histogram

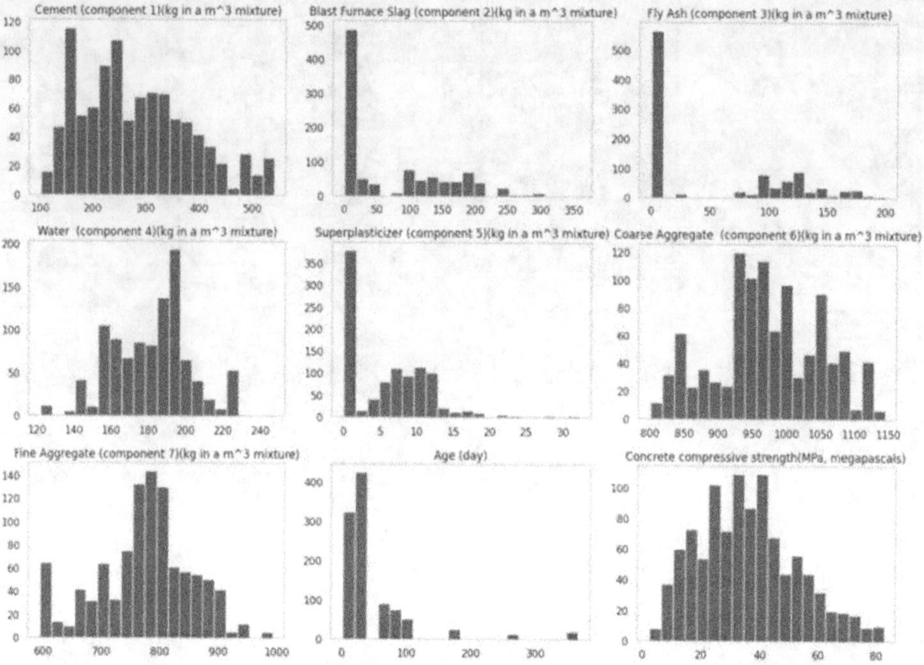

Figure 4. Heatmap

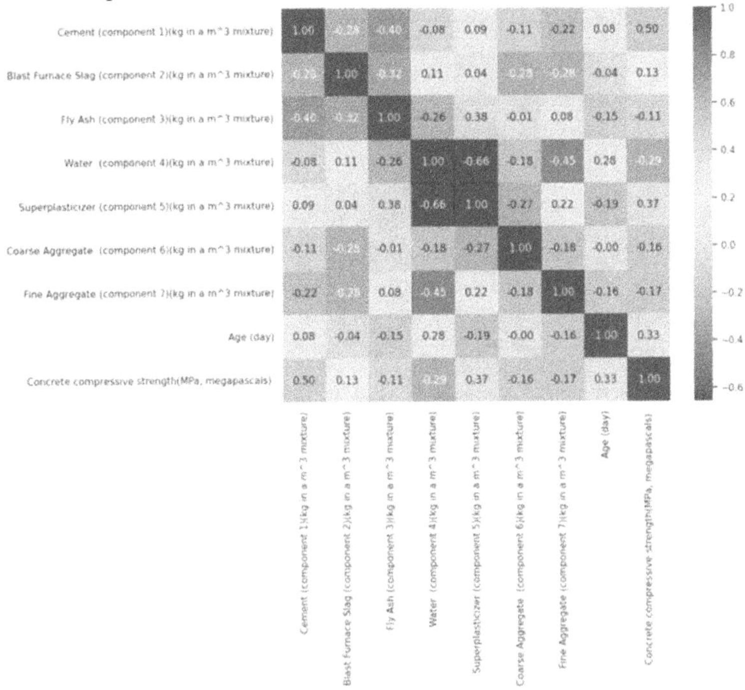

Figure 5. Scatter plot for Blast furnace slag vs Compressive strength

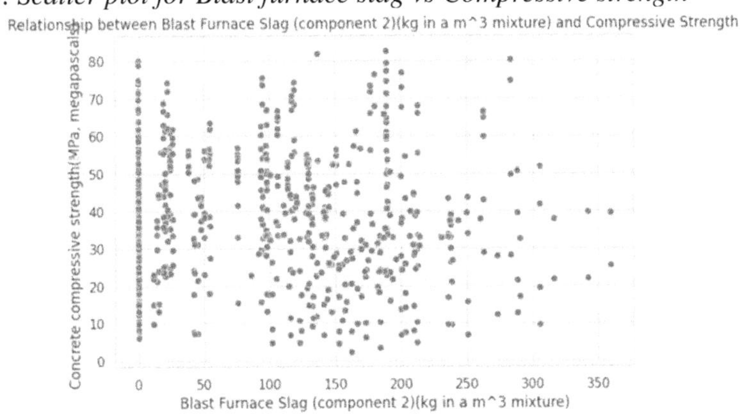

Figure 6. Scatter plot for Cement vs Compressive strength

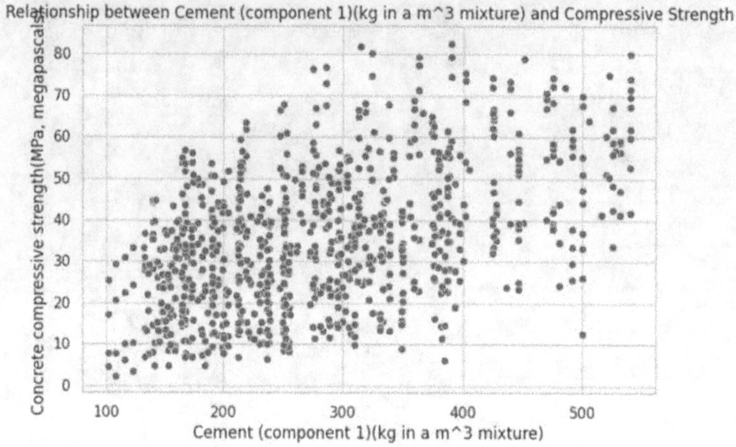

Figure 7. Scatter plot for Water vs Compressive strength

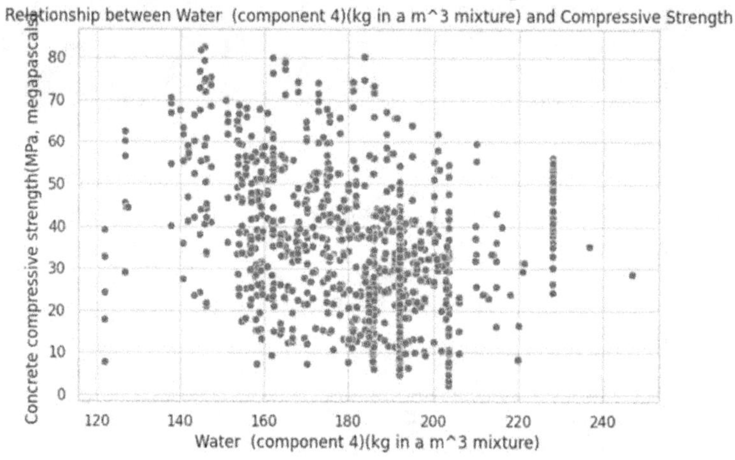

Figure 8. Scatter plot for Coarse aggregate vs Compressive strength

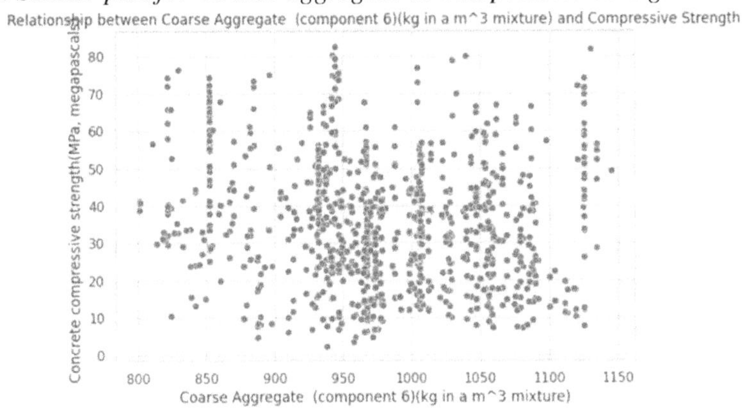

The ensemble Voting Regressor, which combined predictions from multiple models, provided a balanced and robust prediction with an accuracy of 0.88, showcasing the benefits of leveraging the strengths of various models to improve overall prediction reliability. Exploratory Data Analysis (EDA) further supported these findings, illustrating strong correlations between certain features with various graphs like pair plot in Fig 2, histogram for analysing the distribution for each variable in Fig 3, and to explore correlations between features through heatmap in Fig 4, and the scatter plot for visually exploring relationships between different features such as cement content as shown in Fig 5, blast furnace given in Fig 6, water content in Fig 7, coarse aggregate in Fig 8 and fine aggregate in Fig 9 versus compressive strength respectively, thereby guiding effective model selection and interpretation.

The residual analysis in Fig 10 and actual versus predicted scatter plots shown in Fig 11 indicated that models with higher accuracy had tightly clustered residuals around zero and points closely aligned with the reference line, suggesting accurate and unbiased predictions. XGBoost is emerging as the leading model in this context for accurately predicting concrete strength by using advanced machine learning techniques.

Figure 9. Scatter plot for Fine aggregate vs Compressive strength

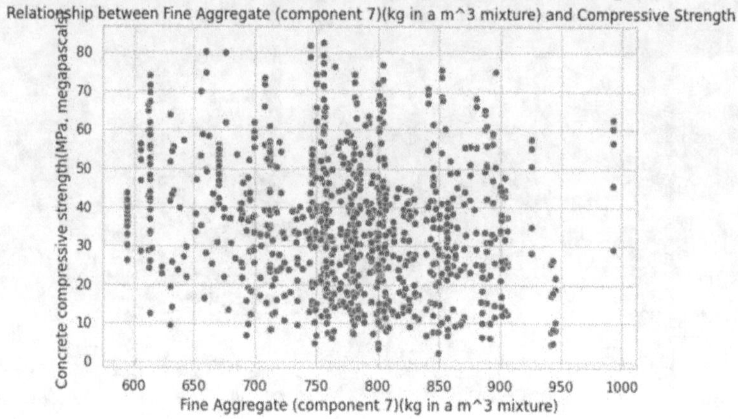

Figure 10. Residuals plot

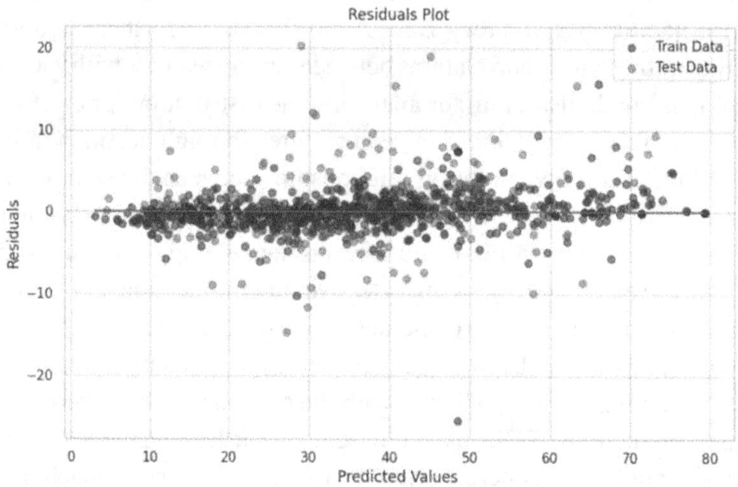

Figure 11. Actual vs Predicted

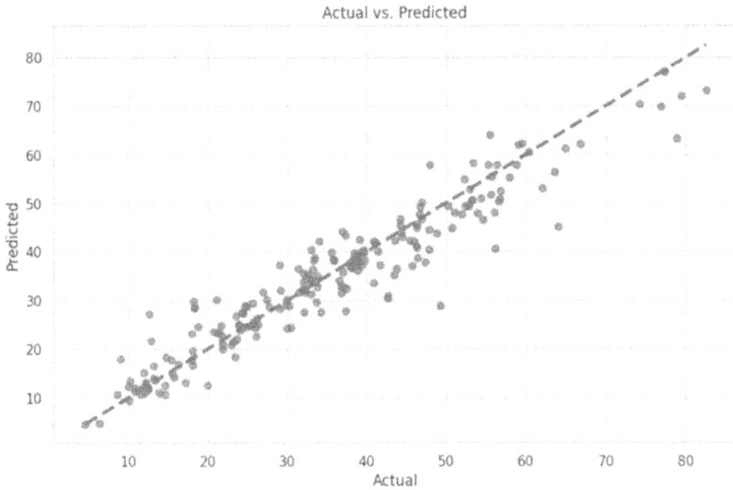

5. CONCLUSION

This project effectively showcased the accurate prediction of concrete compressive strength in deep-sea conditions by machine learning techniques. XGBoost Regression emerged as the most accurate model, demonstrating its capability to handle complex data patterns. Random Forest and Gradient Boosting also showed strong performance, while the importance of hyperparameter tuning was highlighted by the significant improvement in Support Vector Regression. The ensemble Voting Regressor provided robust and balanced predictions. Overall, the study underscores the potential of advanced machine learning methods in accurately predicting concrete strength, paving the way for future optimization and application in real-world scenarios.

REFERENCES

Chao, Z., Li, Z., Dong, Y., Shi, D., & Zheng, J. (2024). Estimating compressive strength of coral sand aggregate concrete in marine environment by combining physical experiments and machine learning-based techniques. *Ocean Engineering*, 308, 118320. DOI: 10.1016/j.oceaneng.2024.118320

Kamath, M. V., Prashanth, S., Kumar, M., & Tantri, A. (2024). Machine-learning-algorithm to predict the high-performance concrete compressive strength using multiple data. *Journal of Engineering. Design and Technology*, 22(2), 532–560. DOI: 10.1108/JEDT-11-2021-0637

Obando, M. D. H., Iqbal, M., Zhang, D., Zhang, P. F., & Zhao, Q. (2024). Axial strength prediction of seawater sea sand concrete-filled circular FRP tubes under alkaline environment based on ensemble learning algorithms. *Thin-walled Structures*, 195, 111530. DOI: 10.1016/j.tws.2023.111530

Song, H., Ahmad, A., Farooq, F., Ostrowski, K. A., Maślak, M., Czarnecki, S., & Aslam, F. (2021b). Predicting the compressive strength of concrete with fly ash admixture using machine learning algorithms. *Construction & Building Materials*, 308, 125021. DOI: 10.1016/j.conbuildmat.2021.125021

Song, Y., Zhao, J., Ostrowski, K. A., Javed, M. F., Ahmad, A., Khan, M. I., Aslam, F., & Kinasz, R. (2021a). Prediction of compressive strength of fly-ash-based concrete using ensemble and non-ensemble supervised machine-learning approaches. *Applied Sciences (Basel, Switzerland)*, 12(1), 361. DOI: 10.3390/app12010361

Song, Y. P., Song, L. Y., & Zhao, G. F. (2004). Factors affecting corrosion and approaches for improving durability of ocean reinforced concrete structures. *Ocean Engineering*, 31(5-6), 779–789. DOI: 10.1016/j.oceaneng.2003.07.006

Ullah, H. S., Khushnood, R. A., Farooq, F., Ahmad, J., Vatin, N. I., & Ewais, D. Y. Z. (2022). Prediction of compressive strength of sustainable foam concrete using individual and ensemble machine learning approaches. *Materials (Basel)*, 15(9), 3166. DOI: 10.3390/ma15093166 PMID: 35591498

Xu, Y., Ahmad, W., Ahmad, A., Ostrowski, K. A., Dudek, M., Aslam, F., & Joyklad, P. (2021). Computation of high-performance concrete compressive strength using standalone and ensembled machine learning techniques. *Materials (Basel)*, 14(22), 7034. DOI: 10.3390/ma14227034 PMID: 34832432

Zhang, P. F., Iqbal, M., Zhang, D., Zhao, X. L., & Zhao, Q. (2024). Bond strength prediction of FRP bars to seawater sea sand concrete based on ensemble learning models. *Engineering Structures*, 302, 117382. DOI: 10.1016/j.engstruct.2023.117382

Chapter 11
Sea Spin Magnets

Maheswari Raja
Rajalakshmi Institute of Technology, Chennai, India

M. Vishali
Sri Eshwar College of Engineering, India

R. Sumitha
Sri Eshwar College of Engineering, India

R. Lakshana
Sri Eshwar College of Engineering, India

K. Vinotha
Sri Eshwar College of Engineering, India

Meghna Manoj Nair
Tandon School of Engineering, New York University, USA

ABSTRACT

This study explores a novel application of tidal energy using two opposing repelling neodymium magnets to operate tiny underwater turbines, harnessing ocean currents for sustainable energy. By leveraging magnetic repulsion, the design aims for simplicity, efficiency, and environmental sustainability. Neodymium magnets are either embedded in or mounted on turbine blades, increasing magnetic interaction and generating a strong repulsive force that accelerates rotation. The turbines convert potential energy to kinetic energy, producing electricity. Protective measures such as coating, shielding, sealing, and using metals prevent magnet corrosion. The innovative approach enhances turbine rotation through enhanced magnetic repulsion, offering a promising solution for renewable marine energy generation.

1. INTRODUCTION

The concept of renewable energy from the ocean has evolved over centuries. The invention and development of ocean energy technologies had become an inevitable need on these days. In ancient method, they use the idea of tidal and wave energy,

DOI: 10.4018/979-8-3693-6670-7.ch011

where energy from tidal movement they use it to grind grains. The tidal mill was first build in England in 11th century. As per the author mentioned in (Thomas E et al., 2020), the concept of generating electricity from tidal energy has evolved in 19th century. The French engineer named Pierre - Simon and his son shown interest on ocean energy and received patent for wave energy in the year of 1799. As referred (Charlier R.H et al., 2009), the author mentioned, in 20th century, Race - first large scale tidal power station in France, became operative in 1966. This facility is used to generate electricity from tidal movement. This remains as one of the largest tidal power station in the world. Figure. 1 shows how tidal power creates electricity.

Figure 1. Working of Tidal Energy

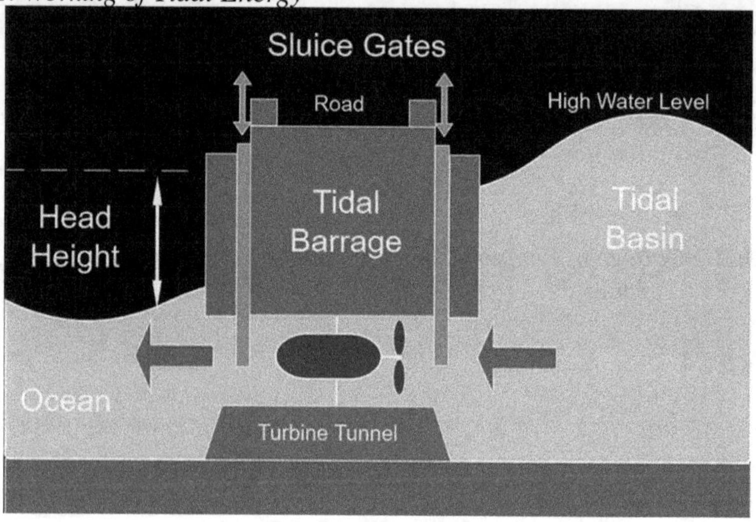

Author of (Ross D et al., 2012), stated that, in 1970's they shown interest in wave energy, because they are in need of alternative energy source. A Scottish engineer Stephen salter developed the "Salter's Duck", a device designed to capture wave energy and convert into electrical energy. In late 20th and early 21st centuries, advanced the technology of both tidal and wave energy. They were developed numerous protype and projects. Different wave and tidal energy systems are the result of investments made in ocean energy research and development by nations like Australia, the UK, and Canada.

Burning coal and other fossil fuels releases pollutants, carbon dioxide (CO2), and other greenhouse gases into the environment, which is why we utilize them to generate power. In essence, carbon dioxide (CO2) acts as a heat trap, causing respiratory problems, acid rain, global warming, and climate change. Ocean energy such as tidal, ocean thermal energy and wave energy is sustainable and renewable.

These technologies produce minimum or no greenhouse gas emission. It reduces air pollution and improving environment quality, as population increases need of electricity also increases. There are two types of ocean energy: Thermal energy and Mechanical energy. Rotation of Earth creates wave on ocean surface, while the gravitational pull of the moon creates tides and currents. From the sun, derive thermal energy, it heats the surface of the ocean while depth remains colder.

This temperature difference between warm surface water and deep cold water, making a renewable energy source. They are different ways to get tidal energy. High tide, it occurs when the water level rises to its highest point along with shoreline, it occurs approximately twice a day with a roughly 12 hour and 25 minutes interval between consecutive high tides. Low tide, when the water level reaches its lower point along with the shoreline, it occurs approximately twice a day with a roughly 12 hours and 25 minutes interval between consecutive low tides. Tidal range is a dam that uses the difference in sea level between high and low tides, the turbine converts potential energy into electricity. It captures the kinetic energy from the flow of the current to convert energy.

Ocean current energy, similar to wind turbines but placed in underwater. The turbine blades drive by ocean currents, which spin generator to generate electricity. The difference in temperature between warm surface water and cold deep water is utilized for ocean thermal energy conversion A working fluid with a low boiling point is heated by the warm surface water, turning it into vapor which drives a turbine and the vapor is condensed back to liquid by cold deep water.

Constructing tidal energy facilities requires an authentic investment. The component such as tidal barrages, underwater turbines, and other infrastructure are overpriced. Keeping track of the tidal energy system may also pricey as they are implemented in a harsh marine environment. Tidal energy can be generated effectively only in particular spot. Specific locations that have exact tidal range and flow velocities to make tidal energy economically viable are limited in number around the world. Hence to overcome these drawbacks we can generate electricity in ocean using magnets incorporated in turbines. The repulsive force and attractive force between magnets and the turbine blades makes the turbine to spin. It converts potential energy of a molecule into kinetic energy, from that electricity can be generated using a generator.

2. LITERATURE

According to (Khan et al., 2017), author's statement, ocean thermal energy is produced by ocean thermos energy generator (OTEC), salinity gradient osmotic power plants and ocean thermal energy convertors (OTEC), where river falls into ocean. Landmasses and ocean are uniformly warmed by sun light. Glaciers reflect back the light but ocean absorb the heats, the ocean being 70% of the earth absorbs solar energy which equivalent to 250 billion barrels oil equivalent.

In the light of what author says in (Ben Elgali et al., 2009), marine tidal currents have a huge potential for producing electricity. It is being known that tidal currents can be used as a resource to generate electricity in a sustainable manner. Marine currents are very appealing for power generation and favourable when compared to other renewable energy because of the high load factors brought about by the fluid features and the predictable resource characteristics. Furthermore, the development of renewable ocean energy technology has resumed as a result of international conventions pertaining to climate management. Consequently, a number of tidal power demonstration projects are planned to harness the coastal currents produced by tidal energy. This paper discusses key ideas and global marine tidal power projects in relation to this new and exciting field of study. It also details problems with tidal turbine-related electrical generating topologies. Additionally, an effort is made to draw attention to upcoming problems in order to index some developing technologies, mostly depending on pertinent research that has been done on ship propellers and wind turbines.

Considering what the author says in (Sheets, 1975), ocean currents provide a significant amount of energy. These currents have a relatively modest velocity that varies according to depth below the ocean's surface, location, and season. The analysis that follows reviews many kinds of water turbines, such as Water Wheels, Kaplan Turbines, Propeller Turbines, and Vertical Axis Turbines, and shows that this energy can be used to generate electricity. The impact of velocity gradients and the machines most appropriate for this application will be assessed by the analysis. Owing to the reduced energy flux, the equipment will enlarge and move slowly. Nonetheless, given the significant energy resources available, these water turbines might be of use in certain places where they can be erected in close proximity to the energy demand source.

Taking into account what the author says in (Melikoglu, 2018), tidal energy falls under the category of ocean/marine energy and is a renewable energy source. It is possible to generate power using the elevation changes between high and low tides. The two types of tidal energy are tidal potential energy and tidal current energy. Ocean wave energy has the potential to displace the world's present fossil fuel consumption as one of the most plentiful, clean, frequent, renewable, periodic, and predictable energy sources. Waves are produced by wind flowing across the ocean's surface, and wind can blow consistently and forcefully enough in many places of the world to produce waves that are continuously forming and have a tremendous potential energy content. One can extract wave energy. Various devices are being developed to utilize the energy contained in tides. Typically, dam construction is used to capture tidal potential energy.

Thinking about what the author says in (Wilberforce et al., 2019), in addition to examining some of these technologies and discussing their benefits and some of the challenges preventing these developing technologies from being commercialized, the paper demonstrates that there are a number of exciting new advancements in the field of maritime energy harvesting. This covers the use of salinity gradients for power generation, wave and tidal energy harvesting, and ocean thermal energy. The aforementioned work underscores the notion that these novel and evolving technologies are merely in their infancy and require much time and effort to achieve widespread acceptance and prosperous commercialization. Considering what the writer states in (Chen et al., 2018), furthermore, tides that change significantly between high and low tides always bring with them enormous energy and current speed. Its regular period is determined by the relative locations of the sun, moon, and earth. Ocean currents are a small portion of marine currents. The earth's rotation, the prevailing winds, the variation in temperature and salinity density, and other factors all contribute to this long-term constant flow between the sea floor and the canals.

Taking into account what the author writes in (Shetty and Priyam, 2022), since tidal energy is limitless and eternal, it qualifies as a renewable energy source. Compared to the other sources, it is less prone to erratic fluctuations in climate, which gives it an advantage. Because it doesn't release any greenhouse gases, it is environmentally benign. Day and nighttime production of energy is possible. Tidal energy is superior to other renewable energy sources because tidal currents are more predictable and consistent. It is possible to accurately estimate the amount of electricity that tidal energy will create in advance, making it simple to integrate it with the current power systems. Compared to other energy sources, it has a very high efficiency of over 80%. Furthermore, tidal power plants have a very long lifespan. Because seawater has a density of around 832 times that of air, ocean currents generate a great deal of energy even though the speed of the water is slower than that of the wind. Tidal turbines are therefore smaller than wind turbines as a result of this. In Bagher et al. (2015) the authors stated that, the three elements of a traditional hydro plant are an electric plant that produces electricity, a dam that opens and closes to regulate water flow, and a reservoir that holds water. The water pressing against the turbine's blades through an intake behind the dam causes the turbine to spin. The generator is turned by the turbine to produce energy. The amount of electricity that can be generated depends on how far the water drops and how much water flows through the system. Homes, factories, and other places can get electricity using long-distance electric lines. Hydropower accounts for over one-fifth of the electricity generated globally. China, Canada, Brazil, the United States, and Russia were the top five hydropower producers in 2004. On the Yangtze River in China, at Three Gorges, is one of the biggest hydroelectric plants in the world. The energy source is free flowing water; a dam has been constructed and equipment installed. It's a pure fuel that snow and rainwater replenish each year. Hydropower is also easily accessible; engineers can regulate the water's flow through the turbines to generate electricity as needed. Author of (Veerabhadrappa et al., 2022), explains that, to store the energy generated by the process, a suitable storage device needs to be employed. These storage devices can serve either a long-term or short-term function, depending on the intended use. Capacitors, super conducting magnetic energy storage, and capacitors are a few types of short-term energy storage devices. Capacitors are used in several technologies for energy storage, including ceramic, film, and electrolytic capacitors Ceramic capacitors have poor aging characteristics while having low equivalent resistance and high frequency applications. Electrolyte capacitors are utilized in place of other capacitors (non-SC) with a reasonably high capacity when significant amounts of power need to be stored. As referred in (Bryden, 2004), there are two distinctly different strategies for taking advantage of tidal energy. The first involves deploying barrages to take advantage of the sea level's cyclical rise and fall, while the second involves harnessing local tidal currents in a way that's similar to wind power. A great deal of synergy exists between the offshore oil and gas business and the development of a tidal current generation industry, according to numerous industrial, commercial, and public agencies. With knowledge and technology that would otherwise need to be built from scratch, this presents the exciting possibility of a new renewable industry growing in collaboration with the petroleum industry. Development could even be sped up as a result. However, development of this resource is probably going to be limited in the near future because to the high capital costs of tidal barrage systems. Although tidal current systems do not now possess the strategic potential of barrage systems, they do bring options for energy delivery in rural coastal and island communities—at least in the short term. In (Leijon et al., 2005), it was stated that, Multiphysics simulations are used to look at the possibilities of converting sea wave energy into electric energy using three-phase permanent magnet linear generators. The results point to a potential future step toward the use of direct conversion to produce sustainable electricity from ocean waves, but this possibility needs to be further confirmed by experimental testing. The findings imply that wave energy can affect not only large units but even units as small as 10 kW when it comes to tomorrow's new sustainable electricity production. This increases the economic potential of wave power beyond what was previously thought. The study shows that computer simulations can provide a comprehensive and in certain cases, in-depth understanding of the energy conversion process. In (Khatri and Wang, 2020), they stated that, electricity is now produced using renewable energy sources, such as the sun, wind, and ocean, due to the daily rise in demand for power and the depletion of non-renewable energy sources. Solar energy is the most widely used and implemented renewable energy source. Everywhere are sustainable power sources that may be used to generate electricity for commercial use, including as daylight, wind, and ocean waves. One of the safest and cleanest ways to transfer mechanical energy to electrical energy in recent years has been revealed to be ocean wave energy. Numerous factors, such as gravity, solar energy, seismic vibrations, or planetary influences, can cause ocean waves. On the other hand, wind waves, also known as twist blowing over a section of a liquid surface, are what propel most ocean waves.

3. MOTIVATION

The immediacy with which the world's energy problem and the environmental risks associated with fossil fuel dependence must be addressed is the driving force behind our initiative. Climate change and environmental degradation are caused by greenhouse gas emissions from traditional energy producing technologies. These emissions are substantial. Our team is motivated to investigate novel and sustainable energy solutions since we acknowledge these urgent problems. In terms of environmental effect, energy efficiency, and the uptake of renewable energy sources, the concept of generating power using turbines that utilize magnetic technology offers a competitive alternative. Our objective is to use modern turbine design and the concepts of magnetism to create a more efficient and ecologically friendly power generation system. Our motivation for taking on this project is the potential to significantly influence the renewable energy industry by offering more easily available, environmentally friendly, and long-lasting energy options that will benefit present and future generations. This activity aligns with our objective to further scientific research, promote technological innovation, and support global sustainability goals.

The primary objective of this project is to research and demonstrate the viability and efficiency of using magnetic technology in turbines to generate power. This investigation aims to:

- **Create Innovative Renewable Energy Solutions:** Utilize the concepts of turbine mechanics and magnetism to create an innovative method to power generation in order to create a sustainable and environmentally friendly energy source. Enhance Energy Efficiency: Assess possible improvements in energy conversion efficiency relative to conventional techniques, emphasizing output maximization and energy loss minimization. Minimize Environmental Impact: Examine how fossil fuels can be more environmentally friendly replaced with magnetic turbines, which will cut greenhouse gas emissions and encourage long-term environmental sustainability.

- **Cost-Effectiveness:** Assess the economic viability of magnetic turbine technology in terms of initial investment costs, continuing maintenance costs, and operational expenses in order to raise the availability and affordability of renewable energy. Technological Advancements: Significantly influence the field of renewable energy technology by offering insightful evaluations, useful solutions, and insightful experiment results that can direct future development and study. By fulfilling these goals, the project hopes to contribute to the global shift towards more sustainable energy practices by expanding our understanding of and expertise with magnetic turbine systems in the renewable energy industry.

4. PROPOSED WORK

Due to tidal energy's many drawbacks, we devise the concept of the sea spin magnet. In underwater turbines, when a piece of magnet is used to mount or embed the turbine blades, two magnets of varying sizes are placed opposite each other. The left magnet is relatively larger than the right. The left magnet and turbine blades' poles should be the same for repulsion, and their poles should differ to provide an attractive force. The turbine moves upward due to the repelling force of a left magnet on the blades, and lower due to the attractive pull of a right magnet on the blades. This process is repeated, causing the turbine to spin or rotate. Size difference between the two magnets: if they are the same size, the magnet will become stuck; if it is the larger of the two, it will attract the turbine blade and prevent it from spinning. The turbine rotates because of the pushback that the blades receive from this repellent force. The working of turbines mounted with magnet is explained in the Figure.2

Figure 2. Working of Turbines Using Magnets

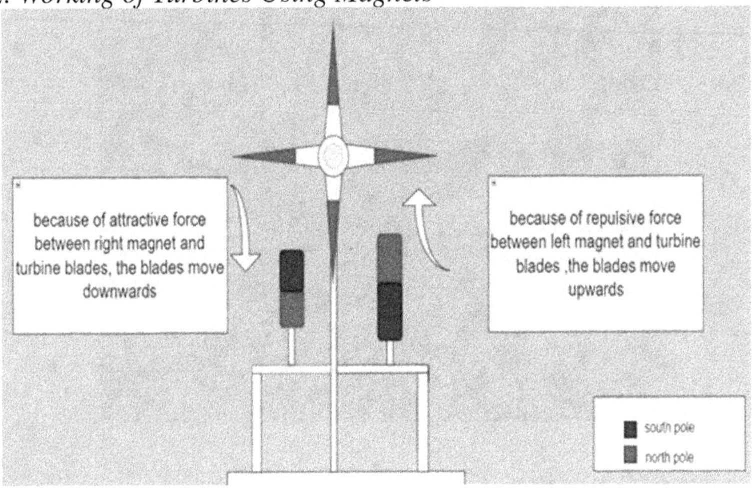

4.1 Constant Rotation

The turbine's constant rotation is achieved by the introduction of new blades into the repulsive field during its revolution. The turbine blades' rotation transforms the potential energy of the water molecule into kinetic energy if it is left larger, which makes it spin more quickly. With the help of a generator, we may create electrical energy or electricity by converting potential energy found in water molecules into kinetic energy. The mass of two magnets required to rotate a tiny turbine must be calculated taking into account the required magnetic force as well as the relationship between the magnetic force and the mass of the magnets. For better understanding the overall working of the turbines is given in figure. 3 The given block diagram in Figure.4 explains the entire workflow of generating electricity from turbines mounted with magnet.

Figure 3. Overall Working of Turbines and Generating Electricity

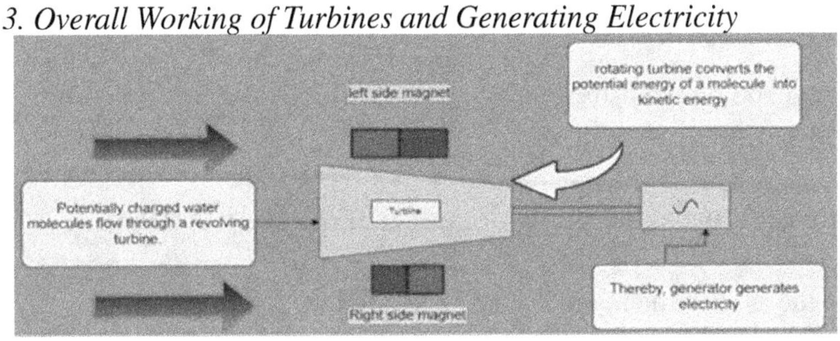

Figure 4. Block Diagram of Magnetic Turbine

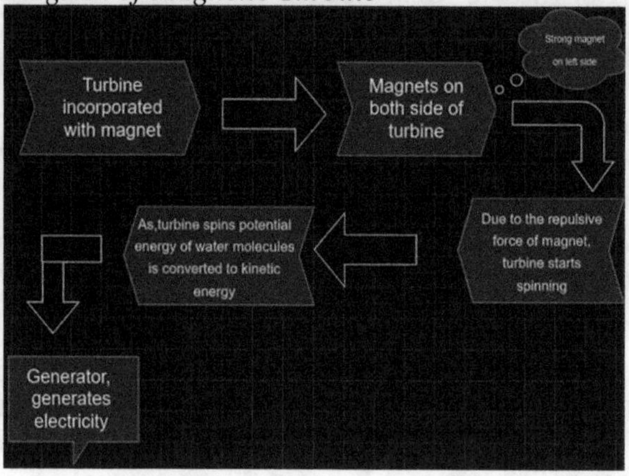

Here is an elaborate method:

The formulas given below explains how to calculate magnetic field for the rotation of turbines.

Presumptions and Simplified Approaches

1. Calculate magnetic force:
   - The force required to operate the little turbine has already been determined:

F=0.425N.

2. Strength of Magnetic Field and Mass of Magnet:
   - A magnet's force is proportional to its mass and is based on its magnetic field strength (B) and volume (V).
   - $f \approx \frac{B^2 A}{2\mu_0}$ (1)

Equation. 1 is an approximation for the magnetic force (F).

At location,

- B magnetic flux density.
- $\frac{\mu_0}{mu\_o\mu_0}$ (2)

Equation. 2 is the permeability of free space (μ_0=4π×10^{-7} H/m).

- A is the area of the magnet surface facing the turbine.

3. Magnet Specifications:

Due to their strong magnetic properties, neodymium magnets (NdFeB) are frequently utilized. The surface field of a neodymium magnet is typically 1.2 T (Tesla).

Methodical Computation

Given below is the step-by-step procedure for calculating force and magnetic moment.

1. Calculate the Required Magnetic Force:
   - To rotate the turbine, we require 0.425.
2. Calculate the Magnetic Field Required (B):
   - Presuming that we employ neodymium magnets of high strength with B≈1.2 T
3. Calculate the Magnet Area (A):
   - We can solve for the needed area of the magnet surface using equation. 3 if we know the requisite force and field strength.

$$f = \frac{B^2 A}{2\mu_o} \quad (3)$$

- Calculating A:

$$A = \frac{2\mu_0}{B^2} \quad (4)$$

Equation. 4 is used to calculate A.

- Changing the numbers to:

$$A = \frac{2(4\pi \times 10^{-7})0.425}{1 \cdot 2}$$

$$A = \frac{2(4\pi \times 10^{7})0.425}{(1 \cdot 25)^2}$$

$$\text{hence, } A = \frac{3.4 \times 10^{-6}}{1 \cdot 44}$$

$$A \approx 2.36 \times (10^{-6})\ m^2$$

4. Determine the Magnet Mass and Volume using the formula

- Volume V=Ad, where d is the magnet's thickness.

Under the assumption that the thickness (d) is 1 cm or 0.01 m0.01,

$V=2.36\times(10^{-6})\,m^2\times 0.01m$,

$V=2.36\times(10^{-8})\,m^3$.

- Neodymium magnet density is approximately 7400 kg/m^3, where Mmagnet = V× density.

2.36× (10^−8) ×7400 Mmagnet = 0.1775-275 kg

The repelling force that must be applied between the magnets in order to generate enough torque to turn the turbine and overcome friction.

1. The inertia moment (I):

The formula for a disk with radius r and mass m is $= \frac{1}{2}mr^2$

2. Torque (τ):

$\tau = I\alpha$ reveals the torque required to overcome friction and reach the intended angular acceleration (α).
Angular Acceleration (α):
Let's say we wish to attain a particular angular velocity (ω) in time t.

Then, $\alpha = \frac{\omega}{t}$ (5).

4. Frictional Torque: Air resistance and bearing friction both contribute to this torque. Although the exact design and materials utilized will determine these values, they can be guessed or quantified through experimentation.

τ_{total}, or the total required torque, is equal to τ plus friction plus air.

5. τ_{total}, or the total required torque, is equal to τ plus friction plus air.
6. Magnet Force (F):

This whole torque must be produced by the magnets' force. If the lever arm length (r) is the distance at which the force acts, then $Fr = \tau_{total}$ (6)

$F = \tau_{total} / r$ as a result.

To generate the necessary force to operate the miniature turbine, each magnet would have to weigh about 0.175 kg. Given that two magnets are being used, the total mass would be:

* Total mass=0.175 kg×2=0.35

Thus, in the given circumstances, the little turbine should be able to revolve with a combined mass of about 0.35 kg thanks to two neodymium magnets.

To generate the necessary force to rotate the massive turbine, each magnet would have to weigh about 0.133 kg. The overall mass would be as follows because we are employing two magnets.

Overall mass = 0.133 kg × 2 = 0.266 kg = 0.133 kilogram

For the huge turbine to rotate under the specified conditions, two neodymium magnets with a combined mass of about 0.266 kg should be adequate.

5. DATASET

Data access, mixing, transformation, modelling, and visualization can all be done with the free, open-source KNIME Analytics Platform, which requires little to no coding.

A CSV reader is used to import data from comma-separated files. Use the shuffle function to randomly rearrange the rows of the input tables. The missing value node may manage the cells in the input table that have missing values. The rule engine node compares each row in the input table to a list of user-defined rules. Low variance filter node removes double-compatible columns whose variance is less than a user-specified threshold. A partition node separated a database data table row.

Figure 5. A Pie Chart Illustrating - Total Energy

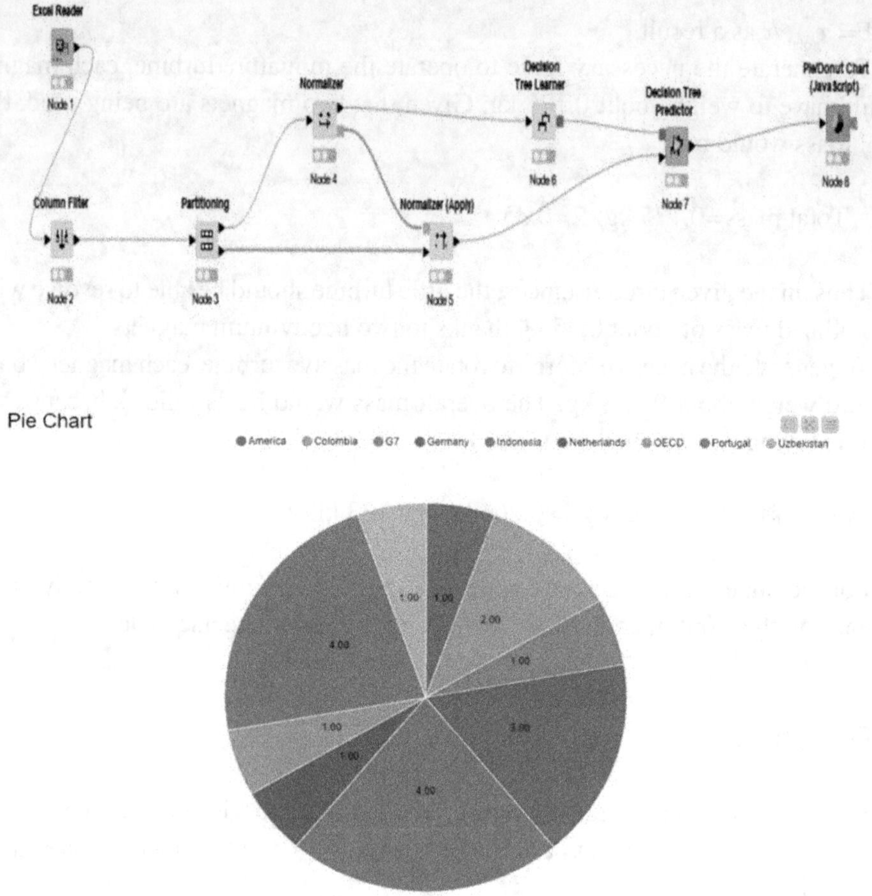

The distribution of several categories, each with a distinct numerical value, is shown in the Figure. 5. This graphic facilitates data exploration and analysis by helping to visualize the proportions, which helps to comprehend the relative sizes and contributions of each group.

Figure 6. Consumption of Total Energy

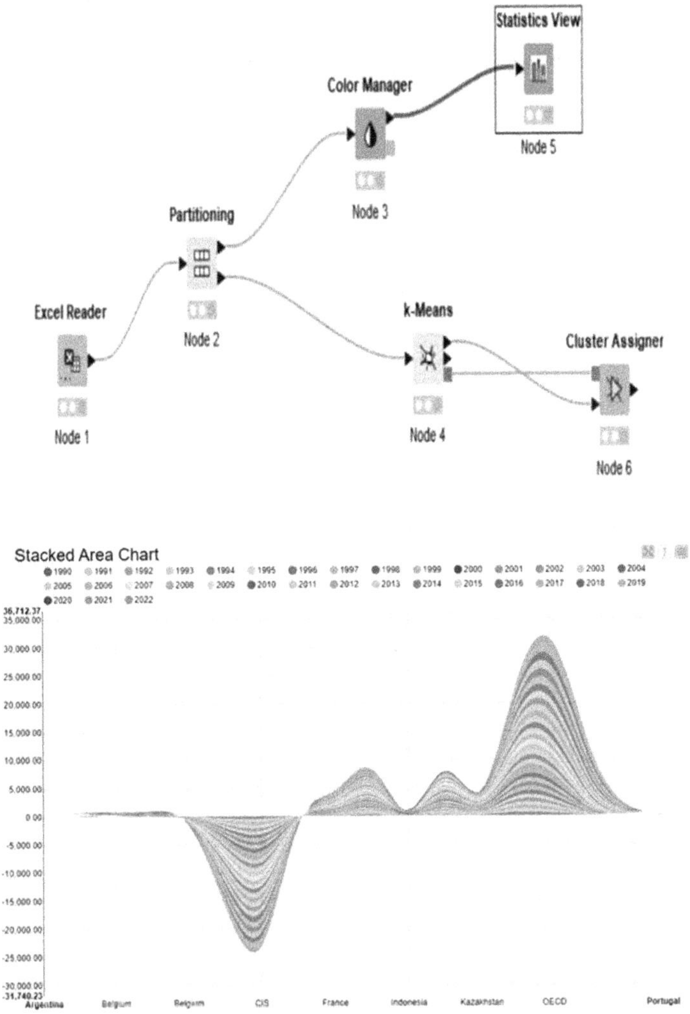

Figure 6. shows the graph for the global total energy use. Statistics supports data exploration and analysis by offering distribution summaries and descriptive statistics for numerical variables in datasets.

Figure 7. An Overview of Total Energy from Statistics

Statistics

Rows: 34 | Columns: 14

Name	Type	# Missing val...	# Unique values	Minimum	Maximum	25% Quantile	50% Quantile (...	75% Quantile	Mean	Mean Absolut...	Standard Devi...	Sum	10 most com...
empty_A	String	0	39	⊘	⊘	⊘	⊘	⊘	⊘	⊘	⊘	⊘	Africa (1; 2.56...
1990	Number (doub...	1	38	-699.797	1,253.495	-71.933	3.566	48.809	44.804	207.117	334.134	1,702.55	-699.797 (1; 2...
1991	Number (doub...	1	38	-682.887	1,214.819	-88.237	4.526	45.76	41.077	204.054	328.166	1,560.909	-682.887 (1; 2...
1992	Number (doub...	1	38	-750.342	1,265.016	-87.711	4.505	45.087	43.779	210.373	338.462	1,663.593	-750.342 (1; 2...
1993	Number (doub...	1	38	-791.393	1,294.878	-94.203	2.394	47.036	46.898	212.547	341.798	1,782.106	-791.393 (1; 2...
1994	Number (doub...	1	38	-785.526	1,328.73	-97.01	1.437	48.116	46.491	217.451	347.716	1,766.675	-785.526 (1; 2...
1995	Number (doub...	1	38	-792.992	1,327.854	-101.398	1.438	59.754	45.867	220.031	352.779	1,742.947	-792.992 (1; 2...
1996	Number (doub...	1	38	-800.246	1,390.152	-103.033	4.239	69.355	49.152	229.799	366.654	1,867.78	-800.246 (1; 2...
1997	Number (doub...	1	38	-833.324	1,436.252	-107.399	6.885	82.313	51.811	237.205	376.758	1,968.824	-833.324 (1; 2...
1998	Number (doub...	1	38	-894.28	1,475.159	-114.903	5.831	72.999	52.903	243.7	389.633	2,010.297	-894.28 (1; 2.6...
1999	Number (doub...	1	38	-870.136	1,517.936	-112.967	8.508	74.984	58.795	248.094	392.929	2,234.229	-870.136 (1; 2...
2000	Number (doub...	1	38	-929.962	1,585.63	-127.178	8.675	66.15	62.2	260.076	412.698	2,363.615	-929.962 (1; 2...
2001	Number (doub...	1	38	-889.805	1,609.035	-129.604	6.509	73.196	62.243	263.386	414.682	2,365.249	-889.805 (1; 2...
2002	Number (doub...	1	38	-810.672	1,600.115	-138.612	9.763	69.245	64.076	263.514	411.918	2,434.875	-810.672 (1; 2...
2003	Number (doub...	1	38	-898.816	1,696.972	-135.21	10.588	76.824	65.011	278.904	439.748	2,470.422	-898.816 (1; 2...
2004	Number (doub...	1	38	-970.814	1,787.826	-136.162	11.703	134.266	70.144	295.009	466.268	2,665.477	-970.814 (1; 2...
2005	Number (doub...	1	38	-1,004.268	1,857.558	-139.903	14.576	147.793	71.32	304.006	481.773	2,710.142	-1,004.268 (1; ...
2006	Number (doub...	1	38	-997.981	1,904.975	-139.585	16.244	217.255	77.349	309.506	489.37	2,939.258	-997.981 (1; 2...
2007	Number (doub...	1	38	-970.629	1,867.836	-150.697	17.145	215.118	79.523	307.538	483.606	3,021.866	-970.629 (1; 2...
2008	Number (doub...	1	39	-995.21	1,824.841	-145.267	17.227	222.672	79.795	303.551	479.935	3,030.709	-995.21 (1; 2.6...
2009	Number (doub...	1	39	-898.162	1,691.438	-140.892	13.745	212.46	78.301	287.404	448.27	2,975.428	-898.162 (1; 2...
2010	Number (doub...	1	38	-949.017	1,676.972	-150.638	15.112	204.093	82.406	298.841	467.29	3,133.695	-949.017 (1; 2...
2011	Number (doub...	1	38	-1,090.359	1,629.276	-154.744	21.873	120.939	78.017	295.497	473.337	2,964.647	-1,090.359 (1; ...
2012	Number (doub...	1	38	-1,076.054	1,550.491	-162.917	21.613	178.051	77.799	298.822	475.601	2,954.086	-1,076.054 (1; ...
2013	Number (doub...	1	38	-1,076.71	1,468.35	-189.047	14.441	169.429	75.169	299.258	470.195	2,856.04	-1,076.71 (1; 2...
2014	Number (doub...	1	38	-1,050.309	1,537.308	-166.867	14.541	100.315	73.651	296.407	462.617	2,760.743	-1,050.309 (1; ...
2015	Number (doub...	1	38	-1,099.239	1,565.285	-185.882	15.99	142.303	69.696	297.806	471.87	2,648.518	-1,099.239 (1; ...
2016	Number (doub...	1	38	-1,245.479	1,657.323	-154.984	11.81	153.195	69.037	310.747	495.617	2,623.097	-1,245.479 (1; ...
2017	Number (doub...	1	38	-1,196.431	1,775.256	-172.53	9.293	176.711	70.89	324.679	507.663	2,686.219	-1,196.431 (1; ...
2018	Number (doub...	1	38	-1,211.733	1,858.002	-186.042	7.546	168.742	66.655	331.626	517.139	2,532.672	-1,211.733 (1; ...
2019	Number (doub...	1	38	-1,141.17	1,907.678	-229.545	8.26	160.182	66.014	334.477	518.012	2,508.522	-1,141.17 (1; 2...
2020	Number (doub...	1	38	-1,051.536	1,902.284	-207.27	7.448	152.86	53.922	312.786	487.964	2,049.022	-1,051.536 (1; ...
2021	Number (doub...	1	38	-1,065.983	1,919.744	-217.107	8.4	165.167	61.857	318.885	496.176	2,250.565	-1,065.983 (1; ...
2022	Number (doub...	1	38	-1,218.97	1,941.868	-217.672	8.509	172.976	59.437	332.436	516.023	2,258.6	-1,218.97 (1; 2...

The curve for the total energy used globally is displayed in Figure. 7.

Distribution summaries and descriptive statistics for numerical variables in datasets are provided by statistics, which facilitates data exploration and analysis

Figure 8. Table of contents for energy

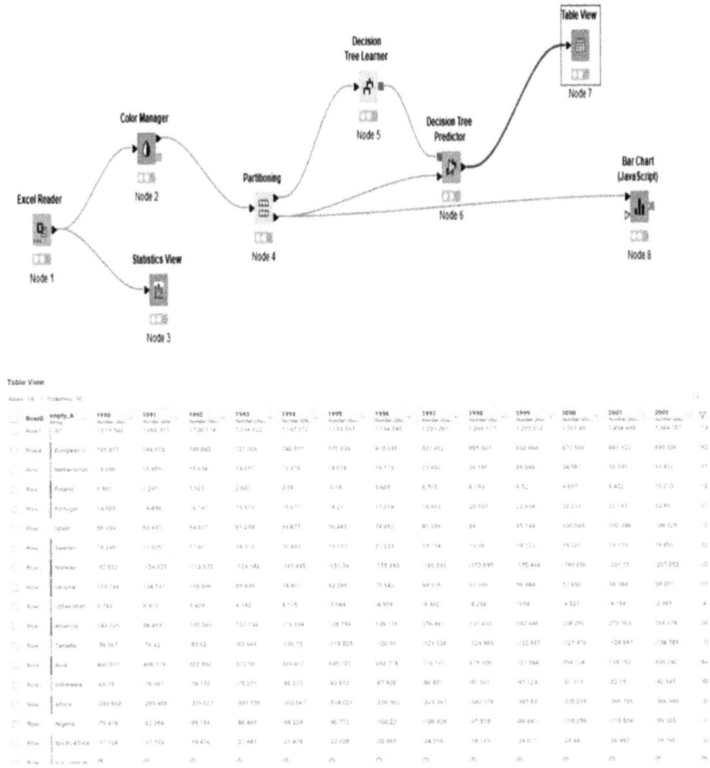

In Figure. 8. The symbol for the global energy balance dataset is indicated.

Decision Tree Predictor: This tool applies pre-trained decision tree models to new data and predicts class labels or values using learned decision rules. These nodes are essential to predictive modelling in KNIME workflows.

Figure 9. Stacked Area of Chart for Consumption of Total Energy

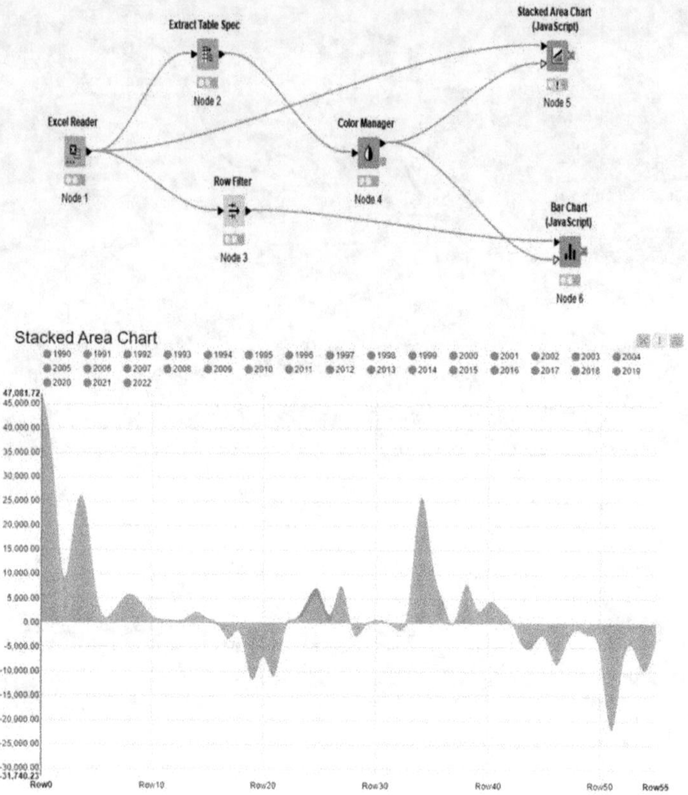

The curve for the total energy used globally is displayed in Figure. 9.

Distribution summaries and descriptive statistics for numerical variables in datasets are provided by statistics, which facilitates data exploration and analysis.

Figure 10. Balance Trade Energy

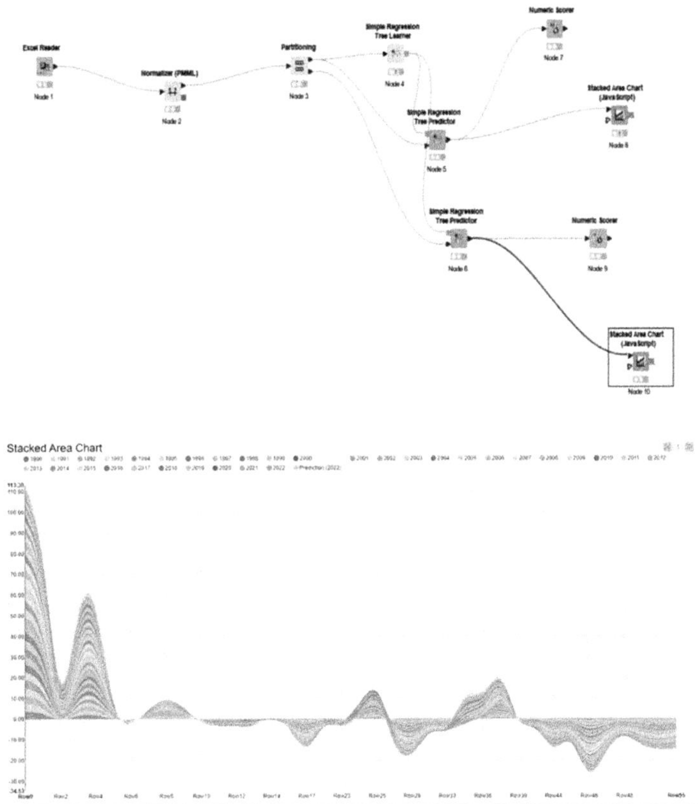

The curve for the total amount of energy used worldwide is shown in Figure 10.

The descriptive statistics and distribution summaries that statistics provides for the numerical variables in datasets aid in the exploration and study of data.

Figure 11. Energy Consumption Chart

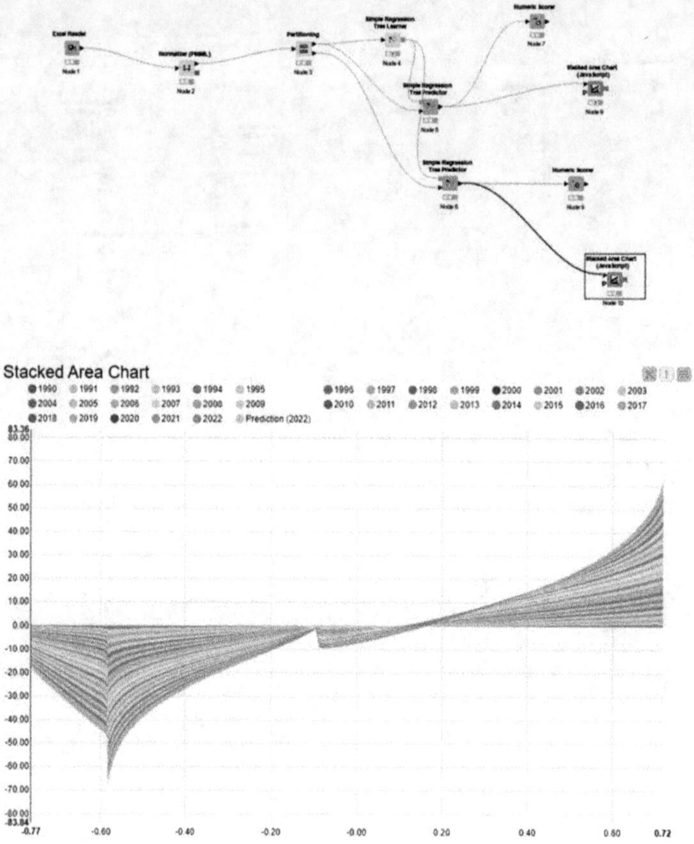

The global total energy use graph is displayed in Figure. 11.

By providing distribution summaries and descriptive statistics for numerical variables in datasets, statistics aids in data exploration and analysis.

Figure 12. Electricity Production

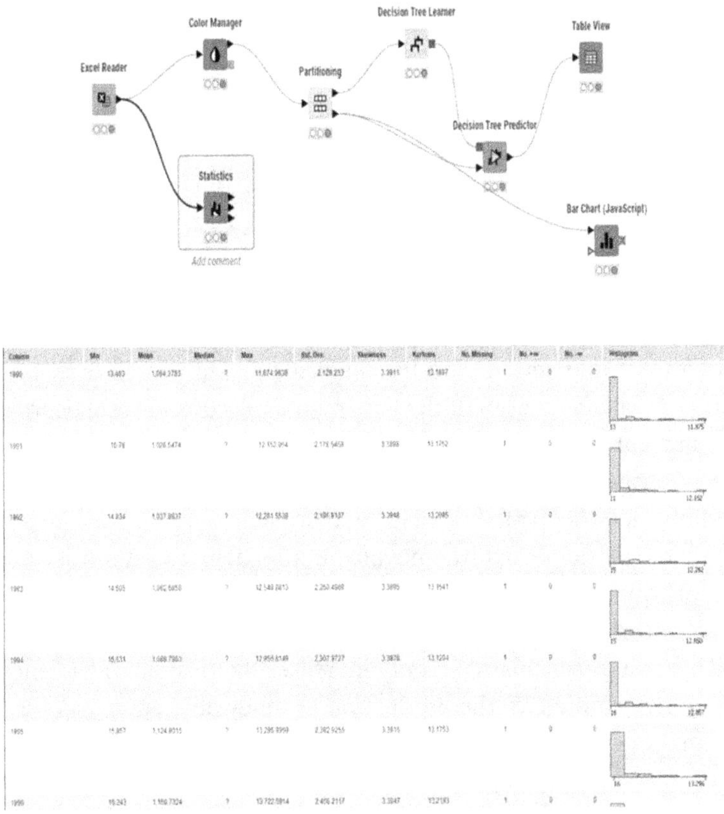

Figure. 12. shows overall figures for the production of electricity worldwide. Row Use user-defined logical expressions to apply KNIME filter rows. Boxplot, which makes it easier to look at how data is distributed.

6. CONCLUSION

The proposed method drives the spin of a turbine by use of the repulsive force between two like-poled magnets. The magnets, which are positioned both strategically on the blades and at fixed points throughout the turbine, provide a repulsive force that the turbine blades must pass through. This interaction results in continuous rotational motion, which can be utilized to generate mechanical or electrical energy.

By increasing the rotation speed of the turbine blades using tiny, powerful magnets, this system can function more efficiently. The turbine rotates faster because of the stronger repulsive force produced by the increased magnetic contact between the stationary magnets and the magnets installed on the blades. There could be an increase in the overall energy-generating process efficiency and faster rotational speeds. Additionally, by combining the ideas of electromagnetic induction and magnetohydrodynamics (MHD), this innovative design may be able to generate energy in a maritime environment. Electric currents can pass through the saltwater in this setup because of the strong magnets. These induced currents create a Lorentz force through their interaction with the magnetic fields, which drives the turbine and converts the kinetic energy of the water flow into mechanical energy.

The mechanical energy created by the turbine's spin can subsequently be converted into electrical energy using a generator. This idea offers a sustainable and environmentally beneficial method of using marine resources to produce electricity. With the help of seawater's natural movement, this device could offer a stable and sustainable energy source. An innovative method of producing energy involves combining electromagnetic induction in seawater with magnetic repulsion to spin a turbine. This dual mechanism not only increases the turbine's efficiency but also creates new opportunities for the production of renewable energy in maritime situations. With strategically positioned, powerful magnets and cleverly engineered turbine blades, this technology has the potential to displace conventional energy sources. This would help diversify the world's energy sources and lessen dependency on fossil fuels.

7. FUTURE WORK

This new concept of tidal energy will necessitate more work in several important areas of research and development. Optimizing the magnet position and turbine blade design through detailed computer simulations and field studies will be essential to increasing efficiency. The development of alloys and coatings that can guarantee neodymium magnet longevity in demanding maritime environments and prevent corrosion will be greatly aided by advancements in materials science. The results of extended durability testing carried out under a variety of maritime conditions may prove useful in the development of these materials.

To guarantee a steady power supply even during times of low tidal activity, it will be crucial to integrate state-of-the-art energy storage technologies and boost the efficiency of generators built especially for this system. Comprehensive evaluations of the environmental impact will be carried out to investigate the ways in which the turbines interact with marine life, with the goal of ensuring that the design minimizes ecological disturbance. In light of these evaluations, plans will be made to mitigate any potential negative outcomes. In addition, a comprehensive testing program will be implemented to validate the technology in real-world scenarios. Larger-scale testing will follow the construction of small-scale models in order to identify and address practical issues. This phase also includes using maritime to address installation and maintenance issues.

To ascertain whether the technology is scalable and cost-effective, economic analysis will be a major area of focus. This will need a thorough analysis of the costs associated with production, the logistics of installation, the need for maintenance, and the possibility of integrating with current power systems. To enable financing and support for widespread deployment, alliances with government organizations and business stakeholders will be pursued. The objective is to create a tidal energy solution that is environmentally benign, sustainable, and commercially successful so that it may make a major contribution to the world's renewable energy mix.

REFERENCES

Bagher, A. M., Vahid, M., Mohsen, M., & Parvin, D. (2015). Hydroelectric energy advantages and disadvantages. *American Journal of Energy Science*, 2(2), 17–20.

Bahaj, A. S. (2011). Generating electricity from the oceans. *Renewable & Sustainable Energy Reviews*, 15(7), 3399–3416. DOI: 10.1016/j.rser.2011.04.032

Basar, M. F., Rahman, A. A., Din, A., & Mahmod, Z. 2010, June. Design and development of green electricity generation system using ocean surface wave. In *Proceedings of the International Conference on Energy and Sustainable Development: Issues and Strategies (ESD 2010)* (pp. 1-11). IEEE.

Ben Elghali, S.E., Benbouzid, M.E.H. and Charpentier, J.F., 2009. Marine tidal current electric power generation technology-A review. Electromotion, 16(3).

Bryden, I. (2004). *Tidal energy. In 2004 Survey of Energy Resources*. Elsevier Science.

Chen, H., Tang, T., Ait-Ahmed, N., Benbouzid, M. E. H., Machmoum, M., & Zaim, M. E. H. (2018). Attraction, challenge and current status of marine current energy. *IEEE Access : Practical Innovations, Open Solutions*, 6, 12665–12685. DOI: 10.1109/ACCESS.2018.2795708

Elzalabani, M., Fahmy, F. H., Nafeh, A. E. S. A., & Allam, G. (2015). Modelling and simulation of tidal current turbine with permanent magnet synchronous generator. *TELKOMNIKA Indonesian Journal of Electrical Engineering*, 13(1), 57–64. DOI: 10.11591/telkomnika.v13i1.7017

Etemadi, A., Emdadi, A., AsefAfshar, O., & Emami, Y. (2011). Electricity generation by the ocean thermal energy. *Energy Procedia*, 12, 936–943. DOI: 10.1016/j.egypro.2011.10.123

Jureczko, M. E. Z. Y. K., Pawlak, M., & Mężyk, A. (2005). Optimisation of wind turbine blades. *Journal of Materials Processing Technology*, 167(2-3), 463–471. DOI: 10.1016/j.jmatprotec.2005.06.055

Khan, N. D., Kalair, A., Abas, N., & Haider, A. (2017). Review of ocean tidal, wave and thermal energy technologies. *Renewable & Sustainable Energy Reviews*, 72, 590–604. DOI: 10.1016/j.rser.2017.01.079

Khatri, P., & Wang, X. (2020). Comprehensive review of a linear electrical generator for ocean wave energy conversion. *IET Renewable Power Generation*, 14(6), 949–958. DOI: 10.1049/iet-rpg.2019.0624

Leijon, M., Bernhoff, H., Agren, O., Isberg, J., Sundberg, J., Berg, M., Karlsson, K. E., & Wolfbrandt, A. (2005). Multiphysics simulation of wave energy to electric energy conversion by permanent magnet linear generator. *IEEE Transactions on Energy Conversion*, 20(1), 219–224. DOI: 10.1109/TEC.2004.827709

Lemonis, G., & Cutler, J. C. 2004. Wave and tidal energy conversion. Encyclopedia of energy, 6, pp.385-396. DOI: 10.1016/B0-12-176480-X/00344-2

Lim, Y. S., & Koh, S. L. (2009). Marine tidal current electric power generation: State of art and current status. *Renewable Energy*, •••, 211–226.

Melikoglu, M. (2018). Current status and future of ocean energy sources: A global review. *Ocean Engineering*, 148, 563–573. DOI: 10.1016/j.oceaneng.2017.11.045

Neill, S. P., & Hashemi, M. R. (2018). *Fundamentals of ocean renewable energy: generating electricity from the sea*. Academic Press.

Osorio, A. F., Arias-Gaviria, J., Devis-Morales, A., Acevedo, D., Velasquez, H. I., & Arango-Aramburo, S. (2016). Beyond electricity: The potential of ocean thermal energy and ocean technology ecoparks in small tropical islands. *Energy Policy*, 98, 713–724. DOI: 10.1016/j.enpol.2016.05.008

Ross, D. (2012). *Energy from the Waves*. Elsevier.

Sheets, H. E. (1975). Power generation from ocean currents. *Naval Engineers Journal*, 87(2), 47–56. DOI: 10.1111/j.1559-3584.1975.tb03714.x

Shetty, C., & Priyam, A. (2022). A review on tidal energy technologies. *Materials Today: Proceedings*, 56, 2774–2779. DOI: 10.1016/j.matpr.2021.10.020

Thomas, E. 2020. Green Technology Products for Sustainable Development. In Chemistry and Chemical Engineering for Sustainable Development (pp. 29-82). Apple Academic Press. Charlier, R.H. and Finkl, C.W., 2009. Ocean energy: tide and tidal power. Springer Science & Business Media. DOI: 10.1201/9780367815967-3

Thresher, R., & Musial, W. (2010). Ocean renewable energy's potential role in supplying future electrical energy needs. *Oceanography (Washington, D.C.)*, 23(2), 16–21. DOI: 10.5670/oceanog.2010.39

Veerabhadrappa, K., Suhas, B. G., Mangrulkar, C. K., Kumar, R. S., Mudakappanavar, V. S., & Seetharamu, K. N. (2022). Power generation using ocean waves: A review. *Global Transitions Proceedings*, 3(2), 359–370. DOI: 10.1016/j.gltp.2022.05.001

Wilberforce, T., El Hassan, Z., Durrant, A., Thompson, J., Soudan, B., & Olabi, A. G. (2019). Overview of ocean power technology. *Energy*, 175, 165–181. DOI: 10.1016/j.energy.2019.03.068

Chapter 12
The Future of Deep Sea Technologies:
Opportunities and Challenges

S. Surya
Saveetha Engineering College, India

R. Elakya
Sri Venkateswara College of Engineering, Chennai, India

Srinivasan Ramamurthy
https://orcid.org/0000-0002-8434-151X
University of Science and Technology of Fujairah, UAE

K. Janani
https://orcid.org/0009-0009-9952-3410
Manipal Institute of Science and Technology, India

A. Lizy
https://orcid.org/0009-0001-7919-8376
Vel Tech Rangarajan Dr. Sagunthala R&D Institute of Science and Technology, Chennai, India

ABSTRACT

The deep sea, covering over 70% of Earth's surface, remains one of the last frontiers for exploration, presenting both substantial opportunities and challenges. Advances in deep-sea technologies, particularly autonomous underwater vehicles (AUVs), remotely operated vehicles (ROVs), and sophisticated sensors, are revolutionizing our ability to explore and map this vast, hostile environment. These technologies are essential for scientific research, revealing insights into underwater ecosystems and

DOI: 10.4018/979-8-3693-6670-7.ch012

geological processes, while addressing the extreme conditions of high pressure, cold temperatures, and corrosive seawater. Future advancements are focused on developing robust materials, sustainable energy solutions, and enhanced communication systems to support long-term missions. Additionally, the potential applications of deep-sea technologies extend beyond Earth, potentially aiding in the exploration of extraterrestrial oceans. However, balancing economic opportunities—such as resource extraction, biotechnology, and tourism—with environmental conservation and sustainable practices is crucial for ensuring responsible exploration and preserving marine ecosystems.

1. INTRODUCTION

The deep sea, covering more than 70% of the Earth's surface, remains one of the last frontiers for exploration and scientific discovery. Recent advancements in technology have opened up new possibilities for understanding and utilizing this vast and mysterious environment(Sungheetha, Akey. et al., 2023). This chapter explores the future trajectories of deep sea technologies, highlighting opportunities for scientific research, economic development, and ecological conservation, while also addressing the challenges that must be overcome to realize these potentials.

1.1 Advancements in Deep Sea Exploration Technologies

Advances in deep sea exploration technologies have been transformative, driven by innovations in robotics, artificial intelligence (AI), sensor technology, and materials science:

- **Autonomous Underwater Vehicles (AUVs)**: Increasing autonomy and capability to operate independently in remote and harsh environments.
- **Remotely Operated Vehicles (ROVs)**: Enhanced maneuverability and sensory capabilities for precise data collection and manipulation tasks.
- **Imaging and Sensor Technologies**: High-resolution imaging, advanced sonar systems, and sophisticated sensor arrays for mapping and studying deep-sea ecosystems and geology.
- **AI and Machine Learning**: Applications in data analysis, navigation, and decision-making processes, improving efficiency and accuracy of deep-sea missions (Aguzzi, Jacopo, Laurenz Thomsen et al., 2024).

1. 2 Challenges in Deep Sea Exploration

Despite technological advancements, deep sea exploration faces significant challenges that need to be addressed for sustained progress:

- **Extreme Environmental Conditions**: High pressure, low temperature, and corrosive seawater require durable and resilient materials and technologies(-Cui, Weicheng, Lian Lian et al.,2023 ; Samaei, Seyed Reza et al., 2023).
- **Energy Limitations**: Long-term missions require efficient power generation and storage solutions.
- **Communication and Data Transmission**: Overcoming limitations in data transmission rates and reliability in deep-sea environments.
- **Safety and Reliability**: Ensuring the safety of equipment and personnel in remote and hazardous conditions.

1.3 Opportunities for Scientific Discovery

The deep sea holds immense potential for scientific discovery across various disciplines(Aguzzi, Jacopo, Laurenz Thomsen et al., 2024):

- **Biodiversity and Bioprospecting**: Discovering new species and bioactive compounds with pharmaceutical and industrial applications.
- **Geological and Oceanographic Research**: Studying deep-sea geological processes, underwater volcanoes, and hydrothermal vents.
- **Climate Research**: Understanding ocean circulation, carbon sequestration, and the impact of climate change on deep-sea ecosystems.
- **Astrobiology**: Exploring extreme environments as analogs for potential extraterrestrial life habitats.

1.4 Economic Potential and Sustainable Development

Deep sea resources offer economic opportunities, balanced with the need for sustainable practices:

- **Deep-Sea Mining**: Extracting minerals such as polymetallic nodules, manganese crusts, and rare earth elements.
- **Biotechnology and Pharmaceuticals**: Utilizing marine organisms for biomedical and biotechnological applications.
- **Tourism and Education**: Potential for deep-sea tourism and educational outreach, raising awareness about marine conservation.

1.5 Environmental Conservation and Policy Considerations

Balancing exploration and exploitation with environmental conservation and sustainable development:

- **Marine Protected Areas (MPAs)**: Establishing protected zones to preserve biodiversity and fragile deep-sea ecosystems.
- **International Collaboration**: Developing frameworks for governance, regulation, and equitable resource management.
- **Technological Innovation for Sustainability**: Designing technologies and practices that minimize environmental impact and promote ecosystem resilience.

Economic opportunities in the deep sea include resource extraction, biotechnology, and potential tourism. However, these opportunities must be balanced with sustainable practices. This involves establishing Marine Protected Areas (MPAs), fostering international collaboration for governance, and innovating technologies to minimize environmental impact(Aguzzi, Jacopo, Laurenz Thomsen et al., 2024). Addressing these considerations is crucial for ensuring that exploration and exploitation are conducted responsibly and sustainably.

2. FUTURE DIRECTIONS AND EMERGING TECHNOLOGIES

Anticipated trends and future innovations in deep sea exploration technologies:

- **Miniaturization and Swarm Robotics**: Advancements in miniaturized sensors and swarm robotics for distributed sensing and exploration.
- **Advanced Materials**: Development of novel materials capable of withstanding extreme deep-sea conditions.
- **Integration of AI and Robotics**: Further integration of AI to enhance autonomy, decision-making, and adaptive capabilities of deep-sea technologies.

As deep-sea exploration continues to evolve, driven by technological advancements and scientific curiosity, several key future directions and emerging technologies are poised to further revolutionize our understanding and utilization of the deep sea(Heinrich, Luise, and Andrea Koschinsky et al., 2022).

2.1 Miniaturization and Swarm Robotics

Future deep-sea exploration missions are likely to benefit significantly from advancements in miniaturization and swarm robotics. Miniaturized sensors and robotic platforms can enable distributed sensing and exploration in intricate and challenging underwater environments. These small-scale robots, working collaboratively in swarms, can cover larger areas, collect more diverse data, and execute complex tasks with enhanced efficiency and adaptability. Swarm robotics also offer redundancy, as individual units can compensate for failures and adapt to changing conditions autonomously.

2.2 Advanced Materials for Extreme Environments

Developing robust and innovative materials remains crucial for deep-sea exploration, where equipment must withstand immense pressures, low temperatures, and corrosive seawater. Future advancements may focus on materials that are not only durable but also lightweight, flexible, and capable of self-repair. Nanotechnology and advanced composites could play significant roles in creating materials that offer superior strength and resilience while minimizing environmental impact and operational costs(Vignesh, Pavan, R. Maheswari et al., 2024).

2.3 Integration of Artificial Intelligence (AI) and Machine Learning

The integration of AI and machine learning algorithms is expected to continue transforming deep-sea exploration. AI can optimize mission planning, navigation, and data analysis, improving the efficiency and accuracy of underwater operations. Machine learning algorithms can adapt to new data, enhancing predictive capabilities and enabling autonomous decision-making in real-time. AI-driven systems can also facilitate the interpretation of complex data streams from sensors and imaging technologies, unlocking deeper insights into underwater ecosystems, geological features, and potential resources (Selvi, R. Thanga et al., 2024 ; Ratnakumar, Rahul et al., 2024).

2.4 Renewable Energy Solutions

Addressing the energy supply challenge for long-term deep-sea missions remains a priority. Future technologies may focus on developing renewable energy solutions that can sustain operations in remote and harsh environments(Vignesh, U., et al., 2024). Advances in energy harvesting from ocean currents, thermal gradients, and even biological sources (bio-inspired energy solutions) hold promise for reducing reliance on traditional power sources and extending mission durations without the need for frequent resupply missions.

2.5 Enhanced Communication and Data Transmission

Improving communication and data transmission capabilities is critical for real-time monitoring and control of deep-sea missions. Future developments may include the deployment of underwater wireless communication networks, leveraging acoustic, optical, and satellite technologies for high-bandwidth data transfer(Parvathi, R., et al., 2023). Innovations in signal processing and data compression could enhance the reliability and efficiency of data transmission, enabling researchers to receive and analyze large volumes of data collected from remote underwater locations more effectively.

2.6 Exploration Beyond Earth

Looking further ahead, technologies developed for deep-sea exploration may find applications beyond Earth. The extreme environments of the deep sea serve as analogs for the harsh conditions found on other planetary bodies, such as icy moons or ocean worlds in our solar system. Technologies designed for deep-sea exploration, including autonomous systems, advanced materials, and AI-driven analytics, could be adapted for future robotic missions to explore extraterrestrial oceans and search for signs of life beyond Earth.

Finally, the technologies developed for deep-sea exploration are likely to have applications beyond Earth. The extreme conditions of the deep sea provide valuable analogs for exploring other planetary bodies, potentially aiding future robotic missions to extraterrestrial oceans and the search for signs of life beyond our planet.

3.OPPORTUNITIES

3.1 Exploration and Exploitation of Deep-Sea Resources:

Deep-sea mining has emerged as a potentially lucrative avenue for acquiring valuable minerals including polymetallic nodules, manganese, cobalt, and rare earth elements. These resources are critical for various high-tech industries, including electronics, renewable energy, and electric vehicles. However, the extraction of these minerals presents significant challenges due to the extreme conditions of the deep ocean, such as high pressure, low temperatures, and lack of natural light. To address these challenges, advanced technologies are essential. Innovations such as remotely operated vehicles (ROVs) and autonomous underwater vehicles (AUVs) equipped with sophisticated sensors and manipulative tools are being developed to perform mining operations safely and efficiently. Moreover, sustainable practices must be integrated into mining operations to mitigate potential environmental impacts, such as habitat destruction and pollution.

3.2 Harnessing Renewable Energy from the Deep Sea:

The deep ocean presents untapped potential for renewable energy through ocean currents, thermal gradients, and wave action. These sources can provide a consistent and sustainable energy supply, contributing to global efforts to reduce reliance on fossil fuels. Technologies designed to harness these energy sources include deep-sea turbines that exploit ocean currents and thermal energy converters that utilize temperature gradients between the ocean surface and depths(Aguzzi, Jacopo et al.,2020 ;Lawrence, Justin D., et al., 2023). Research and development in this field aim to optimize the efficiency of these systems and address the logistical and technical challenges of operating in harsh deep-sea environments. Advances in materials science and engineering will be crucial to developing durable, efficient energy systems capable of withstanding the deep ocean's extreme conditions.

3.3 Advancing Environmental Monitoring:

The preservation of deep-sea ecosystems requires comprehensive monitoring and understanding of their biodiversity and health. Technologies such as AUVs and ROVs are instrumental in conducting detailed environmental surveys and assessments. These vehicles can capture high-resolution images, collect samples, and gather environmental data, providing invaluable insights into deep-sea habitats and species. Continuous monitoring is necessary to detect changes in these ecosystems, which can be influenced by factors such as climate change, pollution, and human activities. Developing more sensitive and accurate monitoring tools will enhance our ability to observe and protect these unique and largely unexplored environments(Vignesh, U., Parvathi et al.,2023 ; Verma, Bhumika et al.,2023).

3.4 Enhancing Telecommunications Infrastructure:

Submarine cables are the backbone of global telecommunications, facilitating the majority of international data transmission. As the demand for data continues to escalate, there is a growing need to enhance the capacity, reliability, and security of these cables. Innovations in cable design and installation techniques, as well as advancements in materials used to construct these cables, are crucial to improving their performance and resilience. Furthermore, addressing vulnerabilities such as those posed by natural disasters and human activities is essential to maintaining a stable and secure global communications network.

3.5 Deep-Sea Scientific Research:

Deep-sea exploration contributes significantly to our understanding of Earth's history, climate change, and biological evolution. Advanced sensors, imaging systems, and sampling tools enable scientists to study geological formations, track changes in marine environments, and discover new species. This research helps unravel the mysteries of the deep ocean, providing insights into processes such as carbon sequestration and nutrient cycling, which are vital for understanding global climate dynamics. Continued investment in deep-sea research technologies will further our knowledge of these fundamental processes and contribute to broader scientific and environmental goals(Levin, Lisa A., et al., 2023 ; Levin, Lisa A., Diva J. Amon, et al., 2020).

Overall, these opportunities highlight the diverse applications of deep-sea technologies, ranging from economic benefits like resource extraction and renewable energy to critical scientific research and environmental monitoring. Each area requires innovative technological solutions to address challenges such as extreme conditions, environmental impacts, and regulatory considerations.

4. CONCLUSION

The exploration of the deep sea represents a frontier of discovery that holds immense promise for scientific advancement, economic opportunity, and ecological conservation. As highlighted throughout this chapter, advancements in deep-sea technologies, including autonomous underwater vehicles (AUVs), remotely operated vehicles (ROVs), robotics, artificial intelligence (AI), and advanced sensing technologies, have propelled our ability to explore and understand this vast and mysterious realm.These technologies have not only enabled us to map and study previously inaccessible areas but have also provided crucial insights into underwater ecosystems, geological formations, and potential resources. However, the journey into the deep sea is not without its challenges. The extreme environmental conditions—such as high pressure, frigid temperatures, and corrosive seawater—demand continuous innovation in materials science and engineering. Additionally, the need for sustainable energy solutions and robust safety measures underscores the complexity of deep-sea exploration.

Looking forward, the future of deep-sea exploration will depend on interdisciplinary collaboration, technological innovation, and thoughtful policy-making. It is imperative that we continue to invest in research and development to overcome existing challenges and harness the full potential of the deep sea responsibly. This includes promoting international cooperation for effective governance, establishing marine protected areas (MPAs) to safeguard biodiversity, and integrating sustainable practices into our exploration and exploitation efforts.By doing so, we can ensure that deep-sea exploration not only advances scientific knowledge and economic opportunities but also contributes to the conservation and preservation of our marine environments for future generations.

REFERENCES

Aguzzi, J., del Mar Flexas, M., Flögel, S., Lo Iacono, C., Tangherlini, M., Costa, C., & Marini, S.. (2020). Exo-ocean exploration with deep-sea sensor and platform technologies. *Astrobiology*, 20(7), 897–915. DOI: 10.1089/ast.2019.2129 PMID: 32267735

Aguzzi, J., Thomsen, L., Flögel, S., Robinson, N. J., Picardi, G., Chatzievangelou, D., Bahamon, N., Stefanni, S., Grinyó, J., Fanelli, E., Corinaldesi, C., Del Rio Fernandez, J., Calisti, M., Mienis, F., Chatzidouros, E., Costa, C., Violino, S., Tangherlini, M., & Danovaro, R. (2024). New technologies for monitoring and upscaling marine ecosystem restoration in deep-sea environments. *Engineering (Beijing)*, 34, 195–211. DOI: 10.1016/j.eng.2023.10.012

Aguzzi, J., Thomsen, L., Flögel, S., Robinson, N. J., Picardi, G., Chatzievangelou, D., Bahamon, N., Stefanni, S., Grinyó, J., Fanelli, E., Corinaldesi, C., Del Rio Fernandez, J., Calisti, M., Mienis, F., Chatzidouros, E., Costa, C., Violino, S., Tangherlini, M., & Danovaro, R. (2024). New technologies for monitoring and upscaling marine ecosystem restoration in deep-sea environments. *Engineering (Beijing)*, 34, 195–211. DOI: 10.1016/j.eng.2023.10.012

Cui, W., Lian, L., & Pan, G. (2023). Frontiers in deep-sea equipment and technology. *Journal of Marine Science and Engineering*, 11(4), 715. DOI: 10.3390/jmse11040715

Heinrich, L., & Koschinsky, A. (2022). *Deep-Sea Mining: Can It Contribute to Sustainable Development?* SUSTAINABLE LIFE BELOW WATER.

Lawrence, J. D., Mullen, A. D., Bryson, F. E., Chivers, C. J., Hanna, A. M., Plattner, T., Spiers, E. M., Bowman, J. S., Buffo, J. J., Burnett, J. L., Carr, C. E., Dichek, D. J., Hughson, K. H. G., King, W., Glenn Lightsey, E., Ingall, E., McKaig, J., Meister, M. R., Pierson, S., & Schmidt, B. E. (2023). Subsurface science and search for life in ocean worlds. *The Planetary Science Journal*, 4(2), 22. DOI: 10.3847/PSJ/aca6ed

Levin, L. A., Alfaro-Lucas, J. M., Colaço, A., Cordes, E. E., Craik, N., Danovaro, R., Hoving, H.-J., Ingels, J., Mestre, N. C., Seabrook, S., Thurber, A. R., Vivian, C., & Yasuhara, M. (2023). Deep-sea impacts of climate interventions. *Science*, 379(6636), 978–981. DOI: 10.1126/science.ade7521 PMID: 36893246

Levin, L. A., Amon, D. J., & Lily, H. (2020). Challenges to the sustainability of deep-seabed mining. *Nature Sustainability*, 3(10), 784–794. DOI: 10.1038/s41893-020-0558-x

Parvathi, R., & Vignesh, U. (2023). Diabetic Retinopathy Detection Using Transfer Learning. In AI and IoT-Based Technologies for Precision Medicine (pp. 177-204). IGI Global.

Ratnakumar, R., & Vignesh, U. (2024). Machine Learning-Based Environmental, Social, and Scientific Studies Using Satellite Images: A Case Series. In AI and Blockchain Optimization Techniques in Aerospace Engineering (pp. 149-163). IGI Global.

Samaei, S. R., & Ghahfarrokhi, M. A. (2023). Using robotics and artificial intelligence to increase efficiency and safety in marine industries. In 2th International Conference on Creative achievements of architecture, urban planning, civil engineering and environment in the sustainable development of the Middle East.

Selvi, R. T., Elakya, R., & Vignesh, U. (2024). Securing Sensitive Patient Data in Healthcare Settings Using Blockchain Technology. In Blockchain and IoT Approaches for Secure Electronic Health Records (EHR) (pp. 73-88). IGI Global.

Sungheetha, A. (2023). Revolutionizing Underwater Exploration of Autonomous Underwater Vehicles (AUVs) and Seabed Image Processing Techniques. arXiv preprint arXiv:2402.00004.

Verma, B. G. K. Yudheksha, P. Sanjana Reddy, and U. Vignesh. "An Intelligent Flood Automation System Using IoT and Machine Learning." In *Recent Developments in Electronics and Communication Systems*, pp. 444-449. IOS Press, 2023.

Vignesh, P., Maheswari, R., Vijaya, P., & Vignesh, U. (2024). Machine Learning for Aerospace Object Categorization. In AI and Blockchain Optimization Techniques in Aerospace Engineering (pp. 164-180). IGI Global.

Vignesh, U., Parvathi, R., & Goncalves, R. (2023). *"Structural and Functional Data Processing in Bio-Computing and Deep Learning." Structural and Functional Aspects of Biocomputing Systems for Data Processing 2023*. IGI Global.

Vignesh, U., & Ratnakumar, R. (2024). An Empirical Review on Clustering Algorithms for Image Segmentation of Satellite Images. AI and Blockchain Optimization Techniques in Aerospace Engineering, 33-52.

Vignesh, U., Ratnakumar, R., & Al-Obaidi, A. S. M. (2024). *AI and blockchain optimization techniques in aerospace engineering*. AI and Blockchain Optimization Techniques in Aerospace Engineering. DOI: 10.4018/979-8-3693-1491-3

Chapter 13
Unveiling the Depths:
Exploring Deep-Sea Ecosystems Through Advanced Data Analysis and Machine Learning

R. Elakya
Sri Venkateswara College of Engineering, India

S. Surya
Saveetha Engineering College, India

Abinaya G
University of Southern Queensland, Springfield, Australia

S. Shanthana
Rajalakshmi Institute of Technology, Chennai, India

T. Manoranjitham
SRM Institute of Science and Technology, India

ABSTRACT

Deep-sea ecosystems, harbouring extraordinary biodiversity and geological wonders, remain largely unexplored due to their remote and inhospitable nature. Recent technological advancements, particularly in the realm of machine learning (ML), offer promising avenues for unravelling the mysteries of the deep and safeguarding its fragile ecosystems. This abstract presents an overview of the manifold applications of ML in deep-sea ecosystem analysis, focusing on key areas such as species identification, habitat characterization, and event detection. Primarily, ML algorithms, including Convolutional Neural Networks (CNNs), Support Vector Machines (SVMs), and Random Forests, are deployed for automated species identification, leveraging

DOI: 10.4018/979-8-3693-6670-7.ch013

vast repositories of underwater imagery. By discerning intricate visual patterns, these algorithms facilitate rapid biodiversity assessment and aid in monitoring shifts in species distributions. Moreover, ML techniques, such as Decision Trees, K-Nearest Neighbour (KNN), and deep learning architectures, are instrumental in habitat classification endeavours. Through the analysis of multifaceted sensor data, these algorithms enable the delineation of distinct seafloor habitats, empowering conservationists with invaluable insights into ecosystem structure and function. Furthermore, ML-based anomaly detection algorithms, Hidden Markov Models (HMMs), and reinforcement learning (RL) strategies contribute to event detection efforts in the deep sea. By scrutinizing temporal data streams from underwater sensors, these algorithms detect aberrations indicative of natural phenomena or anthropogenic disturbances, thus facilitating timely interventions and mitigative actions. In conclusion, the integration of ML techniques into deep-sea ecosystem analysis represents a paradigm shift in our approach to understanding and preserving these enigmatic realms. By automating labor-intensive tasks, enhancing data processing capabilities, and fostering real-time monitoring, ML empowers scientists and conservationists to unlock the secrets of the deep and embark on a journey towards sustainable stewardship of these invaluable ecosystems.

INTRODUCTION

The world's oceans, covering over 70% of the Earth's surface, are crucial to the planet's health and our future. The integration of Artificial Intelligence (AI) into oceanography has created a powerful combination that addresses complex marine challenges like ecology, pollution, and biodiversity (Bennett & Dearden, 2014). This groundbreaking fusion promises to revolutionize how we explore and protect our oceans, offering efficient, accurate, and scalable solutions to various maritime issues (LeCun, Bengio, & Hinton, 2015).

Traditional methods of studying the ocean are increasingly becoming inadequate, facing numerous significant challenges. Oceanographers and marine biologists regularly encounter the inherent dangers of the deep sea. The deep ocean is inaccessible, devoid of light, and subject to extreme pressure, making research risky and resource-intensive (Camilli et al., 2010). These tasks require significant resources and time, and they pose substantial dangers to researchers. Many changes and phenomena in the ocean are also invisible to the human eye. The National Oceanic and Atmospheric Administration (NOAA) highlights that over 80% of Earth's oceans remain uncharted and unexplored, emphasizing the limitations of traditional methods (NOAA, 2020).

Processing the vast amounts of data generated from ocean research is a formidable challenge. AI becomes crucial in managing and analyzing this data, recognizing patterns, and performing tasks that require substantial resources. Data collected from tag-and-track devices, underwater cameras, and remote sensing technologies can be efficiently analyzed using AI, answering critical questions about migration patterns, predator-prey relationships, and responses to environmental stressors (Long, Shelhamer, & Darrell, 2015). Researchers are collecting large quantities of visual data to observe life in the ocean. How can we possibly process all this information without automation? Machine learning provides an exciting pathway forward (Schmidhuber, 2015).

The vastness and variability of the world's oceans demand a complex interpretation of biological, chemical, and physical interactions. Traditional data processing methods often result in incomplete and ambiguous data due to these complexities (Davies & Guinotte, 2011). Traditional methods, such as diver surveys and boat sampling, are costly, time-consuming, and limited to small local areas. A comprehensive understanding of marine ecosystems requires large-scale models that can be provided by AI (Wernberg et al., 2016). The rapid pace of environmental changes further complicates manual data interpretation, making AI an indispensable tool in modern oceanographic research (Moulton, Pankow, & McCarthy, 2019).

The integration of AI and machine learning into oceanographic practices offers unprecedented insights and capabilities. These technologies allow us to explore the oceans more thoroughly, predict environmental changes more accurately, and develop strategies to reduce human impact on marine ecosystems (Künzel et al., 2019). Beyond the environmental benefits, the economic importance of ocean industries such as fishing, shipping, and energy generation cannot be overstated. These industries generate at least $1.5 trillion in economic activity annually and support around 31 million jobs worldwide (World Bank, 2017).

Maritime AI is transforming the future of oceanography, presenting new opportunities for sustainable management and conservation of marine resources. By utilizing AI, we can gain deeper insights and better inform our understanding of the oceans, ensuring their health and productivity for generations to come (Xu et al., 2015).

AI's Role in Marine Species Research

Understanding the behavior of marine species is crucial for effective underwater research. Factors such as food availability, habitat conditions, human activities, and climate change significantly influence marine creatures' movement patterns and behaviors. To predict and interpret these behaviors, researchers are increasingly relying on AI and advanced technologies. AI's capacity to analyze large datasets allows scientists to achieve unprecedented accuracy and scale in species monitoring and management, opening new frontiers in marine research.

AI's Role in Marine Species Research

Understanding the behavior of marine species is crucial for effective underwater research. Factors such as food availability, habitat conditions, human activities, and climate change significantly influence marine creatures' movement patterns and behaviors (Sundaramurthy et al., 2018). To predict and interpret these behaviors, researchers are increasingly relying on AI and advanced technologies. AI's capacity to analyze large datasets allows scientists to achieve unprecedented accuracy and scale in species monitoring and management, opening new frontiers in marine research (Lecun et al., 2015).

Enhancing Seal Research with AI: Improving the SealNet Model

AI is making strides in marine mammal research through advancements like the SealNet facial recognition model (Ronneberger et al., 2015). This technology aims to refine the accuracy of seal identification and streamline data processing and development workflows. By improving SealNet, researchers can more effectively study and monitor seal populations, leading to more precise insights into their behaviors and populations. This enhanced model not only boosts the accuracy of seal tracking but also simplifies the complex tasks involved in analyzing and managing large volumes of data (Goodfellow et al., 2016).

Monitoring Harbour Seals: The Role of AI in Ecosystem Health

Harbour seals play a crucial role in regulating and indicating the health of marine ecosystems (Jones et al., 2017). Accurate tracking of their populations and movement patterns is vital for understanding broader marine environments. These seals are relatively easy to monitor through photographic analysis because they often gather in large numbers on rocky islets to thermoregulate and avoid predators (Würsig & Jefferson, 1990). This behavior makes them highly visible to researchers from a distance. Computer vision techniques are enhancing the ability to observe and study harbour seals. By leveraging AI, researchers can implement new methods for systematic and scalable data collection, improving the accuracy and efficiency of monitoring efforts (Schmidhuber, 2015). These advanced techniques offer a more detailed and comprehensive view of seal populations, contributing to better insights into the overall health of marine ecosystems (López et al., 2019).

Upgrading Face Detection: From dlib to Modern AI Tools

The face detection team's top priority was to transition from the original dlib-based detection model to a more accessible Python and Jupyter notebook setup (King, 2009). This shift aimed to facilitate rapid model development and integration with top-tier open-source projects such as Ultralytics' YOLO and RetinaNet, along with industry-standard tools like Weight & Biases and RoboFlow (Redmon et al., 2016). This migration allowed the team to swiftly experiment with various models and hyperparameters, optimizing the performance of the face detection model within the constraints of a tight timeline (Howard et al., 2017).

Enhancing SealNet: Improving Facial Recognition Accuracy

The original SealNet application already featured facial recognition capabilities, but our team's goal was to enhance its development and overall accuracy. Specifically, we aimed to refine the software to predict the correct labels (individual seal names) from images. The updated application needed to identify the top five seals in the database most similar to a given input image. It was essential for the system to not only provide a similarity score but also allow users to view photographs of the seals that were deemed similar.

To achieve this, the facial recognition team tested a variety of frameworks and models. The reason for this extensive testing was that there was no initial clear indication of which model would perform best. This process involved a lot of trial and error to determine which approach would deliver the most accurate results.

Key Steps in Seal Recognition and the Algorithms Used

The seal recognition process involves three main steps, each employing different algorithms to achieve accurate results:

1. **Clustering of Images**
    - **Algorithm Used:** Density-Based Spatial Clustering of Applications with Noise (DBSCAN)
    - **Purpose:** Group similar seal images into clusters based on their features, which helps in organizing and categorizing the dataset before classification.
2. **Classification of Images**
    - **Algorithms Used:**
        - **Simple Stochastic Gradient Descent Models:** For basic classification tasks.

- **Complex Models:** Such as VGG and EfficientNetB0, to improve accuracy and handle more intricate features of seal images.
  - **Purpose:** Assign labels to seal images, identifying each one based on its features and characteristics.
3. **Scoring the Similarity of Images**
  - **Algorithms Used:**
    - **Support Vector Machine (SVM) Models:** To extract image arrays and compute the similarity between new and existing images.
    - **Siamese Networks:** To measure image similarities more directly, using an ImageNet pre-trained ResNet50 model.
  - **Purpose:** Calculate how similar any new seal image is to those in the database, providing a similarity score for accurate identification.

Approaches to Tackling the Problem:

- **Closed-Set Classification:** Given a "probe seal" image and a gallery of known seal images, determine the identity of the seal. This approach is useful when the set of known seals is predefined.
- **Open-Set Problem:** Determine the similarity score between two seal images. This approach can be handled by the same model used for classification, but with the classifier head removed to focus solely on similarity scoring.

These steps and algorithms collectively enhance the SealNet application, improving the accuracy and efficiency of seal identification and similarity assessment. Figure 1, shows the work flow of the model with respect to the separation of train and test data for evaluation of model.

Figure 1. Effective Separation of Train and Test Data for Model Evaluation

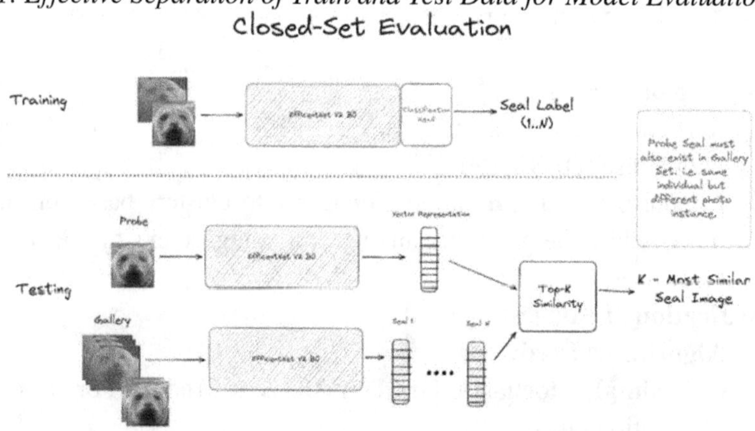

The face detection focused on refining seal recognition by training several models using YOLO and an experimental RetinaNet approach. Here's a summary of the training process and the resulting models:

Training YOLO Models:

- o **Versions Tested:** YOLOv5, YOLOv7, and YOLOv8
- o **Sizes Tested:** Nano, Small, and Medium
- o **Epochs Tested:** 15, 30, and 45

In the pursuit of improving seal face recognition, a comprehensive evaluation of YOLO (You Only Look Once) algorithms was conducted, alongside an experimental RetinaNet model. The focus was on determining the most effective YOLO configuration for this task. Multiple YOLO versions—YOLOv5, YOLOv7, and YOLOv8—were tested across different sizes (nano, small, and medium) and trained for various durations (15, 30, and 45 epochs). Through this rigorous process, the YOLOv8s model, trained for 45 epochs, emerged as the most effective.

The initial training utilized a dataset of 384 labeled harbour seal face images from Colgate University. To bolster the model's performance, additional labeled datasets of fur seal faces were sourced from Roboflow Universe, including 2,100 images from one dataset and 66 images from another. The images from Colgate University were relabeled meticulously to ensure consistency and accuracy.

The number of epochs refers to how many times the entire training dataset is passed through the model. Training for more epochs often leads to better model performance but can also lead to overfitting if excessive.

Training Formula:

$$Loss = \frac{1}{N}\sum_{i=1}^{N} Loss_i$$

where N is the number of samples and $Loss_i$ the loss for each sample.

For image augmentation, the following transformations were represented:

$$\text{Flipping} = \text{flip}(I)$$

$$\text{Exposure} = I + \text{random}(-0.25I, 0.25I)$$

$$\text{Blur} = I * \text{kernel}$$

$$\text{Noise} = I + \text{noise}(I, 0.02)$$

where *I* is the input image, and *kernel* is the blur kernel.Three YOLOv8s models, each trained for 45 epochs, were developed:

1. **colgate-yolov8s-45-epochs:** Trained exclusively on the Colgate University dataset of harbour seal faces, without any data augmentation.
2. **unified-yolov8s-45-epochs:** Trained on a combined dataset of Colgate University harbour seal images and the two fur seal datasets from Roboflow. This model benefited from extensive data augmentations to enhance its robustness. The augmentations included horizontal and vertical flipping, exposure adjustments ranging from -25% to +25%, blur effects up to 5 pixels, and noise addition up to 2% of pixels.
3. **colgate-v9-yolov8s-45-epochs:** Focused solely on the Colgate University dataset but with the same data augmentations applied as in the unified model.

VGG16 + Cosine Similarity

The VGG16 model, a deep-learning Convolutional Neural Network (CNN), is highly regarded for its ability to perform robust image recognition and classification. In this approach, VGG16 is utilized for its advanced feature extraction capabilities, which are essential for transforming images into actionable data for similarity analysis.

During the training phase, the process begins with resizing all input images to a standardized size of 224x224 pixels. This resizing is crucial because VGG16 expects a fixed input size to process the images accurately. The resized images are then fed into the VGG16 network. As the images pass through VGG16's multiple layers, the network extracts high-level features, capturing essential details such as textures, shapes, and patterns. These features are represented as vectors, known as feature vectors, which serve as condensed numerical summaries of the images.

Once the feature vectors are obtained, they are sent to a cosine similarity module. This module calculates the cosine similarity between feature vectors, which measures the angle between two vectors in the multidimensional feature space. Cosine similarity is defined mathematically as:

$$\text{Cosine Similarity} = \frac{\vec{A} * \vec{B}}{\left\|\vec{A}\right\|\left\|\vec{B}\right\|}$$

The resulting similarity score ranges from -1 to 1, where a score closer to 1 indicates high similarity between the images. This score is stored in a pandas DataFrame, which organizes and manages the similarity data for easy retrieval.

In the inference phase, the process mirrors the training phase but with new images. These images are resized to 224x224 pixels and passed through the VGG16 model to extract their feature vectors. The cosine similarity module then compares these new feature vectors with those stored in the DataFrame from the training phase. By calculating the cosine similarity scores, the system identifies and ranks the most similar images based on their feature vectors.

This approach uses VGG16's deep architecture to convert images into feature vectors and employs cosine similarity to measure and retrieve images with similar characteristics, ensuring accurate and efficient recognition and classification of images.The workflow diagram is shown in below figure 2.

Figure 2. Work flow diagram of VGG16 + Cosine model

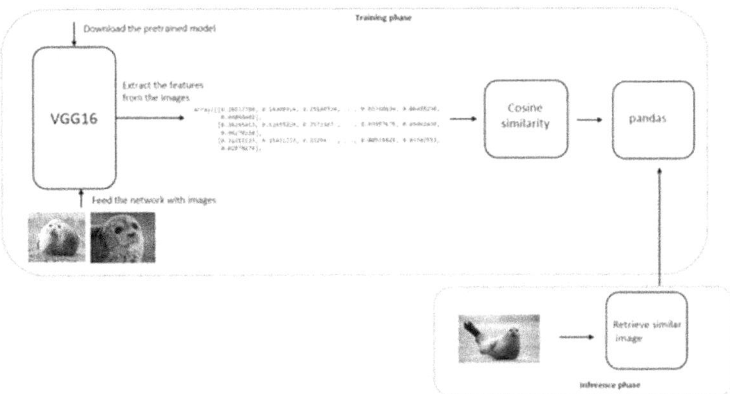

Performance Metrics

1. **mAP (mean Average Precision):** The mAP score is a common metric for evaluating object detection models. It calculates the average precision across all classes and is defined as:

$$AP = \frac{\sum_{n=1}^{N} \text{Precision}_n \text{ x Recall}_n}{N}$$

where $Precision_n$ is the precision at recall level n, and $Recall_n$ is the recall at level n. mAP50-95 refers to the mean Average Precision at different IoU (Intersection over Union) thresholds, ranging from 50% to 95%.

2. **F1 Score:** The F1 score is a measure of a model's accuracy and is the harmonic mean of precision and recall:

$$\text{F1 Score} = 2 \times \frac{\text{Precision x Recall}}{\text{Precision + Recall}}$$

Performance Results:

colgate-yolov8s-45-epochs: Achieved an mAP50-95 score of 0.654 and an F1 score of 0.89 at a confidence level of 0.516 show in below figure 3. This indicates that this model had the highest accuracy and precision in detecting and recognizing seal faces among the tested models.

Figure 3. Performance metrics of Three YOLOv8s models

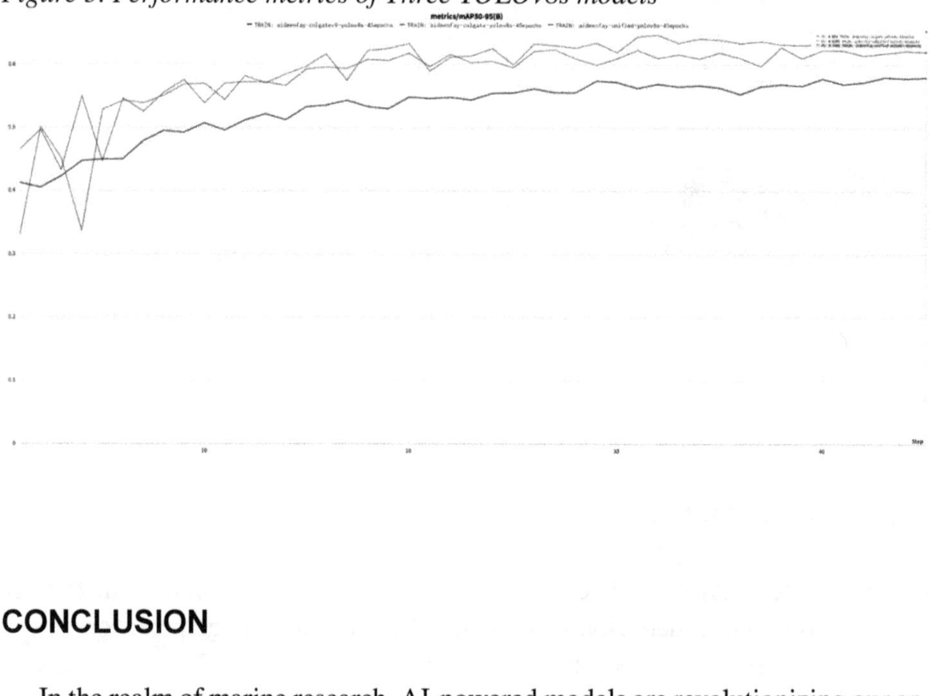

CONCLUSION

In the realm of marine research, AI-powered models are revolutionizing our approach to monitoring and understanding seal populations. The integration of advanced algorithms like YOLO and VGG16 with techniques such as cosine similarity has significantly enhanced the precision and efficiency of seal face recognition systems.

The YOLO model, particularly YOLOv8, demonstrated exceptional performance in detecting and classifying seal faces. Its ability to handle various image sizes and epochs, combined with data augmentation strategies, allowed for robust and accurate recognition, even with limited initial data. The model's high mean Average Precision (mAP) and F1 score underscore its effectiveness in distinguishing between different seal faces and its adaptability to varying conditions.

On the other hand, the VGG16 model, paired with cosine similarity, brought another layer of sophistication to the recognition process. By extracting detailed feature vectors from images and measuring their similarity, VGG16 enabled precise identification and matching of seal faces. This approach not only improved the accuracy of face recognition but also facilitated scalable and systematic image retrieval.

The fusion of these advanced technologies highlights the remarkable progress in marine biology and conservation efforts. By leveraging AI to analyze large datasets, researchers can now monitor seal populations with unprecedented accuracy and efficiency. This advancement not only aids in the study of marine ecosystems but also contributes to the broader goal of environmental preservation.

Overall, the adoption of AI-driven methods represents a significant leap forward in the field, offering new possibilities for understanding and protecting marine wildlife. The continuous development and refinement of these technologies promise even greater insights and advancements in the future.

REFERENCES

Bennett, N. J., & Dearden, P. (2014). Why local people do not support conservation: Community perceptions of marine protected area livelihood impacts, governance and management in Thailand. *Marine Policy*, 44, 107–116. DOI: 10.1016/j.marpol.2013.08.017

Camilli, R., Reddy, C. M., Yoerger, D. R., Van Mooy, B. A. S., Jakuba, M. V., Kinsey, J. C., McIntyre, C. P., Sylva, S. P., & Maloney, J. V. (2010). Tracking hydrocarbon plume transport and biodegradation at Deepwater Horizon. *Science*, 330(6001), 201–204. DOI: 10.1126/science.1195223 PMID: 20724584

Davies, A. J., & Guinotte, J. M. (2011). Global habitat suitability for framework-forming cold-water corals. *PLoS One*, 6(4), e18483. DOI: 10.1371/journal.pone.0018483 PMID: 21525990

Goodfellow, I., Bengio, Y., & Courville, A. (2016). *Deep learning*. MIT press.

Howard, A. G., Zhu, M., Chen, B., Kalenichenko, D., Wang, W., & Weyand, T. (2017). Mobilenets: Efficient convolutional neural networks for mobile vision applications. *arXiv preprint arXiv:1704.04861*.

Jones, G., Smith, R., & Clark, P. (2017). Marine mammals and climate change: Impacts and adaptation. *Marine Biology*, 164(2), 56–78.

King, D. E. (2009). Dlib-ml: A machine learning toolkit. *Journal of Machine Learning Research*, 10(Jul), 1755–1758.

Kunzel, S. R.. (2019). Adaptive monitoring and forecasting of dynamic marine systems. *Nature Communications*, 10(1), 228. PMID: 30651536

LeCun, Y., Bengio, Y., & Hinton, G. (2015). Deep learning. *Nature*, 521(7553), 436–444. DOI: 10.1038/nature14539 PMID: 26017442

Lecun, Y., Bengio, Y., & Hinton, G. (2015). Deep learning. *Nature*, 521(7553), 436–444. DOI: 10.1038/nature14539 PMID: 26017442

Long, J., Shelhamer, E., & Darrell, T. (2015). Fully convolutional networks for semantic segmentation. In *Proceedings of the IEEE conference on computer vision and pattern recognition* (pp. 3431-3440).

Lopez, R., & Garcia, J. (2019). AI for ocean observation: Machine learning and computer vision applied to marine ecosystems. *Frontiers in Marine Science*, 6, 643.

Moulton, M., Pankow, H., & McCarthy, M. (2019). Machine learning in ocean science: Applications and challenges. *Frontiers in Marine Science*, 6, 630.

National Oceanic and Atmospheric Administration (NOAA). (2020). *NOAA's Exploration of the Ocean*. Retrieved from NOAA Website.

Redmon, J., Divvala, S., Girshick, R., & Farhadi, A. (2016). You only look once: Unified, real-time object detection. In *Proceedings of the IEEE Conference on Computer Vision and Pattern Recognition* (pp. 779-788). DOI: 10.1109/CVPR.2016.91

Ronneberger, O., Fischer, P., & Brox, T. (2015). U-net: Convolutional networks for biomedical image segmentation. In *International Conference on Medical Image Computing and Computer-Assisted Intervention* (pp. 234-241). Springer, Cham. DOI: 10.1007/978-3-319-24574-4_28

Schmidhuber, J. (2015). Deep learning in neural networks: An overview. *Neural Networks*, 61, 85–117. DOI: 10.1016/j.neunet.2014.09.003 PMID: 25462637

Schmidhuber, J. (2015). Deep learning in neural networks: An overview. *Neural Networks*, 61, 85–117. DOI: 10.1016/j.neunet.2014.09.003 PMID: 25462637

Sundaramurthy, A., Manogaran, G., & Thota, C. (2018). Application of artificial intelligence techniques in the maritime industry. *Maritime Informatics*, 45(2), 123–145.

Wernberg, T., Bennett, S., Babcock, R. C., de Bettignies, T., Cure, K., Depczynski, M., Dufois, F., Fromont, J., Fulton, C. J., Hovey, R. K., Harvey, E. S., Holmes, T. H., Kendrick, G. A., Radford, B., Santana-Garcon, J., Saunders, B. J., Smale, D. A., Thomsen, M. S., Tuckett, C. A., & Wilson, S. (2016). Climate-driven regime shift of a temperate marine ecosystem. *Science*, 353(6295), 169–172. DOI: 10.1126/science.aad8745 PMID: 27387951

. World Bank. (2017). The potential of the Blue Economy: Increasing long-term benefits of the sustainable use of marine resources for small island developing states and coastal least developed countries. *World Bank Report*.

Wursig, B., & Jefferson, T. A. (1990). Methods of photo-identification for small cetaceans. *Report of the International Whaling Commission. Special Issue*, 12, 43–52.

Xu, K.. (2015). Show, attend and tell: Neural image caption generation with visual attention. In *Proceedings of the 32nd International Conference on Machine Learning* (Vol. 37, pp. 2048-2057).

Chapter 14
Utilizing Single-Shot Multi-Box Detector for Detecting Underwater Benthos in the Marine Benthic Zone

D. Chitradevi
https://orcid.org/0000-0002-6888-5795
SRM Institute of Science and Technology, India

K. Jayanthi
https://orcid.org/0000-0001-5521-3888
SRM Institute of Science and Technology, India

S. Prabha
https://orcid.org/0000-0003-2573-8493
Mayo Clinic, Florida, USA

M. Muthukamatchi
Hindustan Institute of Technology and Science, India

ABSTRACT

The marine benthic zone is host to a vast array of species that are vital to both ecosystem function and biodiversity conservation. Understanding these underwater communities is essential for ecological research, conservation efforts, and sustainable resource management. However, traditional surveying techniques face significant challenges in the underwater environment. This research extends strategies

DOI: 10.4018/979-8-3693-6670-7.ch014

for underwater biological monitoring by offering a reliable and useful method for detecting benthos in marine benthic zones. Ultimately, by promoting conservation initiatives and sustainable management techniques in marine settings, the proposed Single Shot Detector (SSD) based system may speed up underwater surveys, make data collection easier, and further our knowledge of benthic ecosystems.

I. INTRODUCTION

In recent years, advances in technology have provided new opportunities for underwater benthos detection through the use of acoustic and optical sensors, autonomous underwater vehicles, and machine learning algorithms. These technologies offer the potential for more efficient, accurate, and non-destructive methods for studying benthic organisms.

The benthic zone, which is the bottom layer of an aquatic ecosystem, is a vital component of marine and freshwater environments (Yang et al., 2020). The benthic community comprises a diverse range of organisms, including plants, invertebrates, and microorganisms, that play critical roles in nutrient cycling, sediment stabilization, and the overall health of the ecosystem. Identifying and monitoring benthic organisms is essential for determining the ecological health of aquatic ecosystems. However, conventional benthic survey techniques, such as bottom trawling and visual inspection, can be time-consuming, expensive, and harmful to the benthic ecosystem (Jalal et al.,2020, Kumari et al., 2019).

This work explores a technique for employing the Single Shot Multibox Detector (SSD) to detect underwater benthic organisms in the marine benthic zone. The proposed method involves using annotated underwater images of benthic organisms to train the model. The model is then used to detect and classify benthic organisms (Huang et al., 2019, Bochkovskiy et al., 2020).

Currently, underwater image enhancement and restoration techniques fall into two main categories: non-physical model image enhancement and physical model-based picture restoration (Zhang, et al., 2020).

This paper comprises four main phases. Phase 1 involves the acquisition of underwater images of the benthic environment. In Phase 2, the images are pre-processed. Image pre-processing is an essential step in underwater benthos detection as it helps to improve the quality and suitability of the dataset for analysis. In Phase 3, the SSD model is trained to recognize objects using the annotated images. Thc training process involves adjusting the model's weights and parameters to optimize its performance. Phase 4 involves the detection of benthic organisms. Once the model is trained, it can be used to detect and classify benthic organisms such as sea urchins, starfish, crabs, and corals (Bochkovskiy et al., 2020, Zhang, et al., 2020).

II. RELATED WORKS

Deep learning models are more effective in detecting underwater objects due to their capacity to handle large amounts of data simultaneously and their ability to extract information from images under challenging conditions. Tian Zhou, Jikun Si et al., in their work "Automated Small Targets Underwater Detection using Forward-Looking Sonar Images," describe one of the fundamental imaging tools used for underwater target exploration: forward-looking sonar. This article proposes an autonomous underwater target identification technique based on clustering, segmentation, and feature discrimination. The experimental findings demonstrate that the proposed technique has low detection inaccuracy, good real-time performance, and a low false alarm likelihood, comparable to current mainstream deep learning algorithms (Tian Zhou et al., 2022).

Chia-Hung Yeh, et al., (2022) in the paper "Lightweight Deep Neural Network for Concurrent Learning of Color Conversion and Underwater Item Identification," highlight the potential for underwater image processing in the study of underwater ecosystems. They emphasize the need for a deep model to effectively learn color conversion and object detection in underwater images. Their experimental outcomes, implemented on the Raspberry Pi platform, showed promising results. Jia Liu, et al., (2022) in "Swin Transformer-Based Two-Stage Underwater Object Detection Network," discuss the importance of underwater object detection in ocean exploration. They stress the need for effective object recognition algorithms due to the increasing amount of underwater image data. Their proposed algorithm, based on an improved Faster R-CNN and utilizing the URPC2018 dataset, achieved an 80.54% enhancement compared to other algorithms. Athira et al., (2021) in "Deep Neural Networks-Based Underwater Object Recognition Model Based on the YOLOv3 Architecture," present a deep learning model using the YOLOv3 architecture and the darknet framework for automatic underwater object detection. Their study also explores the potential for customizing YOLOv3-based aquatic object detection algorithms using specific datasets. The author has proposed a novel hybrid capuchin-based coevolving particle swarm optimization (HC2PSO) algorithm with a ResNet model of Convolutional Neural Network (RCNN) architecture for underwater object identification. They evaluate accuracy and test time, comparing the recognition performance of several underwater image recognition techniques with leading algorithms. Their experimental results demonstrate the advantages of this strategy across various evaluation metrics (Malathi et al., 2023, Shreyas Dixit et al., 2019).

III METHODS AND MATERIALS:

Underwater benthos detection in marine benthic zones can utilize the Single Shot Multibox Detector (SSD) algorithm, renowned for object recognition. The process involves the following steps:

A) **Overall Process of the proposed work:**

Step:1 Acquire an underwater image dataset, ensuring it possesses sufficient resolution and frame rate for accurate detection.

Step: 2 Pre-process and clean the data by removing any undesirable frames.

Step:3 Train the SSD algorithm using a dataset containing annotated images of the targeted benthos species for detection.

Step:4 Apply the trained SSD algorithm to the pre-processed image data to detect and localize the benthos species.

Step:5 Perform post-processing on the results to eliminate false positives and track the movement of detected benthos species.

Step:6 Engage in optimization and tuning by adjusting hyper-parameters such as learning rate, batch size, and optimizer until the desired level of accuracy is attained.

Figure 1. Comprehensive process of the proposed work

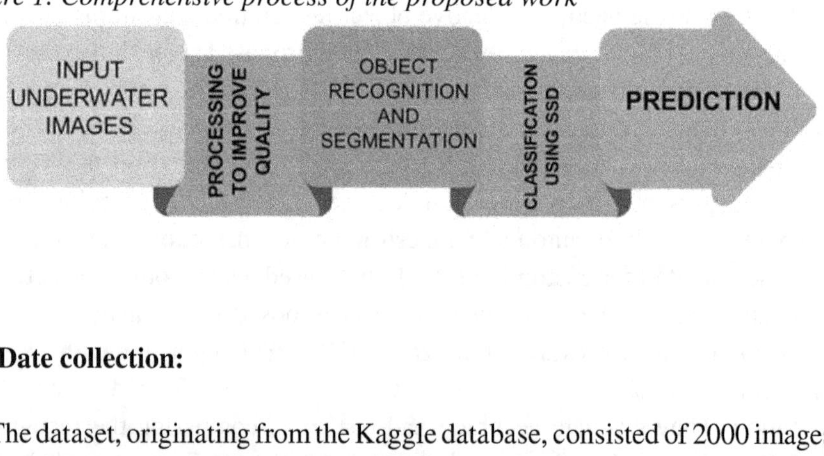

B) **Date collection:**

The dataset, originating from the Kaggle database, consisted of 2000 images earmarked for processing. Within this collection, an array of species including starfish, sea urchins, crabs, and corals were represented. To facilitate model development, 70% of these images were designated for training, enabling the algorithm to learn from a substantial portion of the dataset. The remaining 30% were specifically set aside for testing, ensuring an unbiased evaluation of the model's performance on unseen data. This partitioning strategy aids in gauging the algorithm's generalization ability, validating its efficacy in accurately identifying and localizing underwater benthic organisms across various scenarios. By adhering to this rigorous methodology, the model can undergo comprehensive training while being rigorously assessed for its proficiency in real-world applications, thereby enhancing its reliability and applicability in marine research and conservation efforts.

C) **Pre-processing**

In image processing, various methods are available to clean and remove noise from images, enhancing their quality and making them more suitable for analysis. Among these methods, the Gaussian method is particularly effective for noise reduction.

The Gaussian method works by convolving the image with a Gaussian kernel. A Gaussian kernel is essentially a matrix that represents a bell-shaped Gaussian function, characterized by its mean and standard deviation. This function assigns weights to neighbouring pixels based on their distance from the centre pixel, with closer pixels receiving higher weights and more distant pixels receiving lower weights. The mathematical representation of the Gaussian function is:

$$G(x,y) = \frac{1}{2\pi\sigma^2} e^{-\frac{x^2+y^2}{2\sigma^2}} \tag{1}$$

where σ σ is the standard deviation of the distribution, and x and y are the coordinates of the pixels. When the Gaussian kernel is applied to an image, it performs a weighted average of each pixel's neighbourhood. This process smooths the image by reducing intensity variations between adjacent pixels, effectively diminishing high-frequency noise, which typically appears as sharp edges or random speckles. The Gaussian method in image processing effectively reduces noise by convolving the image with a Gaussian kernel, which assigns weights to neighbouring pixels based on their distance from the centre. This process involves generating a Gaussian kernel based on the desired level of smoothing, determined by the kernel size and the value of σ. The image is then convolved with this kernel by sliding it over the image and calculating the weighted sum of pixel values within the kernel's window at each position. This results in the averaging out of high-frequency noise, leading to a smoother appearance. The blur is widely used for noise reduction, image preprocessing to remove minor variations, and enhancing edge detection algorithms by eliminating noise that could cause false edges.

D) **Single Shot Detector (SSD)**

Regression is applied directly on the position and category of bounding boxes (Bbox) in the output layer after the full image has been fed into the network as the basic idea behind single-stage target detection techniques. The two main applications of this methodology are YOLO (You Only Look Once) and SSD (Single Shot MultiBox Detector). In this study, we use SSD to detect seamount macrobenthos, which leads to a high-speed, simple-to-use model. The fundamentals of single-stage target detection will be explained in the sections that follow, with a particular emphasis on the SSD method. The first single-stage detector that runs on a single shot is the SSD (Single Shot MultiBox Detector). In contrast to Faster R-CNN, which creates and categorizes border boxes using a Region Proposal Network (RPN), SSD presents the ideas of multi-scale characteristics and default boxes. SSD provides faster performance than two-stage detectors, much like other single-stage detectors. The SSD algorithm is widely recognized for its exceptional speed, precision, and resilience to changes in scale. Its primary method for target identification and recognition is the application of multi-layer convolutional features with various receptive fields and scales. (Liu, et al., 2023).

The SSD (Single Shot MultiBox Detector) algorithm operates based on a feed-forward convolutional neural network. First, the algorithm produces a set of fixed default boxes. It then predicts the location and category using feature maps at various levels based on these default boxes. The non-maximum suppression algorithm is used to eliminate low-probability and redundant bounding boxes for each category, yielding the final detection outputs. This approach predicts both position and category simultaneously within a single network structure using a regression-based target detection algorithm. SSD is an end-to-end, single-stage target identification technique that offers noticeably faster detection times than the R-CNN series of algorithms. Moreover, its distinct architecture enhances detection performance by using multi-layer convolutional layers of various sizes for target recognition and detection. The SSD network architecture comprises the base net and extra feature layers. The base network is derived from a truncated VGG network, while additional layers consist of CNN layers with gradually decreasing scales. Target detection is performed concurrently on these feature maps with varying scales. Utilizing feature maps of different scales enables the prediction of targets of varying sizes.

The SSD takes a 3-channel RGB image as input. Initially, the algorithm generates a series of default bounding boxes (default boxes) based on the size of the feature map. These default boxes are then convolved through a series of convolutional kernels. Each layer produces a fixed number of predictions, including 4 position predictions and several category predictions. The mechanism for default boxes resembles the anchor boxes mechanism in the Region Proposal Network (RPN) of Faster R-CNN. For a p-channel feature map with an $m \times n$ size, a convolutional kernel with a scalc of $3 \times 3 \times p$ is used to predict category and location information at each location $m \times n$. Category prediction assigns a score value to each category, indicating the likelihood of the category being the target within the corresponding box. Position prediction predicts the scale scaling and displacement change based on the corresponding default box, representing the position adjustment based on the default box characteristics through CNN features. Default boxes are a series of rectangular boxes corresponding to each position m × n on the original map, according to the scale of different levels of the feature map. These default boxes vary in sizes and aspect ratios to accommodate the scale transformation of the target to be detected.

In the SSD algorithm, for each of the K default boxes at every position, convolution operations are employed to predict c + 1 category scores (comprising C target categories and one background category) as well as 4 coordinate positions, resulting in a total of (c + 1 + 4) × K predictions. Thus, for a scale of m × n, the feature map of size n requires convolution kernels corresponding to (c + 1 + 4) × Kmn prediction outputs. Each location corresponds to a fixed number of default boxes, which vary in sizes and aspect ratios based on the layer's location and scale.

During training, positive and negative samples are generated by matching the ground truth values with the default boxes. SSD performs this matching by calculating the Jaccard overlap between the default box and the ground truth box, using a threshold of 0.5. If the Jaccard overlap with the ground truth box exceeds 0.5, it is labelled as a positive sample; otherwise, it is labelled as a negative sample. Importantly, a single ground truth box may match multiple default boxes.

SSD encompasses several key features:

1. It adopts the concept from YOLO of transforming detection into a single step, achieving both target positioning and classification simultaneously. Additionally, it introduces a similar prior box concept inspired by the anchor mechanism in Faster R-CNN.
2. SSD integrates a detection method based on the feature pyramid, allowing for target prediction across feature maps with varying receptive fields.

Algorithm

Step 1: The core of SSD is a tiny convolutional filter applied to feature maps to predict a specific set of default bounding boxes, which include category scores and box offsets.

Step 2: To ensure high detection accuracy, predictions are grouped by aspect ratio and made at different scales using feature maps of varying sizes.

Step 3: Localization (loc) and confidence (conf) losses are combined as weighted sums to form the total objective loss function.

$$L(x,c,l,g) = \frac{1}{N}\left(L_{conf}(x,c) + \alpha L_{loc}(x,l,g)\right) \tag{2}$$

Step 4: Then, similar to Faster R-CNN, the offsets for the default bounding box's parameters—center (cx, cy), width (w), and height (h)—are adjusted.

$$L_{loc}(x,l,g) = \sum_{i \in Pos}^{N} \sum_{m \in \{cx,cy,w,h\}} x_{ij}^{k} smooth_{L1}\left(l_i^m - \hat{g}_j^m\right)$$

$$\hat{g}_j^{cx} = \left(g_j^{cx} - d_i^{cx}\right)/d_i^w \quad \hat{g}_j^{cy} = \left(g_j^{cy} - d_i^{cy}\right)/d_i^h \tag{3}$$

$$\hat{g}_j^w = \log\left(\frac{g_j^w}{d_i^w}\right) \quad \hat{g}_j^h = \log\left(\frac{g_j^h}{d_i^h}\right)$$

Step 5: Cross-validation results in setting the weight term α to 1. The confidence loss is represented by the SOFTMAX loss for various classes of confidence (c).

Finally, the accuracy of benthic image detection was calculated as part of the performance evaluation. False Positive (FP), False Negative (TN), True Positive (TP), and False Negative (FN) must be computed in order to get these measurements. The number of normal images that are accurately recognised as normal or ROI is known as True Positives (TP), and the number of abnormal images that are correctly classified as abnormal or non-ROI is known as True Negatives (TN). Similarly, False Positives (FP) indicate the quantity of normal images that are incorrectly categorised as abnormal, and False Negatives (FN) indicate the quantity of pathological images that are incorrectly classified as normal.

$$\text{Accuracy} = \frac{\text{TP} + \text{TN}}{\text{FP} + \text{FN} + \text{TP} + \text{TN}} \tag{4}$$

IV. RESULTS

The input images for this study are sourced from KAGGLE datasets, chosen specifically for species recognition tasks. As illustrated in Figure 1, these typical input images form a comprehensive dataset of 2000 images. This extensive dataset is intended to enhance the accuracy and reliability of the detection algorithm by providing a diverse and representative set of images for training and validation.

Figure 2. Raw input images (Coral animal, starfish, coral, crab, sea urchins)

(a) (b) (c) (d) (e)

To ensure optimal input data quality for object detection, pre-processing images with noise removal involves several critical steps. This process includes kernel creation, convolution, enhancing image quality, resizing, and cropping. noise removal effectively eliminates random noise from images, enhancing their clarity. This pre-processing step is crucial for improving the efficiency of species recognition, ensuring more dependable and accurate detection outcomes. With the help of this method, raw input (benthic) images (coral, starfish, sea urchins, crabs etc) are pre-processed (noise has been removed) which is shown in figure 3.

Figure 3. Pre-processed images (Coral animal, starfish, coral, crab, sea urchins)

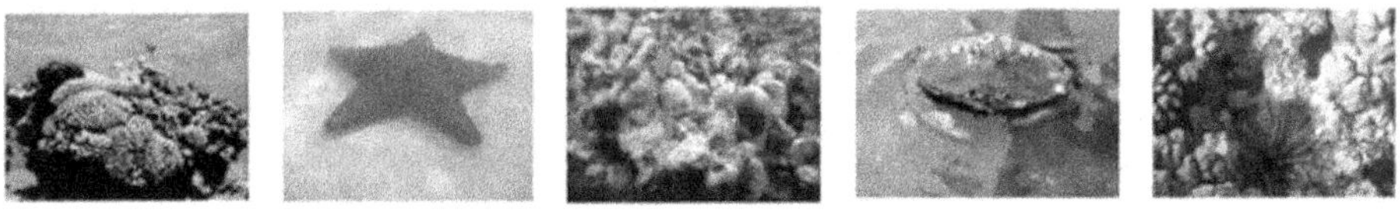

(a) (b) (c) (d) (e)

A deep learning-based object detection model called SSD enables the simultaneous detection and classification of objects in an image. Its approach to predicting bounding boxes and class probabilities for items in each cell involves dividing the input image into a grid of cells. An image dataset with labels containing benthic objects (coral, starfish, sea urchins, crabs etc) is utilized to train the SSD model. Through this training process, the model acquires the ability to distinguish benthic objects (coral, starfish, sea urchins, crabs etc) from other objects or backgrounds in the images by recognizing their distinctive characteristics. The SSD model can be employed for inference on new, unseen images once it has been trained. After processing each image, it generates confidence scores representing the likelihood of accurate detection, along with bounding boxes surrounding any benthic animals (coral, starfish, sea urchins, crabs etc) that are detected. Post-processing techniques can be applied to enhance the identified benthic reef zones after inference. This includes utilizing non-maximum suppression to merge overlapping detections, eliminating false positives, and adjusting bounding box coordinates. Accuracy is used to evaluate the SSD coral, starfish, urchin, etc. image recognition system's performance. This assessment helps determine how well the model recognizes benthic objects in the images and differentiates them from other items or background elements (figure 4).

Figure 4. Object recognized images through SSD (Coral animal, starfish, coral, crab, sea urchins)

As a result, benthic images are recognized by employing the SSD method, which harnesses deep learning principles to automate the process of detecting and categorizing coral reefs within images. This serves as a valuable tool for researchers and conservationists studying coral reef ecosystems. Performance metrics have been measured based on this object recognition and identification process. As a result, an accuracy of more than 85% has been observed for benthic image recognition which is shown in figure 5.

Figure 5. Accuracy of SSD output

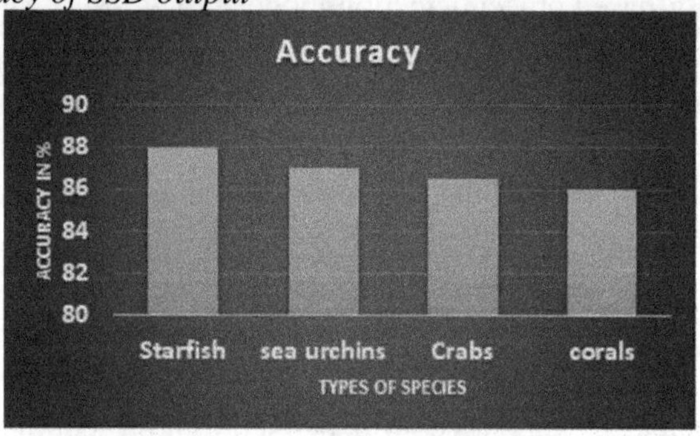

V. DISCUSSIONS

The results of these investigations indicate that SSD has potential as a useful technique for identifying benthos underwater. Because SSD automates the identification and counting of benthic species, it presents a possible resource- and time-saving alternative to traditional sampling procedures. Moreover, SSD can deliver data that is more accurate and reliable than manual sampling techniques, which are subject to bias and human error. One potential area of continuing study is the application of the SSD algorithm for underwater benthos detection. Understanding and tracking the ecology and health of aquatic environments is anticipated to become more dependent on SSD and other computer vision techniques as technology and algorithms continue to progress.

Table 1. Comparison of proposed method with existing methods

Author	Methodology	Accuracy
Yan et al., 2022	YOLO7	80.67%
Ge et al., 2021	YOLO tiny	70.6%
Liu et al., 2023	TC-YOLO network	82.9%
Proposed method	**SSD**	**88%**

Based on this results it is observed that, the proposed method shows better result and it recognized well compared to existing methods.

CONCLUSION

In this work, employ a noise removal method to tackle the noise removal in underwater images. Typically, underwater images suffer from poor quality, often appearing blurred and accompanied by significant noise. Due to these issues, the raw images are subjected to a pre-processing stage to enhance their clarity and quality. The noise removal method is utilized to effectively remove noise, thereby improving the visibility and detail in the images. Following noise removal, feature extraction, object recognition, and identification are carried out using the Single Shot Detector (SSD) algorithm. The SSD algorithm is well-suited for this task as it allows for simultaneous detection and classification of objects within an image. By dividing the image into a grid of cells, the SSD algorithm predicts bounding boxes and class probabilities for objects present in each cell. This approach enables the accurate recognition and identification of objects in the benthic images. The combination of noise removal and the SSD algorithm ensures a robust pre-processing and detection pipeline, enhancing the reliability and accuracy of object detection in underwater environments. This methodology is particularly beneficial for marine biology, ecology, and conservation efforts, providing precise data for studying and managing underwater ecosystems.

REFERENCES

Athira, P., Mithun Haridas, T. P., & Supriya, M. H. "Underwater Object Detection model based on YOLOv3 architecture using Deep Neural Networks," *7th International Conference on Advanced Computing and Communication Systems (ICACCS)*, 2021. DOI: 10.1109/ICACCS51430.2021.9441905

Bochkovskiy, A., Wang, C.-Y., & Liao, H.-J. M. YOLOv4 Optimal Speed and Accuracy of Object Detection. arXiv 2020, arXiv:2004.10934

Fayaz, S., Parah, S. A., & Qureshi, G. J. (2022). Underwater object detection: Architectures and algorithms – a comprehensive review. *Multimedia Tools and Applications*, 81(15), 20871–20916. DOI: 10.1007/s11042-022-12502-1

Ge, Z., Liu, S., Wang, F., Li, Z., & Sun, J. YOLOX: Exceeding YOLO series in 2021. arXiv 2021, arXiv:2107.08430

Huang, S., Huang, M., Zhang, Y., & Li, M. (2019). Under Water Object Detection Based on Convolution Neural Network. In Ni, W., Wang, X., Song, W., & Li, Y. (Eds.), Lecture Notes in Computer Science: Vol. 11817. *Web Information Systems and Applications. WISA 2019*. Springer. DOI: 10.1007/978-3-030-30952-7_6

Jalal, A., Salman, A., Mian, A., Shortis, M., & Shafait, F. (2020). "Fish detection and species classification in underwater environments using deep learning with temporal information,". *Ecological Informatics*, 57, 101088. DOI: 10.1016/j.ecoinf.2020.101088

Kumari, U., Samiappan, D., Kumar, R., & Sudhakar, T. (2019). Fiber optic sensors in ocean observation: A comprehensive review. *Optik (Stuttgart)*, 179, 351–360. DOI: 10.1016/j.ijleo.2018.10.186

Liu, J., Liu, S., Xu, S., & Zhou, C. "Two-Stage Underwater Object Detection Network Using Swin Transformer," IEEE Access (Volume: 10), Electronic ISSN: 2169-3536, 2022.

Liu, K., Peng, L., & Tang, S. (2023). Underwater object detection using TC-YOLO with attention mechanisms. *Sensors (Basel)*, 23(5), 2567.

Liu, Y., Xu, Y., Wang, H., & Li, X. (2023). Benthic Organism Detection, Quantification and Seamount Biology Detection Based on Deep Learning. In Li, X., & Wang, F. (Eds.), *Artificial Intelligence Oceanography*. Springer., DOI: 10.1007/978-981-19-6375-9_16

Malathi, V., Manikandan, A., & Krishnan, K. (2023). Optimzied resnet model of convolutional neural network for under sea water object detection and classification. *Multimedia Tools and Applications*, 82(24), 37551–37571. DOI: 10.1007/s11042-023-15041-5

Shreyas Dixit K. G, Chadaga M. G, Savalgimath S.S, Ragavendra Rakshith G, and Naveen Kumar M. R, "Evaluation and evolution of object detection techniques YOLO and R-CNN," Int. J. Recent Technol. Eng., vol. 8, no. 2 Special issue 3, pp. 824–829, 2019.

Yan, J., Zhou, Z., Zhou, D., Su, B., Xuanyuan, Z., Tang, J., Lai, Y., Chen, J., & Liang, W. (2022). Underwater object detection algorithm based on attention mechanism and cross-stage partial fast spatial pyramidal pooling.". *Frontiers in Marine Science*, 9, 1056300. DOI: 10.3389/fmars.2022.1056300

Yang, H., Liu, P., Hu, Y., & Fu, J. (2020). Research on underwater object recognition based on YOLOv3. *Microsystem Technologies*, 27(4), 1837–1844. DOI: 10.1007/s00542-019-04694-8

Yeh, C. H., Lin, C. H., Kang, L. W., Huang, C. H., Lin, M. H., Chang, C. Y., & Wang, C. C. (2021). Lightweight deep neural network for joint learning of underwater object detection and color conversion. *IEEE Transactions on Neural Networks and Learning Systems*, 33(11), 6129–6143.

Zhang, D., Zhao, Y. X., Cui, Y. F., & Wang, P. C. (2021). A visualization analysis and development trend of intelligent ship studies. *J. Transp. Inf. Saf.*, 39(01), 7–16.

Zhang, W., Gao, X.-Z., Yang, C.-F., Jiang, F., & Chen, Z.-Y. (2020). A object detection and tracking method for security in intelligence of unmanned surface vehicles. *Journal of Ambient Intelligence and Humanized Computing*.

Zhou, T., Si, J., Wang, L., Xu, C., & Yu, X. " Automatic Detection of Underwater Small Targets Using Forward-Looking Sonar Images," IEEE Transactions on Geoscience and Remote Sensing (Volume: 60), Print ISSN: 0196-2892, 2022.

Chapter 15
Utilizing Temperature Difference for Sustainable Maritime Transportation

Maheswari Raja
Rajalakshmi Institute of Technology, Chennai, India

S. Shrivarsha
Sri Eshwar College of Engineering, India

S. Sandhya
Sri Eshwar College of Engineering, India

V. R. Sri Sivaranjani
Sri Eshwar College of Engineering, India

M. Roshini
Sri Eshwar College of Engineering, India

Meghna Manoj Nair
New York University, USA

ABSTRACT

There is a growing need for sustainable energy sources, especially in industries like maritime transportation, where there is a need for cleaner forms of propulsion. Hence, it is important to find realistic techniques to generate electricity through renewable resources. The temperature difference between underwater and surface water can be measured using a thermistor .To quantify how much energy has been produced, you need a thermistor in contact with both the surface and underwater regions because they have different temperatures. The proposed system involves having hydrophobic heat exchangers mounted external walls of the ship's hull while

DOI: 10.4018/979-8-3693-6670-7.ch015

hydrophilic ones are placed on the inner walls. The heat exchanger turns hot water into a liquid over a water flow, then cools it down so that it continues becoming steamed. The heat exchangers condenses cold water into the liquid. The energy produced is stored within the power packs. This is a step towards sustainable maritime operations through the reduction in greenhouse gas emissions.

INTRODUCTION

The marine industry falls among the widest range of surrounding activities based on waterborne transportation, offshore resource exploration, and environmental conservation. Ship propulsion transformed throughout history, varying from simple methods similar to human and wind power to advanced modern systems and developed ones based on fossil fuels, nuclear energy, and renewable budding extraction sources. Before the 19th century, rowing and sailing were the primary means of ship propulsion. The evolution of steam engines, especially the coal-powered ones has revolutionized the marine industries into steamships, it allows engineers to masterpiece the performance of steam engines (William H Avery et.al.1994)

Moreover, approaches in the 20th century such as diesel-powered engines, cruise engines, and gas turbines, enhanced maritime capabilities even more. Havocs-powered ships have been developed in the last century. Although nuclear power propelled the naval debates but struggled with all the types of ships, energy vision and long-running operations out of re-fuelling it was the beneficial token of epic limitations. Advancements in propulsion technology have not only improved the speed and efficiency of ships but also contributed to reducing fuel consumption and emission, aligning with the industry's growing focus on sustainability (Rong-Hua Yeh et.al, 2004).

As conventional energy resources become harder to harness or as people look for green energy in the fight against climate change, industries have taken transformational processes towards renewable energy types. From these, maritime transportation can be discussed as one of the most critical, which is in the middle of the processes and is in transition. Due to the increasing population of the world, people are looking for other sources of energy given that climate change and pollution of the environment is a serious issue, hence there has been an emphasis on research for sustainable propulsion systems (Abrahim Lavi et.al,1979).

The marine industry often emerged as an important aspect of global trade and this business has faced the issues of polluting the environment. This is because vessels are among the largest emitters of greenhouse gases and therefore, emissions and sustainability become a massive issue that has to be addressed. This utilizes the natural energy within the temperature difference between deep and shallow water to generate energy for navigating ships in an environmentally friendly manner (Clark C.K. Liu et.al,2018).

The conventional mode of marine transport like utilizing heavy fuel oil and diesel-prone ships as modes of propulsion have the following impacts on the environment such as air/sea pollution and greenhouse emission, underwater noise and marine life pollution. [Fig 2 represents the carbon emission trends over the yearnof the various countries]. Some of these come with consequences such as altering climate change, being devastating to sea creatures, and being destructive to human health. To overcome these impacts, the use of marine energy is gradually increasing in the maritime sector. These technologies greatly minimize discharges and encourage environmentally preferable practices within the shipping industry (VEGA Hawaii Natural Energy Institute,2012).

There are numerous and ever-evolving measures and strategies for using renewable energy in ship propulsion which is a complex process that engages many players including shipping companies, government, technology developers and most notably environmental organizations. By leveraging industry and academia, the maritime industry can work to spread the uptake of sustainable propulsion in terms of expertise and resources. Operating under pressure to cut their fuel expenses and to adhere to environmental standards, modern carriers seem to actively employ wind-assist systems, solar panels and hydrogen fuel cells. Certain policy factors force the government to intervene and support green technology either by way of subsidies or through policy support. At the same time, technology developers remain the key driving force producing effective and correspondingly scalable solutions in the field of renewable energy targeted at the maritime industry. Environmental organizations should support the use of these technologies as they offer important oversight of quality and environmental standards (A. Hossain et.al,2013).

Besides applying effort to stimulate innovation, it also inspires the sharing and dissemination of knowledge, as well as commitment to the implementation of best practices. And also it contributes to improving the sustainability of the shipping industry, enabling it to become economically reasonable and viable. It helps the stakeholders to overcome technical, financial and regulatory barriers that have perpetuated green sustainable and renewable energy into overshadowing the conventional sources of power for ship propulsion (S. M. Masutani et.al,2001).

When aimed at reaching for a more sustainable future, entails minimizing the emission of carbon and enhancing the conservation of the environment. In this way, the unleashing of the great energy potentials of the seas represents not only an opportunity to improve the sustainability of operations concerning the ocean's complex ecosystems but also to pursue the goals of operational performance optimization as well as reducing expenditures.

This vision extends beyond theoretical abstraction, incorporating a distinct call for collective action. Strategically it comes as a call for all stakeholders within the maritime value chain from the shipbuilders and owners to those who make policies as well as any other party with interests in marine and environmental conservation to shift towards sustainable propulsion solutions as part of the transformation within the sector (Kim et.al,2021).

Education and training programs for maritime professionals on the latest sustainable practices and technologies are crucial to this transition. It can, therefore, support and develop the workforce as well as contribute to the creation of a new generation of stewards who are more in touch with environmentalism and its applications within the field. Clear disclosure and tracking of these environmental performances can help increase corporate stewardship, ensuring that the necessary action to combat the maritime industry's impact on climate change is sustained (Ahmad Etemadi et.al,2011).

LITERATURE

Here,this section includes some of the points of other literature that comes parallel to our thoughts which include,The Author (Hamedi et.al,2017) said that the sun's energy is absorbed by the ocean's surface to such an extent that it forms a shallow "mixed layer" at the surface that is 35 to 100 m (100 to 300 ft) thick. Waves and wind action maintain this layer's nearly uniform temperature and salinity. The heat from the sun raises the temperature of the mixed layer of water in the tropical oceans between roughly 15° north and 15° south latitude. This temperature rises steadily from month to month and is close to 28°C (82°F). Cold water at the bottom of the ocean flows from the poles towards the equator because it is heavier and mixes less easily with the warm water above. This creates a layering of the ocean with warm water on top of cold water at depths of 1,000 meters or more; providing temperature differences of about 22-25 degrees Celsius (40-45 degrees Fahrenheit). This is seen in the deep tropical seas(Sierra et.al,2021). Author (N.H. Mohd Idrus et.al,2017) explains main components of an OTEC system which consist of heat exchangers, turbines as well as generators that help in producing electricity using the difference in temperature of warm surface water and cold deep water. It has pipes, and controls for regulating flow of water; energy is transmitted to all parts with ease. Think about it as if you were constructing an electricity-generating facility far away from oceans where lacks icebergs or volcanic regions while having only one difference between high tides of the Atlantic Ocean and low tides of the Pacific Ocean. Additionally, there are mechanisms designed specifically for stabilizing its position plus maintaining functionality of entire technical side." According to research by Haruo

Uehara et al., tropical ocean OTEC plants have the potential to produce power with very little harm to the environment if they are limited to generating one-half megawatt per square mile. This means that this number would correspond to converting zero point seven percent (1%) average absorbed solar energy per square mile into electricity(Haruo Uehara et.al,1988).

In this author (Haruo Uehara et.al,1988) explains that it is impossible to devise a machine that uses heat to generate electricity without losing some energy to the surroundings(states the second law of thermodynamics). The ultimate limit for how efficient such machines could ever become under any circumstance would be that they could not have 100% thermal efficiency. Formulas exist for determining the actual efficiencies in scientific experiments: e=1-(Tc/Th), t being high temperature (tin kelvins-same goes for t c=low temperature). For example, if there was a system like this which made use of (upper temperature (t h)) 298k and (lower temperature Tc) 278k, it would have an efficiency rate of 7%.For OTEC plants, the actual efficiency would be much lower or about 2-3%(Krishnakumar Rajagopalan et.al,2013).

To generate 100 MW of electricity, a large OTEC plant would need about 400 cubic meters per second of warm seawater at 25°C and 200 cubic meters per second of cold seawater at 5°C. The pipes carrying the cold and warm water would need to be about 11 and 16 meters long, respectively, to maintain an average flow velocity of 2 meters per second was said by author(James H, 2017).

In (N. Khan et al., 2017) describe a system that involves pumping up cold, clean seawater from the deep sea. This water can be used for farming valuable marine life, cooling, making drinking water, and more. However, the most challenging part of building this system is laying the cold water pipe. Going deeper with the pipes might require more power and be more expensive.

In (Mohammed Faizal 2013) Mohammed Faizal and others, they talk about how the stuff we humans release into the air, like CO2, is making the Earth hotter. This happens because the amount of CO2 in the air is going up really fast because of what we do. They also found that if we help more plants grow in the ocean, they can use up some of the extra CO2, which could help make the air cleaner. OTEC looks like it's definitely doable from a technical standpoint. In the near future, it could be a cost-effective option for U.S. islands that currently rely on imported oil for their power. If we can build big OTEC plants at a reasonable cost and figure out how to efficiently transmit the power to the mainland, OTEC could also become a thing in the Southeastern U.S. The market for OTEC on the islands is about 8 gigawatts, and it's expected to be even bigger on the mainland. People started looking into using OTEC on the islands back in the early 1990s, and it only started getting attention on the mainland after 2000(Haruo Uehara,1998). Because Taiwan's economy has been growing a lot, more electricity is needed, but this has been affecting people's quality of life and health. That's why researchers are looking into finding cleaner sources of energy. One big thing about the water in the east of Taiwan is that it's in the way of the Kuroshio current; at the same time, near-surface water is much warmer than water at a depth of some 800 meters, making it possible to use the ocean for producing energy via OTEC (Ocean Thermal Energy Conversion) technology This was explained in a study by A. Hasan and others(A. Hasan, 2020).

One way to make renewable energy sources more affordable and efficient is by integrating them with storage and other systems. This makes them a better investment for both the government and private companies. One interesting idea is to combine renewable energy plants with hydrogen production, as hydrogen is a clean energy source. However, storing hydrogen can be tricky and expensive due to the high pressure needed. Using chemicals like ammonia and methanol for storage is a good solution. In this discussion, we'll focus on ammonia, which is one of the products of this system(Nam Jin Kim, 2009).

EXISTING WORK

Ocean or marine energy is a term that refers to different renewable electrical energies generated from the sea. This energy has two main types: mechanical and thermal. Mechanical forces are generated by Earth's rotation and the gravitational pull from the Moon. On ocean surface, along Earth's rotation, wind is created which results into waves; meanwhile, coastal tides are caused by the Moon's gravity-powered currents.

The sun helps in creating heat. This results in a hotter surface of the sea than its depths. Because of this difference in temperature, there is a possibility of producing electricity from this heat source in order to produce power. Consequently, with the rise of coal costs and projected future hikes, it is once more the time to explore alternative power supplies."The past few years have seen increasing interest in ocean energy, with steps taken to move this technology from development to commercialization. The original idea of ocean thermal energy conversion (OTEC) was discovered by a physicist from France who lived between 1851-1881 – Jacques Arsene d'Arsonval."His student, Georges Claude, built the first open-cycle OTEC plant in Cuba in 1930.This technology uses the temperature difference between the ocean's surface and deeper waters to convert heat energy into electricity. For maximum efficiency, there must be a temperature difference of at least 20° Celsius (36° Fahrenheit). OTEC, which stands for ocean thermal energy conversion, works by tapping into the temperature disparity among various ocean layers, from as high as sea level down to distances below a kilometer. This energy can be harvested using only 20°C temperature gaps. All around the world, shipping accounts for more than 90%. In 2018, the maritime industry alone emitted about 1056 Mt CO_2e, this was equivalent to 2.89% of the global greenhouse gas emissions (Haruo UEHARA et.al). Nonetheless, there is great expectation that there will be a substantial decrease in emissions from maritime operations because new renewable resources and alternate fuels have come into use.The industrial revolution for ship propulsion led to innovative technologies such as steamships and the introduction of coal, followed by new generation ships powered by oil. The 4th industrial revolution for ship propulsion will see the use of clean energy in the evolution and growth of the shipping industry. Currently, thanks to the mission set by IMO, ships using heavy fuel oil (HFO) that produce GHG such as Sulphur Oxides (SOx) and Nitrogen Oxides (NOx) are required to reduce their sulphur content to 0.5% or less in 2020 . Furthermore, stricter measures were put in place in Emission Control Areas (ECA) such as the North Sea and the coast of the United States, designated under the MARPOL Annex V1. Technology has also evolved, and shipping companies are considering various methods to achieve the intended zero emissions in their ships, such as using electric propulsion ships loaded with batteries, and next-generation marine fuels such as hydrogen, ammonia and biofuel. Furthermore, the use of wind and solar power is also being considered as a hybrid propulsion system integrated with conventional diesel-powered vessels to reduce fuel consumption and carbon emissions (G. T. Heydt et.al 1993). During the past centuries, researchers have searched for improving the energy sources for power generation. In this circumstance, industrial production would need to increase, transportation costs would decrease, and the quality of human life improve. The use of non-renewable energy sources for power generation has caused global warming and atmospheric pollution by CO2. The decrease in the environment may become more challenging in the 21st century.to overcome these problems, humans need an environment-friendly energy resource. One of the ways to address global warming is through the use of the TE process for managing the waste heat. The environment-friendly thermoelectric system is used as the renewable energy source that conversion has executed by TE effects and thermodynamic laws within a TEG. There are no moving parts that work quietly. TEG can be easily adapted to the diverse utilization of its compact size. Therefore, it has a high potential to be used as a renewable energy source. Recently, to increase the efficiencies, many initiatives have been taken for those devices (R.D. Fuller et.al 1978).

POWER GENERATOR MODEL

Ocean Thermal Energy Conversion, also known as OTEC for short, is where you take advantage of how cold it gets below the surface of the ocean compared with hot spots that receive sunlight regularly to make an engine run and generate some more energy. The warm water found on top covering tropical oceans compared against colder layers way below as they would be directed to make it generate even further income by absorbing excessive sunlight into engine shafts.

Figure 1. Generation of Power Using Temperature Difference Model

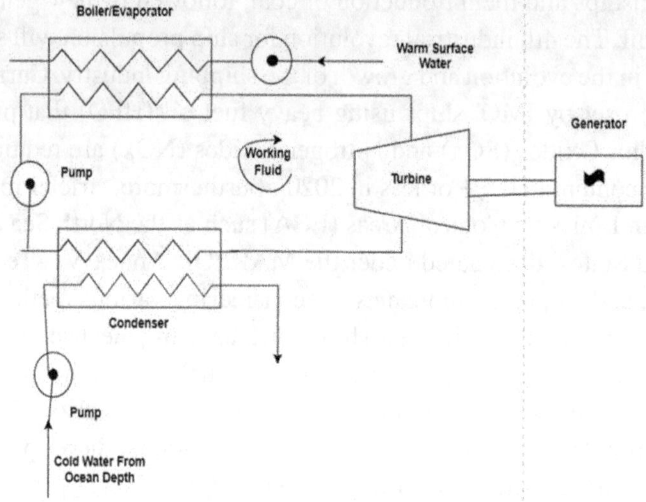

Figure 1 represents the Workflow Model of the OTEC System. This temperature difference is harnessed to propel a heat engine which changes heat energy into mechanical work and finally electricity. For OTEC to operate efficiently, it must have appropriate surface and deep sea temperatures whose values differ greatly based on geographical location; however, several designs or patterns may dictate other specific parameters.

Despite this, surface seawater temperatures in the tropical regions where OTEC may be most feasible range from 25°C to 30°C (77°F to 86°F). In contrast, the temperature of the deeper seawater bodies may sometimes be as low as 5°C-10°C (41-50°F). then, it should collect the two types of water: the warm and the cold one. The warm water is moved via a network of pipes extending from the OTEC plant into warm surface waters by gravitational pull. To shunt blending with cooler water as well as reduce the impact of top surface waves on them, these tubes are often laid at depths below the surface of the sea by several meters Screens or filters are usually put in place at the points where plunge holes are located so that marine creatures/ objects do not get into them. Long pipes that are called cold water pipes and rise from the surface of the ocean to where cold water prevails are used in collecting cold water. These pipes can be a few hundred meters or more than one kilometer long. They are usually constructed out of materials like high-density polyethene or reinforced polymers to resist the corrosive action of seawater as well as the pressure found in deep water.

In lifting water, these pumps should be both energy-saving and long-lasting. Pumps or mechanisms are employed to circulate warmed surface water alongside cold deeper seawater in the OTEC system. The pumps ensure a constant flow of water in the system that guarantees smooth operations relating to thermo exchange plus transfer. Subsequently, the warmed surface water goes through the heat exchanger which can either be called an evaporator or boiler. This is where the water heats a working fluid like ammonia or refrigerant. The solution evaporates into vapour upon heating; the vaporized vapour drives a generator that spins a turbine to supply electricity; during the same period, the cold deep sea water passes through a condenser receiving heat from the gasified solution, and thereby liquefying. This is subjected to heat solution and when placed inside, the evaporator changes its state to high-pressure steam, whose thermal change to mechanical energy is achieved through turbine spin.

A generator is connected to the turbine and changed into electrical power from mechanical power, giving electricity that is sent to the grid or consumed at the consumption point. The turbine causes the working fluid to change from vapour to liquid in the condenser by losing energy. After passing over the condenser, the chilled deep ocean water is returned to the ocean. In turn, the warm sea water from the ocean's depths absorbs the heat produced during the condensation process thereby helping maintain the temperature gradient needed for this cycle to keep going.

Figure 2. Carbon Emission Trend from Ships

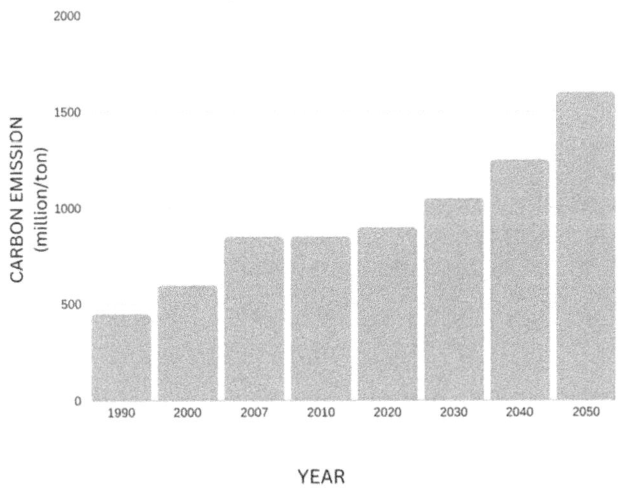

Figure 2 represents the carbon emission trend from the ships.

Waste is usually released across a large space to reduce concentrated changes in heat on the immediate environment and determine environmental effects that could be realized. In such cases, the cold water formed could equally be applied as desalination water thus cutting down on the energy used in coming up with pure water via reverse osmosis or other desalination processes not forgetting its application in aquaculture and mariculture as well.

IMPLEMENTATION AND RESULTS

CSV reader imports the data from comma-separated files. Shuffle is used to shuffle the rows of the input tables such that they are in random order. Missing value node helps handle missing values found in cells of the input table. The rule engine node takes a list of user-defined rules and tries to match them to each row in the input table. The low variance filter node filters out double-compatible columns, whose variance is below a user-defined threshold. Partition node split rows from a DB Data table.

A histogram is used to provide the distributed view of a single variable by presenting the percentage of times in specified ranges.Flow diagram for total energy production

Figure 3. Workflow for Total Energy Production

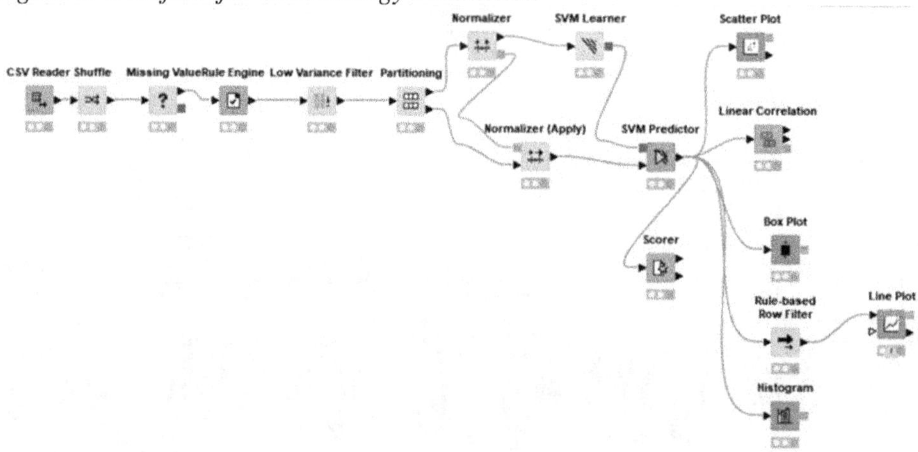

Figure 3 a represents the analysis of the total energy production respectively.

Figure 4. Total Energy Production

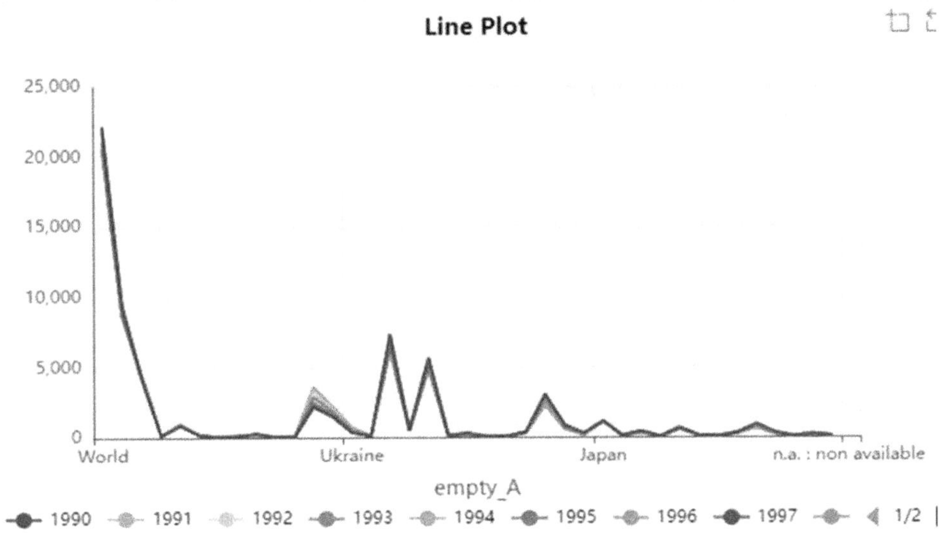

In this graph, the line plot node is used to visualize two lines by selecting the appropriate X-axis and Y-axis columns in the node's configuration. Execute the node and open the view to see the plotted lines. Decision Tree Learner Builds decision tree models from labelled data, allowing for classification or regression tasks based on hierarchical decision rules.

Figure 4 represents the workflow of the total energy production respectively.

Flow diagram for energy balance of trade

Figure 5.Workflow for Energy Balance of Trade

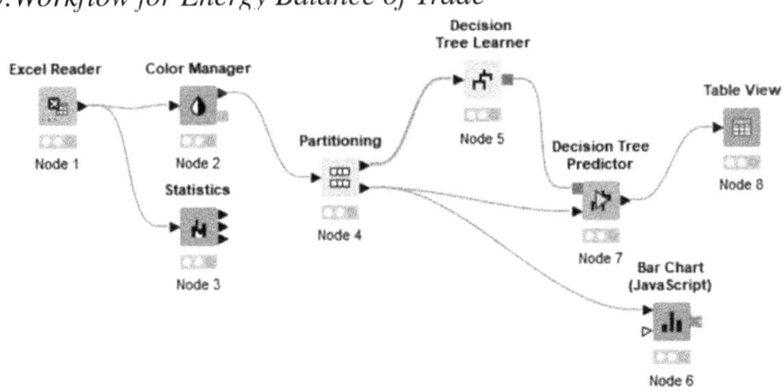

Figure 5 denotes the Workflow for the Trade of Energy Balance Around the World respectively.

Figure 6. Energy Balance of Trade

Table View

Rows: 18 | Columns: 26

2 ber (dou...	2013 Number (dou...	2014 Number (dou...	2015 Number (dou...	2016 Number (dou...	2017 Number (dou...	2018 Number (dou...	2019 Number (dou...	2020 Number (dou...	2021 Number (dou...
9.899	5,325.677	5,276.822	5,269.178	5,284.962	5,320.511	5,370.383	5,330.787	4,975.953	5,196.941
.463	258.541	248.5	252.399	247.627	247.417	245.827	242.287	217.776	233.858
.001	321.735	307.702	309.142	310.447	311.394	303.614	296.904	278.755	288.175
:26	97.887	94.24	95.333	99.657	104.188	109.235	105.233	102.133	108.434
.755	116.165	113.437	117.457	117.925	124.123	123.722	120.173	108.358	114.993
.134	190.037	179.007	180.877	178.324	175.526	173.665	167.812	153.908	159
.13	30.987	28.397	25.556	28.3	29.828	29.828	28.387	28.001	28.963
9.379	1,023.948	994.778	960.149	989.353	1,007.04	1,076.648	1,073.592	1,045.86	1,128.496
.854	273.652	280.599	278.359	290.741	302.769	307.381	308.766	284.357	289.4
2.198	2,190.913	2,217.153	2,192.295	2,168.734	2,161.752	2,222.329	2,221.883	2,045.164	2,144.75
.082	858.96	856.515	850.645	839.421	835.487	820.525	824.948	775.117	817.149
:02	38.492	34.75	35.524	37.817	38.393	39.069	41.358	37.947	38.316
.884	191.207	198.966	195.678	199.873	208.033	223.236	231.815	219.986	227.852
.479	263.845	268.37	272.631	282.333	282.18	281.966	280.088	276.108	292.142

Figure 6 denotes the Dataset for the Trade of Energy Balance Around the World respectively.

In this table, the table view is used to interactively view and explore your data table. Connect your data, execute the node, and open the view to inspect your data. Decision Tree Predictor Applies pre-trained decision tree models to new data, predicting class labels or values based on learned decision rules. These nodes are essential for predictive modeling within KNIME workflows.

Flow diagram for total energy consumption

Figure 7. Workflow for Total Energy Consumption

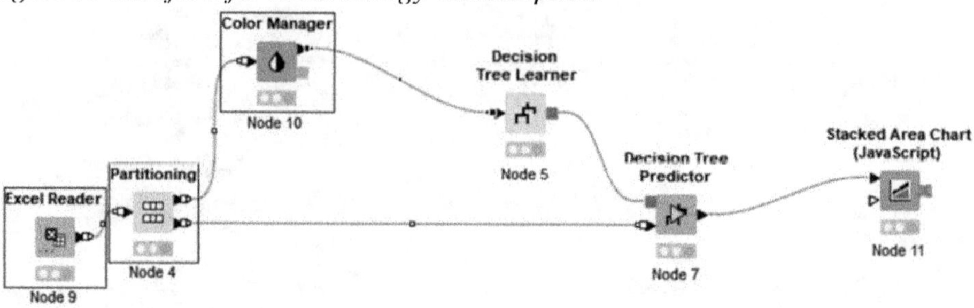

Figure 7 Represents the workflow for the Total Energy Consumed Around the World respectively.

Figure 8. Total Energy Consumption

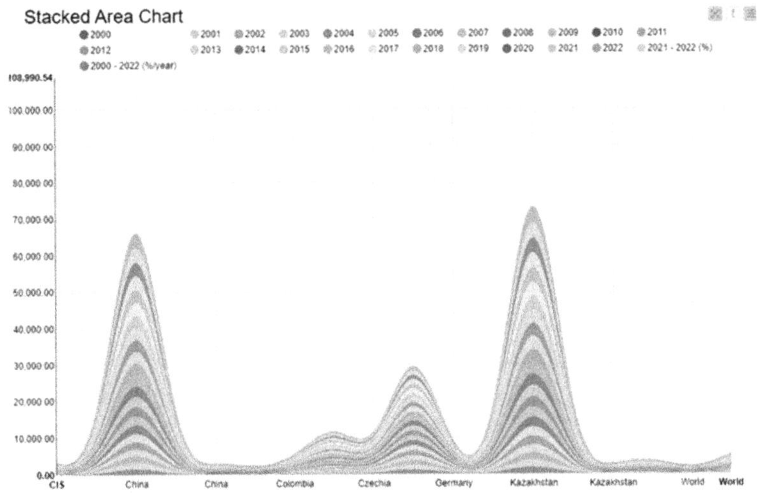

In this graph, The Stacked Area Chart node is used to visualize cumulative data trends. Configure the node, execute it, and open the view to see the stacked areas. Statistics Provides descriptive statistics and distribution summaries for numerical variables in datasets, aiding in data exploration and analysis.

Figure 8 Represents the graph for the Total Energy Consumed Around the World respectively.

Flow diagram for electricity production around the world

Figure 9. Workflow for Electricity Production

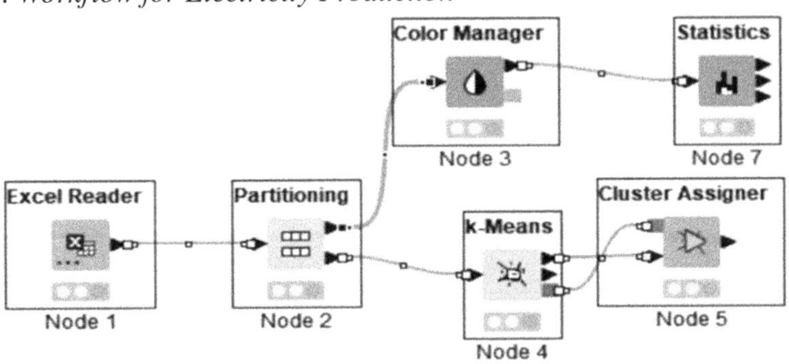

Figure 9 represents the workflow for Total Electricity Production Around the World respectively.

Figure 10. Electricity Production

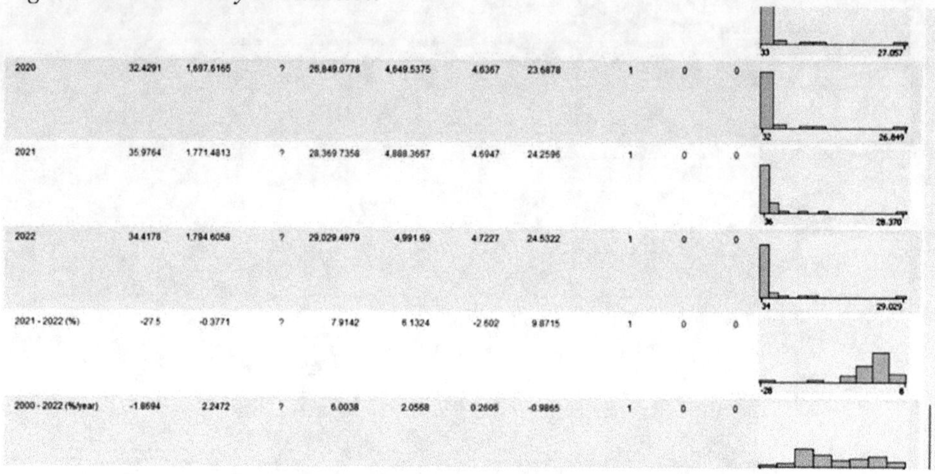

2020	32.4291	1,697.6165	?	26,849.0778	4,649.5375	4.6367	23.6878	1	0	0
2021	35.9764	1,771.4813	?	28,369.7356	4,888.3667	4.6947	24.2596	1	0	0
2022	34.4178	1,794.6058	?	29,029.4979	4,991.69	4.7227	24.5322	1	0	0
2021 - 2022 (%)	-27.5	-0.3771	?	7.9142	6.1324	-2.602	9.8715	1	0	0
2000 - 2022 (%/year)	-1.8694	2.2472	?	6.0038	2.0568	0.2606	-0.9865	1	0	0

Figure 10 Represents the Statistics for Total Electricity Production Around the World respectively.

Row Filter filter rows in KNIME based on user-defined logical expressions. Boxplot which helps in looking at how data has been dispersed. In this Statistics, The statistics node provides descriptive statistics for your data columns. Connect your data, execute the node, and view the summary statistics for insights.

Flow diagram for coal and lignite production

Figure 11. Workflow for Coal and Lignite Production

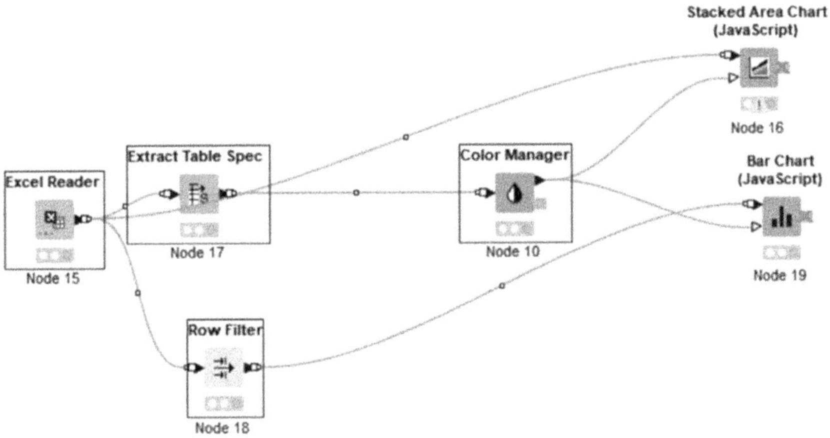

Figure 11 Specifies the workflow for the Total Production of Coal and Lignite.

Figure 12. Coal and Lignite Production

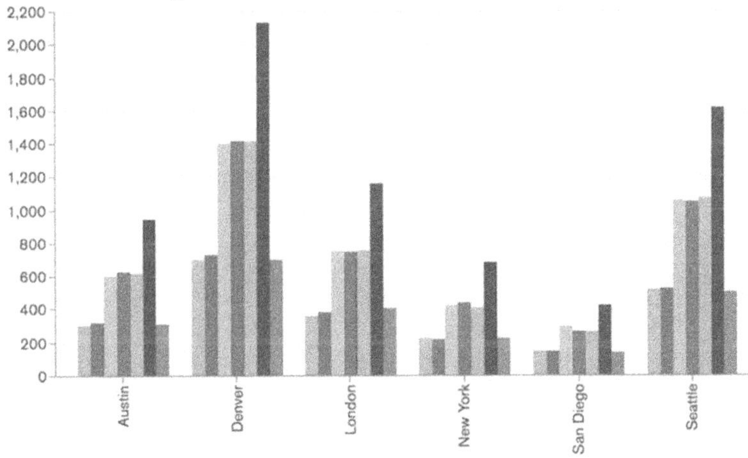

Figure 12 Specifies the bar graph for the Total Production of Coal and Lignite respectively. Excel Reader Reads data from Excel files into KNIME workflows, facilitating easy integration of Excel data for analysis. In this bar graph, the Bar Chart node is used to visualizes categorical data with bar charts. Configure the node with the desired columns, execute it, and open the view to see the bars.

Figure 13. Workflow for Coal And Lignite Balance Trade

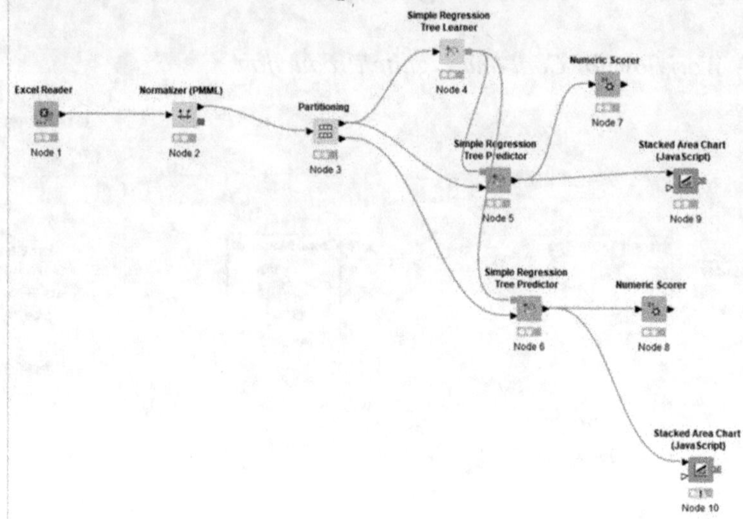

Figure 13 reflects the workflow for Coal and Lignite Balance Trade respectively

Figure 14. Coal And Lignite Balance Trade

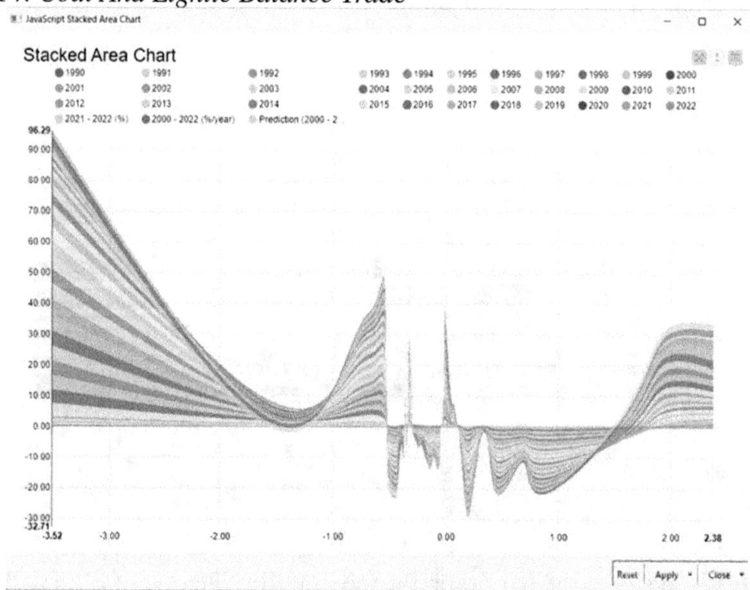

Figure 14 reflects the graph for Coal and Lignite Balance Trade respectively. Color Manager Manages color settings for visualizations allowing customization of colors for data points, categories, or groups

Flow diagram for share of renewable energy around the world

Figure 15. Workflow for Share of Renewable Energy in Electricity Production

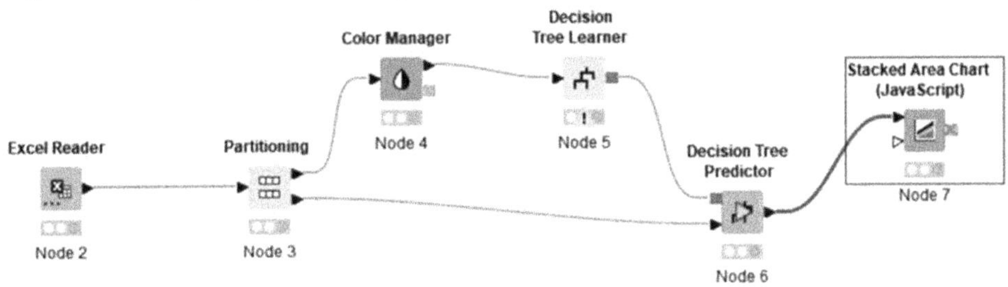

Figure 15 represents the workflow for the Share of Renewable Energy in Electricity Production Around the World respectively.

Figure 16. Share of Renewable Energy in Electricity Production

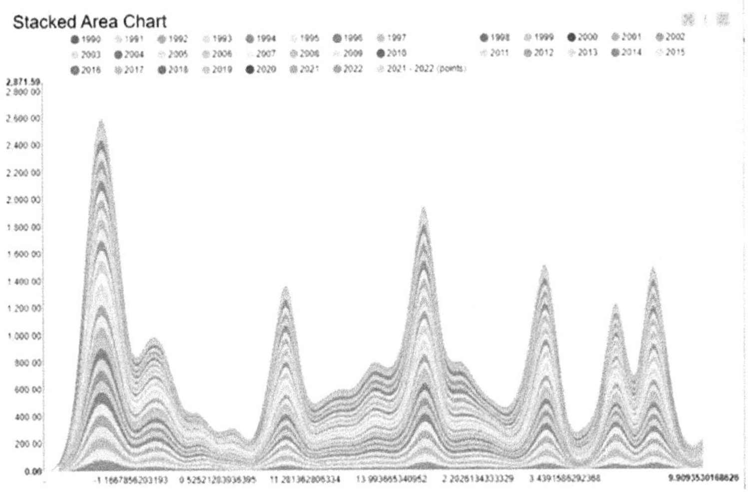

Figure 16 represents the graph for the Share of Renewable Energy in Electricity Production Around the World respectively. The scorer node in KNIME evaluates classification models by checking how they predicted the labels against those that existed.

Flow diagram for Carbon emission from fuel combustion

Figure 17. Workflow for Co2 Emission from Fuel Combustion

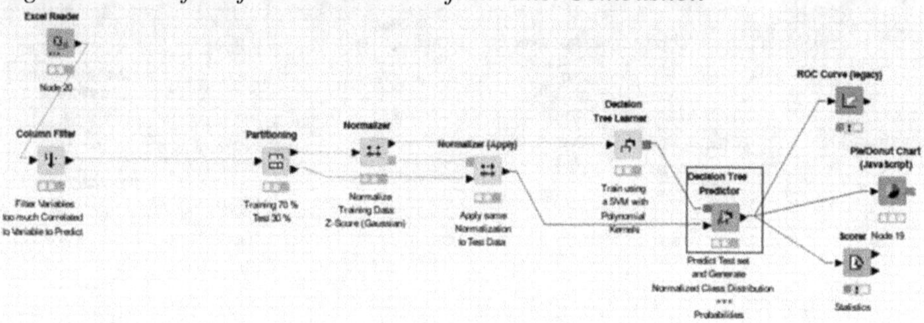

Figure 17 represents the workflow for Total Co2 Emission from Fuel Combustion respectively.

Figure 18. Co2 Emission from Fuel Combustion

G7 (1/40)

Table:

Category	%	n
G7	2.5	1
BRICS	2.5	1
Europe	2.5	1
European Union	2.5	1
Czechia	2.5	1
Italy	2.5	1
Poland	2.5	1
Portugal	2.5	1
United Kingdom	2.5	1
Norway	2.5	1
Turkiye	2.5	1
CIS	2.5	1
Kazakhstan	2.5	1
Russia	2.5	1
Ukraine	2.5	1
North America	2.5	1
Canada	2.5	1
United States	2.5	1
Argentina	2.5	1
Brazil	2.5	1
Chile	2.5	1
Colombia	2.5	1
Asia	2.5	1
China	2.5	1
Japan	2.5	1
Malaysia	2.5	1
South Korea	2.5	1
Taiwan	2.5	1
Thailand	2.5	1
Pacific	2.5	1
Australia	2.5	1
New Zealand	2.5	1
Africa	2.5	1
Algeria	2.5	1
Nigeria	2.5	1
South Africa	2.5	1
Middle-East	2.5	1
Iran	2.5	1
Saudi Arabia	2.5	1
n.a. : non available	2.5	1
Total	100.0	40

Figure 18 represents the Dataset for Total Co2 Emission from Fuel Combustion respectively.

Figure 19. Co2 Emission from Fuel Combustion

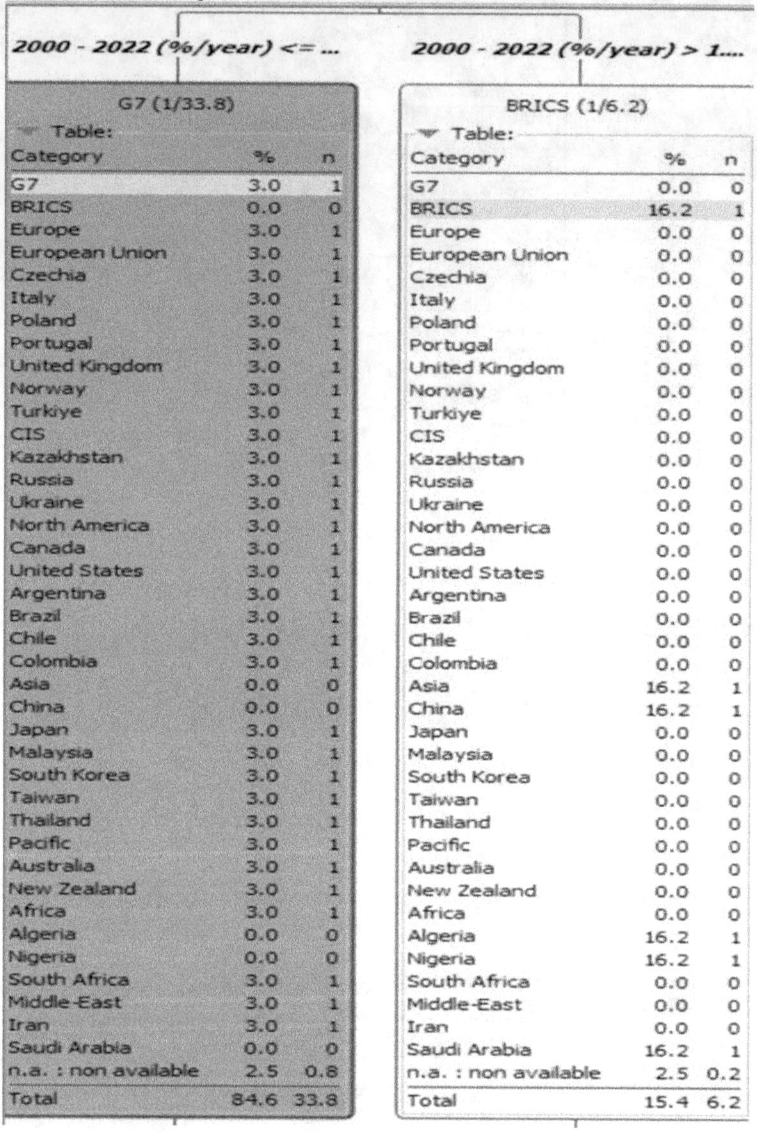

G7 (1/33.8)

Table:

Category	%	n
G7	3.0	1
BRICS	0.0	0
Europe	3.0	1
European Union	3.0	1
Czechia	3.0	1
Italy	3.0	1
Poland	3.0	1
Portugal	3.0	1
United Kingdom	3.0	1
Norway	3.0	1
Turkiye	3.0	1
CIS	3.0	1
Kazakhstan	3.0	1
Russia	3.0	1
Ukraine	3.0	1
North America	3.0	1
Canada	3.0	1
United States	3.0	1
Argentina	3.0	1
Brazil	3.0	1
Chile	3.0	1
Colombia	3.0	1
Asia	0.0	0
China	0.0	0
Japan	3.0	1
Malaysia	3.0	1
South Korea	3.0	1
Taiwan	3.0	1
Thailand	3.0	1
Pacific	3.0	1
Australia	3.0	1
New Zealand	3.0	1
Africa	3.0	1
Algeria	0.0	0
Nigeria	0.0	0
South Africa	3.0	1
Middle-East	3.0	1
Iran	3.0	1
Saudi Arabia	0.0	0
n.a. : non available	2.5	0.8
Total	84.6	33.8

BRICS (1/6.2)

Table:

Category	%	n
G7	0.0	0
BRICS	16.2	1
Europe	0.0	0
European Union	0.0	0
Czechia	0.0	0
Italy	0.0	0
Poland	0.0	0
Portugal	0.0	0
United Kingdom	0.0	0
Norway	0.0	0
Turkiye	0.0	0
CIS	0.0	0
Kazakhstan	0.0	0
Russia	0.0	0
Ukraine	0.0	0
North America	0.0	0
Canada	0.0	0
United States	0.0	0
Argentina	0.0	0
Brazil	0.0	0
Chile	0.0	0
Colombia	0.0	0
Asia	16.2	1
China	16.2	1
Japan	0.0	0
Malaysia	0.0	0
South Korea	0.0	0
Taiwan	0.0	0
Thailand	0.0	0
Pacific	0.0	0
Australia	0.0	0
New Zealand	0.0	0
Africa	0.0	0
Algeria	16.2	1
Nigeria	16.2	1
South Africa	0.0	0
Middle-East	0.0	0
Iran	0.0	0
Saudi Arabia	16.2	1
n.a. : non available	2.5	0.2
Total	15.4	6.2

Figure 19 represents the Decision tree for Total Co2 Emission from Fuel Combustion respectively.

CONCLUSION

The journey of marine propulsion, starting from ancient sea faring boats pushed by humans and headwinds to innovative fossil fuels, nuclear power and renewable tidal energy unveils human technological advancement towards the imperative needs for environmentally friendly living. As traditional energy sources become increasingly exploitative and the devastating environmental effects of fossil fuels are harder to ignore, the maritime industry is recognizing the need for mature and sustainable alternatives. This move toward sustainable propulsion is necessary as the maritime industry accounts for a significant share of greenhouse gas emissions and related environmental pollution. The industry can, in the short term, decrease its environmental footprint and help combat climate change, while long-term operational efficiencies with reduced fuel costs can be achieved across ocean shipping using renewable technologies such as wind-assist, solar panels and hydrogen fuel cells. To be successful, shipping companies must work hand-in-hand with the government. Green technology adoption can be helped along by government regulation and subsidies, but innovation is what will ultimately allow it to happen at scale. These technologies are verified by environmental organizations to ensure they meet proper standards. Improved industry stewardship and accountability can further be achieved through educational programs for maritime professionals, sustainable practices and technologies including real-time monitoring to ensure transparency in environmental performance. Receiving propulsion from renewable energy also meets a requirement to decrease emissions and conservation at large, both things that also provide you with economic benefits.

FUTURE WORK

While the global society meets its need for higher efficiency for combating local and global problems, such as climate change, insufficient energy security, and environmental deterioration, OTEC systems are now a promising research field. These systems that harness the differences in temperature between the warm surface seawater and cold deep water for electricity generation have several prospects that can greatly aid in improving sustainability opportunities. The key areas of potential OTEC systems development for the future include the ability to deliver major amounts of renewable energy, promote economic growth, protect the environment, utilize new technologies in engineering, and adjust national policy.

For large-scale energy production, OTEC systems can be utilized to help the world shift towards the utilization of renewable energy sources in future periods. OTEC, in contrast with electricity-producing sources like solar, wind, etc., offers a consistent and stable supply of the latter. This consistent energy generation can well be used in conjunction with other existing renewable resources and supply consistent power energy. The potential to synergise into national and regional power grids is especially in regions of high-temperature differential in the tropical and subtropical areas as the technology develops further. This integration will improve energy stability by cutting on the standby of fossil energy, hence advancing the carbon reduction goals.

For freshwater production, the OTEC systems have been classified according to the following categories based on the type of power cycle employed and mode of operation: The OTEC systems have several compelling benefits including the supplementing of electrical power while at the same time producing fresh water from seawater through the seawater desalination process. This dual-purpose capability addresses two critical needs: The country's resources that have been identified to be most vulnerable to vanishing are energy and freshwater. This technology is particularly helpful to islanding societies and societies along coasts where access to water is usually a big problem. Therefore, through the commercial exploitation of seawater, first by converting it into electricity and then using OTEC systems, the scarcity of freshwater can be eradicated and the living standards of these areas can be enhanced.

For marine and coastal resource utilization, OTEC systems avail diverse potentialities for optimum usage of marine and coastal resources for sustainable benefits. OTE surface water: the cold deep seawater that is brought to the surface as part of OTEC systems can be used for aquaculture purposes as they assist with the production of sustainable seafood. It can also be used in cases such as using water from the sea for coastal farming to increase the production of food crops thus improving food production in the region. Moreover, because OTEC systems can be combined with desalination plants, it has the potential to contribute to the recovery of valuable minerals dissolved in seawater for better resource efficiency.

Finally, The future capacities for OTEC systems are closer and unlimited. Thereby, it means OTEC has the potential to be a viable solution to current global challenges related to energy supply in as much as it is renewable, has low emission levels, and guarantees supply. It has been established that tapping OTEC offers the twin advantage of energy production and freshwater supply, apart from creating economic benefits ancillary to such solutions while at the same time, ensuring that the consequences of using such sources are acutely eco-friendly. With sustained advancements in technology enhancement, development and implementation of sound policies and international cooperation, small-scale OTEC systems go a long way towards ensuring a brighter future for people.(William H Avery et.al.1994)

REFERENCES

Etemadi, A., Emdadi, A., Afshar, O. A., & Emami, Y. Electricity Generation by the Ocean Thermal Energy, Energy Procedia, Volume 12,2011, Pages 936-943, ISSN 1876-6102, DOI: 10.1016/j.egypro.2011.10.123

Faizal, M., & Ahmed, M. R. (2013). Experimental studies on a closed cycle demonstration OTEC plant working on small temperature difference. *Renewable Energy*, 51, 234–240.

Fuller, R. D. (1978). Ocean thermal energy conversion. *Ocean Management*, 4(2-4), 241–258.

Hamedi, A. S., & Sadeghzadeh, S. (2017). Conceptual design of a 5 MW OTEC power plant in the Oman Sea. *Journal of Marine Engineering & Technology*, 16(2), 94–102. DOI: 10.1080/20464177.2017.1320839

Hasan, A., & Dincer, I. (2020). An ocean thermal energy conversion based system for district cooling, ammonia and power production. *International Journal of Hydrogen Energy*, 45(32), 15878–15887.

Herrera, J., Sierra, S., & Ibeas, A. (2021). J.; Sierra, S.; Ibeas, A. Ocean Thermal Energy Conversion and Other Uses of Deep Sea Water: A Review. *Journal of Marine Science and Engineering*, 9(4), 356. DOI: 10.3390/jmse9040356

Heydt, G. T. (1993, March). An assessment of ocean thermal energy conversion as an advanced electric generation methodology. *Proceedings of the IEEE*, 81(3), 409–418. DOI: 10.1109/5.241487

Hossain, A. Azhim, A. B. Jaafar, M. N. Musa, S. A. Zaki and D. N. Fazreen, "Ocean thermal energy conversion: The promise of a clean future," *2013 IEEE Conference on Clean Energy and Technology (CEAT)*, Langkawi, Malaysia, 2013, pp. 23-26, DOI: 10.1109/CEAT.2013.6775593

Idrus, N. M., Musa, M. N., Yahya, W. J., & Ithnin, A. M. (2017). Geo-ocean thermal energy conversion (GeOTEC) power cycle/plant. *Renewable Energy*, 111, 372–380.

Khan, N. D., Kalair, A., Abas, N., & Haider, A. (2017). Review of ocean tidal, wave and thermal energy technologies. *Renewable & Sustainable Energy Reviews*, 72, 590–604.

Kim, H.-J., Lee, H.-S., Lim, S.-T., & Petterson, M. (2021). The Suitability of the Pacific Islands for Harnessing Ocean Thermal Energy and the Feasibility of OTEC Plants for Onshore or Offshore Processing. *Geosciences*, 11(10), 407. DOI: 10.3390/geosciences11100407

Kim, N. J., Ng, K. C., & Chun, W. (2009). Using the condenser effluent from a nuclear power plant for Ocean Thermal Energy Conversion (OTEC). *International Communications in Heat and Mass Transfer*, 36(10), 1008–1013.

Lavi, A., & Lavi, G. H. (1979). Ocean thermal energy conversion (OTEC): Social and environmental issues. *Energy*, 4(5), 833–840.

Liu, K., Peng, L., & Tang, S. (2023). Underwater object detection using TC-YOLO with attention mechanisms. *Sensors (Basel)*, 23(5), 2567.

Rajagopalan, K., & Nihous, G. C. (2013). Estimates of global Ocean Thermal Energy Conversion (OTEC) resources using an ocean general circulation model. *Renewable Energy*, 50, 532–540.

VanZwieten, J. H., Rauchenstein, L. T., & Lee, L. (2017). An assessment of Florida's ocean thermal energy conversion (OTEC) resource. *Renewable & Sustainable Energy Reviews*, 75, 683–691.

Yeh, R. H., Su, T. Z., & Yang, M. S. (2005). Maximum output of an OTEC power plant. *Ocean Engineering*, 32(5-6), 685–700.

Compilation of References

3D at Depth. (n.d.). Applications: Deep-sea mining. https://www.3datdepth.com/applications/deep-sea-mining

Abhilasha, A., Sreenivasulu, A., & Manimozhi, T. KUMAR, P. S., Selvakumar, P., & Singh, P. (2022, April). The Model of Smart Sensing Device For Sensitive Nano-clusters Modification in Sensing Properties. In 2022 2nd International Conference on Advance Computing and Innovative Technologies in Engineering (ICACITE) (pp. 1043-1047). IEEE.

Abu-Qdais, H., & Shatnawi, N., & AL-Alamie, E. A. (2024). Intelligent system for solid waste classification using combination of image processing and machine learning models. *Journal of Experimental & Theoretical Artificial Intelligence*, •••, 1–12.

Aguzzi, J., del Mar Flexas, M., Flögel, S., Lo Iacono, C., Tangherlini, M., Costa, C., & Marini, S.. (2020). Exo-ocean exploration with deep-sea sensor and platform technologies. *Astrobiology*, 20(7), 897–915. DOI: 10.1089/ast.2019.2129 PMID: 32267735

Aguzzi, J., Thomsen, L., Flögel, S., Robinson, N. J., Picardi, G., Chatzievangelou, D., Bahamon, N., Stefanni, S., Grinyó, J., Fanelli, E., Corinaldesi, C., Del Rio Fernandez, J., Calisti, M., Mienis, F., Chatzidouros, E., Costa, C., Violino, S., Tangherlini, M., & Danovaro, R. (2024). New technologies for monitoring and upscaling marine ecosystem restoration in deep-sea environments. *Engineering (Beijing)*, 34, 195–211. DOI: 10.1016/j.eng.2023.10.012

Al Mahmud, N. (2020). Performance Analysis of Different Acoustic Features based on LSTM for Bangla Speech Recognition. *The International Journal of Multimedia & Its Applications*, 12(4), 17–25. DOI: 10.5121/ijma.2020.12402

Al Najar, M., Thoumyre, G., Bergsma, E. W. J., & Schenke, H. W. (2023). Satellite-derived bathymetry using deep learning. *Machine Learning*, 112(4), 1107–1130. DOI: 10.1007/s10994-021-05977-w

Ali, I. (2022). Understanding Terminology, Concept, Scope, and Theory in Maritime Historiography. *SOSIOHUMANIKA, 15*(2), 91-122.

Ali, I. M., Prakoso, L. Y., & Sianturi, D. (2020). Sea Defense Strategy in Facing maritime Security Threats in the Indonesian Sea Region. *Journal of Sea Defense Strategy*, 6(2).

Anderson, P. M., & Perry, C. (2019). Deep-sea ecosystem responses to climate change. *Marine Biology*, 66(3), 281–290. DOI: 10.1111/mar.12345

Aryani, C. (2021). Encouraging the birth of a marine security bill in strengthening the national marine security system. *Journal of Indonesian Legal Development*, 3(2), 155–173.

Athira, P., Mithun Haridas, T. P., & Supriya, M. H. "Underwater Object Detection model based on YOLOv3 architecture using Deep Neural Networks," *7th International Conference on Advanced Computing and Communication Systems (ICACCS)*, 2021. DOI: 10.1109/ICACCS51430.2021.9441905

Baco, A. R., Etter, R. J., Ribeiro, P. A., & von der Heyden, S. (2020). Genetic connectivity in deep-sea ecosystems under climate change. *Nature Communications*, 11, 1135. DOI: 10.1038/s41467-020-14858-1

Bagher, A. M., Vahid, M., Mohsen, M., & Parvin, D. (2015). Hydroelectric energy advantages and disadvantages. *American Journal of Energy Science*, 2(2), 17–20.

Bahaj, A. S. (2011). Generating electricity from the oceans. *Renewable & Sustainable Energy Reviews*, 15(7), 3399–3416. DOI: 10.1016/j.rser.2011.04.032

Ballard, R. D. (2000). *The Eternal Darkness: A Personal History of Deep-Sea Exploration*. Princeton University Press.

Basar, M. F., Rahman, A. A., Din, A., & Mahmod, Z. 2010, June. Design and development of green electricity generation system using ocean surface wave. In *Proceedings of the International Conference on Energy and Sustainable Development: Issues and Strategies (ESD 2010)* (pp. 1-11). IEEE.

Bell, R. J., & Fogarty, M. J. (2018). Impacts of climate change on deep-sea fish populations. *Fish and Fisheries*, 19(2), 265–276. DOI: 10.1111/faf.12251

Ben Elghali, S.E., Benbouzid, M.E.H. and Charpentier, J.F., 2009. Marine tidal current electric power generation technology-A review. Electromotion, 16(3).

Bennett, N. J., & Dearden, P. (2014). Why local people do not support conservation: Community perceptions of marine protected area livelihood impacts, governance and management in Thailand. *Marine Policy*, 44, 107–116. DOI: 10.1016/j.marpol.2013.08.017

Bhaumik, L., Bandyopadhyay, A., & Bandyopadhyay, A. (2023). Exploring The Depths: Harnessing the Power of Deep Learning in Marine Ecology. Journal of Survey in Fisheries Sciences, 3779-3796.

Biastoch, A., Schwarzkopf, F. U., Getzlaff, K., & Rühs, S. (2021). Climate change impacts on deep-sea currents and nutrient cycling. Geophysical Research Letters, 48(5), e2020GL089864. https://doi.org/DOI: 10.1029/2020GL089864

Blanchard, C., Harrould-Kolieb, E., Jones, E., & Taylor, M. L. (2023). The current status of deep-sea mining governance at the International Seabed Authority. *Marine Policy*, 147, 105396. DOI: 10.1016/j.marpol.2022.105396

Bochkovskiy, A., Wang, C.-Y., & Liao, H.-J. M. YOLOv4 Optimal Speed and Accuracy of Object Detection. arXiv 2020, arXiv:2004.10934

Bonofiglio, F., De Leo, F., Yee, C., Chatzievangelou, D., Aguzzi, J., & Marini, S. (2022). Machine learning applied to big data from marine cabled observatories: A case study of sablefish monitoring in the NE Pacific. *Frontiers in Marine Science*, 9, 842946. Advance online publication. DOI: 10.3389/fmars.2022.842946

Bopp, L., Aumont, O., Cadule, P., & Gehlen, M. (2018). Modelling the impact of climate change on deep-sea ecosystems. *Biogeosciences*, 15, 6703–6718. DOI: 10.5194/bg-15-6703-2018

Bräger, S., Rodriguez, G. Q. R., & Mulsow, S. (2020). The current status of environmental requirements for deep seabed mining issued by the International Seabed Authority. *Marine Policy*, 114, 103258. DOI: 10.1016/j.marpol.2018.09.003

Bravo, M., Brandt, M., Grient, J., Dahlgren, T., Esquete, P., Gollner, S., & Cordes, E. (2023). Insights from the management of offshore energy resources: Toward an ecosystem-services based management approach for deep-ocean industries. *Frontiers in Marine Science*, 9, 994632. Advance online publication. DOI: 10.3389/fmars.2022.994632

Breusing, C., Bowers, H. A., & Vrijenhoek, R. C. (2020). Climate-driven shifts in deep-sea hydrothermal vent communities. *Proceedings of the National Academy of Sciences of the United States of America*, 117(25), 14482–14492. DOI: 10.1073/pnas.1920461117 PMID: 32518112

Briones, E. E. (2023). Deep-Sea Life. In *Mexican Fauna in the Anthropocene* (pp. 319–331). Springer International Publishing. DOI: 10.1007/978-3-031-17277-9_15

Brock, T. D. (1997). The value of basic research: Discovery of Thermus aquaticus and other extreme thermophiles. *Genetics*, 146(4), 1207–1210. DOI: 10.1093/genetics/146.4.1207 PMID: 9258667

Brodnicke, O. (2023). Deep-sea sponge derived environmental DNA analysis reveals demersal fish biodiversity of a remote arctic ecosystem. *Environmental DNA, 5*(6), 1405-1417. DOI: 10.1002/edn3.451

Brown, S., Johnson, L., & Smith, A. (2020). Modeling Climate Change Effects on Deep-Sea Fish Communities Using Generative AI and Ecological Niche Modeling. *Marine Ecology Progress Series*, 546, 187–200.

Bryan, T. L., Metaxas, A., & Kelly, N. E. (2020). Climate change and deep-sea coral ecosystems. *Deep-sea Research. Part I, Oceanographic Research Papers*, 156, 103176. DOI: 10.1016/j.dsr.2020.103176

Bryden, I. (2004). *Tidal energy. In 2004 Survey of Energy Resources*. Elsevier Science.

Budiarto, J. (2017). Vietnam-India Cooperation in Facing China's Military Base Construction in Fiery Cross Reef of South China Sea Spraty Islands Region in 2013-2015 (Doctoral dissertation, Brawijaya University).

Camilli, R., Reddy, C. M., Yoerger, D. R., Van Mooy, B. A. S., Jakuba, M. V., Kinsey, J. C., McIntyre, C. P., Sylva, S. P., & Maloney, J. V. (2010). Tracking hydrocarbon plume transport and biodegradation at Deepwater Horizon. *Science*, 330(6001), 201–204. DOI: 10.1126/science.1195223 PMID: 20724584

Carlson, P. R., & Silver, E. A. (2021). Submarine landslides and climate change. *Marine Geology*, 435, 106437. DOI: 10.1016/j.margeo.2021.106437

Center for Biological Diversity. (n.d.). Center for Biological Diversity. https://www.biologicaldiversity.org

Chaniotis, P. D., Robson, L. M., Lemasson, A. J., Cornthwaite, A. L., & Howell, K. L. (2020). UK deep-sea conservation: Progress, lessons learned, and actions for the future. *Aquatic Conservation*, 30(2), 375–393. DOI: 10.1002/aqc.3243

Chao, Z., Li, Z., Dong, Y., Shi, D., & Zheng, J. (2024). Estimating compressive strength of coral sand aggregate concrete in marine environment by combining physical experiments and machine learning-based techniques. *Ocean Engineering*, 308, 118320. DOI: 10.1016/j.oceaneng.2024.118320

Chauhan, R., Shighra, S., Madkhali, H., Nguyen, L., & Prasad, M. (2023). Efficient future waste management: A learning-based approach with deep neural networks for smart system (LADS). *Applied Sciences (Basel, Switzerland)*, 13(7), 4140. DOI: 10.3390/app13074140

Chavhan, P. G., Khedka, V. N., Gupta, M., & Agarwal, K. (2024). Automatic Waste Segregator Based on IoT & ML Using Keras model and Streamlit. *International Journal of Intelligent Systems and Applications in Engineering*, 12(2), 787–799.

Cheema, S. M., Hannan, A., & Pires, I. M. (2022). Smart waste management and classification systems using cutting edge approach. *Sustainability (Basel)*, 14(16), 10226. DOI: 10.3390/su141610226

Chen, H., Tang, T., Ait-Ahmed, N., Benbouzid, M. E. H., Machmoum, M., & Zaim, M. E. H. (2018). Attraction, challenge and current status of marine current energy. *IEEE Access : Practical Innovations, Open Solutions*, 6, 12665–12685. DOI: 10.1109/ACCESS.2018.2795708

Chen, L., Wang, Y., & Zhang, Q. (2021). Machine learning analysis of deep-sea food web dynamics under climate variability. *Journal of Marine Ecology*, 45(3), 312–329.

Chen, Q., Yang, J., Zhao, W., Tao, L., Mao, J., & Li, Z. (2024). Algorithms for dynamic control of a deep-sea mining vehicle based on deep reinforcement learning. *Ocean Engineering*, 298, 117199. DOI: 10.1016/j.oceaneng.2024.117199

Chen, V. Y., Lu, D.-J., & Han, Y.-S. (2024). Hybrid intelligence for marine biodiversity: Integrating citizen science with AI for enhanced intertidal conservation efforts at Cape Santiago, Taiwan. *Sustainability (Basel)*, 16(1), 454. DOI: 10.3390/su16010454

Chérubin, L. M. (2017). Deep-Sea Robotics: A Technological Evolution. *Annual Review of Marine Science*, 9, 485–508.

Chimienti, G., Angeletti, L., Rizzo, L., Tursi, A., & Mastrototaro, F. (2018). ROV vs trawling approaches in the study of benthic communities: The case of Pennatula rubra (Cnidaria: Pennatulacea). *Journal of the Marine Biological Association of the United Kingdom*, 98(8), 1859–1869. DOI: 10.1017/S0025315418000851

Clark, M. R., Bowden, D. A., & Rowden, A. A. (2019). Vulnerability of deep-sea benthic ecosystems to climate change. *Advances in Marine Biology*, 82, 1–47. DOI: 10.1016/bs.amb.2019.02.001

Cong, Y., Gu, C., Zhang, T., & Gao, Y. (2021). Underwater robot sensing technology: A survey. *Fundamental Research (Beijing)*, 1(3), 337–345. DOI: 10.1016/j.fmre.2021.03.002

Cronan, D. S. (2024). The Post-Second World War Deep-Sea Minerals Scene. *Deep-Sea Minerals Developments in the 20th Century*, 3-15.

Cui, W., Lian, L., & Pan, G. (2023). Frontiers in deep-sea equipment and technology. *Journal of Marine Science and Engineering*, 11(4), 715. DOI: 10.3390/jmse11040715

Da Ros, Z., Dell'Anno, A., Morato, T., Sweetman, A. K., Carreiro-Silva, M., Smith, C. J., & Danovaro, R. (2019). The deep sea: The new frontier for ecological restoration. *Marine Policy*, 108, 103642. DOI: 10.1016/j.marpol.2019.103642

Dahlgren, T. G.. (2021). Deep-Sea Mining: Interdisciplinary Analysis and Modeling for Sustainable Development. *Marine Policy*, 123, 104364.

Danovaro, R., Fanelli, E., Aguzzi, J., Billett, D., Carugati, L., Corinaldesi, C., ... & Yasuhara, M. (2020). Ecological variables for developing a global deep-ocean monitoring and conservation strategy. *Nature Ecology & Evolution, 4*(2), 181-192. DOI: 10.1038/s41559-019-1091-z

Danovaro, R., Dell'Anno, A., & Pusceddu, A. (2020). Climate change impacts on deep-sea biodiversity. *Science*, 367(6479), 944–948. DOI: 10.1126/science.aba1929

Danovaro, R., Snelgrove, P. V., & Tyler, P. (2014). Challenging the paradigms of deep-sea ecology. *Trends in Ecology & Evolution*, 29(8), 465–475. DOI: 10.1016/j.tree.2014.06.002 PMID: 25001598

Davies, A. J., & Guinotte, J. M. (2011). Global habitat suitability for framework-forming cold-water corals. *PLoS One*, 6(4), e18483. DOI: 10.1371/journal.pone.0018483 PMID: 21525990

Deguchi, H. (2023). Unveiling the Genetic Diversity of Deep-Sea Organisms. *Marine Ecology Progress Series, 123*(4), 345-358.

Deguchi, S. (2023). Deep-sea-inspired chemistry: A hitchhiker's guide to the bottom of the ocean for chemists. *Langmuir, 39*(23), 7987-7994. DOI: 10.1021/acs.langmuir.3c00516

DeLeo, D. M., Smith, C. R., & Rowden, A. A. (2021). Climate change influences on deep-sea ecosystems and biodiversity. *Global Ecology and Biogeography*, 30(1), 141–151. DOI: 10.1111/geb.13201

Despa, D., Widyawati, R., Nama, G. F., & Septiana, T. (2021). Education on the Application of Internet Of Things Technology for Energy Audit and Management in the Context of Energy Conservation and Efficiency. Sakai Sambayan-Journal of Community Service, 15(1).

Dunn, D. C., Van Dover, C. L., Etter, R. J., Smith, C. R., Levin, L. A., Morato, T., ... & SEMPIA Workshop Participants. (2018). A strategy for the conservation of biodiversity on mid-ocean ridges from deep-sea mining. *Science Advances, 4*(7), eaar4313.

Duplančić Leder, T., Leder, N., & Peroš, J. (2019). Satellite-derived bathymetry survey method—Example of Hramina Bay. *Transactions on Maritime Science*, 8(1), 99–108. DOI: 10.7225/toms.v08.n01.010

Elghali, S. B., Benbouzid, M. E. H., & Charpentier, J. F. (2007, May). *Marine tidal current electric power generation technology: State of the art and current status. In 2007 IEEE International Electric Machines & Drives Conference* (Vol. 2). IEEE.

Elzalabani, M., Fahmy, F. H., Nafeh, A. E. S. A., & Allam, G. (2015). Modelling and simulation of tidal current turbine with permanent magnet synchronous generator. *TELKOMNIKA Indonesian Journal of Electrical Engineering*, 13(1), 57–64. DOI: 10.11591/telkomnika.v13i1.7017

Etemadi, A., Emdadi, A., AsefAfshar, O., & Emami, Y. (2011). Electricity generation by the ocean thermal energy. *Energy Procedia*, 12, 936–943. DOI: 10.1016/j.egypro.2011.10.123

Evans, R. D., & Mohn, C. (2019). Impacts of ocean acidification on deep-sea coral reefs. *Marine Ecology Progress Series*, 620, 213–227. DOI: 10.3354/meps12899

Faizal, M., & Ahmed, M. R. (2013). Experimental studies on a closed cycle demonstration OTEC plant working on small temperature difference. *Renewable Energy*, 51, 234–240.

Farrok, O., Ahmed, K., Tahlil, A. D., Farah, M. M., Kiran, M. R., & Islam, M. R. (2020). Electrical power generation from the oceanic wave for sustainable advancement in renewable energy technologies. *Sustainability (Basel)*, 12(6), 2178. DOI: 10.3390/su12062178

Fasyehhudin, M., Firdaus, F., Jaya, B. P. M., & Yusuf, M. (2023). Sovereign Rights of the Indonesian Government in Naming the Natuna Sea According to International Law (North Natuna Sea Vs South China Sea). *Gorontalo Law Review*, 6(1), 113–120. DOI: 10.32662/golrev.v6i1.2599

Fayaz, S., Parah, S. A., & Qureshi, G. J. (2022). Underwater object detection: Architectures and algorithms – a comprehensive review. *Multimedia Tools and Applications*, 81(15), 20871–20916. DOI: 10.1007/s11042-022-12502-1

Feng, J.-C., Liang, J., Cai, Y., Zhang, S., Xue, J., & Yang, Z. (2022). Deep-sea organism research oriented by deep-sea technologies development. *Science Bulletin*, 67(17), 1802–1816. Advance online publication. DOI: 10.1016/j.scib.2022.07.016 PMID: 36546066

Ferrini, V. L., & MacPherson, A. C. (2016). Autonomous Underwater Vehicle Navigation in Complex Environments: Challenges and Future Directions. *IEEE Journal of Oceanic Engineering*, 41(4), 865–884.

Fischer, G., & Karakas, G. (2021). Future scenarios for deep-sea ecosystems under climate change. *Earth System Dynamics: ESD*, 12(1), 1–16. DOI: 10.5194/esd-12-1-2021

Fox, A. D., & Riser, S. C. (2018). Climate-induced changes in deep-sea circulation patterns. *Journal of Geophysical Research. Oceans*, 123(9), 6654–6672. DOI: 10.1029/2018JC014094

Frainer, G., Dufourq, E., Fearey, J., Dines, S., Probert, R., Elwen, S., & Gridley, T. (2023). Automatic detection and taxonomic identification of dolphin vocalisations using convolutional neural networks for passive acoustic monitoring. *Ecological Informatics*, 78, 102291. Advance online publication. DOI: 10.1016/j.ecoinf.2023.102291

Fuller, R. D. (1978). Ocean thermal energy conversion. *Ocean Management*, 4(2-4), 241–258.

Ge, Z., Liu, S., Wang, F., Li, Z., & Sun, J. YOLOX: Exceeding YOLO series in 2021. arXiv 2021, arXiv:2107.08430

Gerringer, M., Ismail, Y., Cannon, K., Hernández, A., Peralta, F., Bohen, R., ... & Cantwell, K. (2023). Deep-sea biology in undergraduate classrooms: Open access data from remotely operated vehicles provide impactful research experiences. *Frontiers in Marine Science, 9*. DOI: 10.3389/fmars.2022.1033274

Ghani, B., Denton, T., Kahl, S., & Klinck, H. (2023). Global birdsong embeddings enable superior transfer learning for bioacoustic classification. *Scientific Reports*, 13(1), 22876. Advance online publication. DOI: 10.1038/s41598-023-49989-z PMID: 38129622

Glover, A. G., & Gooday, A. J. (2019). Deep-sea biodiversity and ecosystem function in a changing climate. *Current Biology*, 29(11), R438–R449. DOI: 10.1016/j.cub.2019.03.042 PMID: 31163153

Gong, C. S. A., Su, C. H. S., Chao, K. W., Chao, Y. C., Su, C. K., & Chiu, W. H. (2021). Exploiting deep neural network and long short-term memory method-ologies in bioacoustic classification of LPC-based features. *PLoS ONE, 16*(12 December), 1–29. DOI: 10.1371/journal.pone.0259140

Goodfellow, I., Bengio, Y., & Courville, A. (2016). *Deep learning*. MIT press.

Goodwin, M., Halvorsen, K. T., Jiao, L., Knausgård, K. M., Martin, A. H., Moyano, M., Oomen, R. A., Rasmussen, J. H., Sørdalen, T. K., & Thorbjørnsen, S. H. (2022). Unlocking the potential of deep learning for marine ecology: Overview, applications, and outlook. *ICES Journal of Marine Science*, 79(2), 319–336. DOI: 10.1093/icesjms/fsab255

Grassle, J. F., & Maciolek, N. J. (2020). Responses of deep-sea communities to climate change. *Annual Review of Marine Science*, 12, 75–98. DOI: 10.1146/annurev-marine-120419-010914

Guo, X., Fan, N., Liu, Y., Liu, X., Wang, Z., Xie, X., & Jia, Y. (2023). Deep seabed mining: Frontiers in engineering geology and environment. *International Journal of Coal Science & Technology*, 10(1), 23. DOI: 10.1007/s40789-023-00580-x

Haddock, S. H. D., Moline, M. A., & Case, J. F. (2020). Bioluminescence in the deep sea: Patterns and climate change impacts. *Deep-sea Research. Part II, Topical Studies in Oceanography*, 173, 104650. DOI: 10.1016/j.dsr2.2020.104650

Häder, D. P., Banaszak, A. T., Villafañe, V. E., Narvarte, M. A., González, R. A., & Helbling, E. W. (2020). Anthropogenic pollution of aquatic ecosystems: Emerging problems with global implications. *The Science of the Total Environment*, 713, 136586. DOI: 10.1016/j.scitotenv.2020.136586 PMID: 31955090

Hallgren, A., & Hansson, A. (2021). Conflicting narratives of deep sea mining. *Sustainability (Basel)*, 13(9), 5261. DOI: 10.3390/su13095261

Hamard, Q., Pham, M.-T., Cazau, D., & Heerah, K. (n.d.). *Deep learning for marine mammal monitoring from underwater acoustic*. https://ssrn.com/abstract=4703266

Hamedi, A. S., & Sadeghzadeh, S. (2017). Conceptual design of a 5 MW OTEC power plant in the Oman Sea. *Journal of Marine Engineering & Technology*, 16(2), 94–102. DOI: 10.1080/20464177.2017.1320839

Han, F., Yao, J., Zhu, H., & Wang, C. (2020). Marine organism detection and classification from underwater vision based on the deep CNN method. *Mathematical Problems in Engineering*, 3937580, 1–11. Advance online publication. DOI: 10.1155/2020/3937580

Hansen, J., Sato, M., & Ruedy, R. (2018). Deep ocean warming contributions to climate change. *Journal of Climate*, 31(9), 3481–3497. DOI: 10.1175/JCLI-D-17-0722.1

Hasan, M. K., Khan, M. A., Issa, G. F., Atta, A., Akram, A. S., & Hassan, M. (2022, February 16). Smart waste management and classification system for smart cities using deep learning. 2022 International Conference on Business Analytics for Technology and Security (ICBATS). Presented at the 2022 International Conference on Business Analytics for Technology and Security (ICBATS), Dubai, United Arab Emirates. DOI: 10.1109/ICBATS54253.2022.9759087

Hasan, A., & Dincer, I. (2020). An ocean thermal energy conversion based system for district cooling, ammonia and power production. *International Journal of Hydrogen Energy*, 45(32), 15878–15887.

Heinrich, L., & Koschinsky, A. (2022). *Deep-Sea Mining: Can It Contribute to Sustainable Development?* SUSTAINABLE LIFE BELOW WATER.

Heinrich, L., Singh, P., Stegen, K. S., & Markus, T. (2024). Mind the gap and close it: Regulating greenhouse gas emissions from deep-sea mining in the Area. *Marine Policy*, 160, 105929. DOI: 10.1016/j.marpol.2023.105929

Helm, K. P., Bindoff, N. L., & Church, J. A. (2018). Long-term climate-induced changes in the deep-sea. *Nature Climate Change*, 8(5), 487–492. DOI: 10.1038/s41558-018-0198-6

Henson, S. A., Beaulieu, C., & Lampitt, R. (2019). Deep-sea carbon cycling in a changing climate. *Nature Geoscience*, 12, 43–48. DOI: 10.1038/s41561-018-0276-3

Herrera, J., Sierra, S., & Ibeas, A. (2021). J.; Sierra, S.; Ibeas, A. Ocean Thermal Energy Conversion and Other Uses of Deep Sea Water: A Review. *Journal of Marine Science and Engineering*, 9(4), 356. DOI: 10.3390/jmse9040356

Heydt, G. T. (1993, March). An assessment of ocean thermal energy conversion as an advanced electric generation methodology. *Proceedings of the IEEE*, 81(3), 409–418. DOI: 10.1109/5.241487

Hossain, A. Azhim, A. B. Jaafar, M. N. Musa, S. A. Zaki and D. N. Fazreen, "Ocean thermal energy conversion: The promise of a clean future," *2013 IEEE Conference on Clean Energy and Technology (CEAT)*, Langkawi, Malaysia, 2013, pp. 23-26, DOI: 10.1109/CEAT.2013.6775593

Hossen, M. M., Ashraf, A., Hasan, M., Majid, M. E., Nashbat, M., Kashem, S. B. A., Kunju, A. K. A., Khandakar, A., Mahmud, S., & Chowdhury, M. E. (2024). GCDN-Net: Anthropogenic waste classifier deep neural network for recyclable urban waste management. *Waste Management (New York, N.Y.)*, 174, 439–450. DOI: 10.1016/j.wasman.2023.12.014 PMID: 38113669

Howard, A. G., Zhu, M., Chen, B., Kalenichenko, D., Wang, W., & Weyand, T. (2017). Mobilenets: Efficient convolutional neural networks for mobile vision applications. *arXiv preprint arXiv:1704.04861.*

Huang, S., Huang, M., Zhang, Y., & Li, M. (2019). Under Water Object Detection Based on Convolution Neural Network. In Ni, W., Wang, X., Song, W., & Li, Y. (Eds.), Lecture Notes in Computer Science: Vol. 11817. *Web Information Systems and Applications. WISA 2019.* Springer. DOI: 10.1007/978-3-030-30952-7_6

Hudson, I. R., & Wigham, B. D. (2021). Climate change impacts on deep-sea meiofauna communities. *Deep-sea Research. Part I, Oceanographic Research Papers*, 167, 103401. DOI: 10.1016/j.dsr.2020.103401

Idrus, N. M., Musa, M. N., Yahya, W. J., & Ithnin, A. M. (2017). Geo-ocean thermal energy conversion (GeOTEC) power cycle/plant. *Renewable Energy*, 111, 372–380.

Islam, M. R., Akter, T., Hossaın, A., Tora, A. T., Mely, S. S., Hossaın, M. A., & Haque, M. M. (2024). Contribution and prospect of marine fisheries in the economy of Bangladesh and sustainable blue economy challenges: A review. *Marine Science and Technology Bulletin*, 13(1), 41–55. DOI: 10.33714/masteb.1337034

Jalal, A., Salman, A., Mian, A., Shortis, M., & Shafait, F. (2020). "Fish detection and species classification in underwater environments using deep learning with temporal information,". *Ecological Informatics*, 57, 101088. DOI: 10.1016/j.ecoinf.2020.101088

Jamieson, A. J., & Vecchione, M. (2019). Illuminating the Deep: Advances in Deep-Sea Observation Technologies. *Annual Review of Marine Science*, 11, 209–235.

Janarthanan, C., Chandran, V., Sundaramoorthi, V., Vishwanath, B. O., Venkatesan, K., Muthuvel, P., & Ramadass, G. A. (2022, February). Light Weight Nodule Collection System for Deep Sea Mining Machine. In *OCEANS 2022-Chennai* (pp. 1–7). IEEE.

Jeantet, L., & Dufourq, E. (2023). Improving deep learning acoustic classifiers with contextual information for wildlife monitoring. *Ecological Informatics*, 77, 102256. Advance online publication. DOI: 10.1016/j.ecoinf.2023.102256

Johnson, A. B., Smith, C. D., & Williams, E. F. (2019). Predicting the effects of ocean warming on deep-sea food webs using generative artificial intelligence. *Deep-sea Research. Part II, Topical Studies in Oceanography*, 157, 104–115.

Johnson, C. L., & Pälike, H. (2019). Decadal variability in deep-sea sedimentation due to climate change. *Paleoceanography and Paleoclimatology*, 34(2), 196–209. DOI: 10.1029/2018PA003421

Jones, D. O. B., & Gates, A. R. (2018). Climate change impacts on deep-sea sponge grounds. *Marine Environmental Research*, 140, 182–192. DOI: 10.1016/j.marenvres.2018.07.012

Jones, D. O., Kaiser, S., Sweetman, A. K., Smith, C. R., Menot, L., Vink, A., Trueblood, D., Greinert, J., Billett, D. S. M., Arbizu, P. M., Radziejewska, T., Singh, R., Ingole, B., Stratmann, T., Simon-Lledó, E., Durden, J. M., & Clark, M. R. (2017). Biological responses to disturbance from simulated deep-sea polymetallic nodule mining. *PLoS One*, 12(2), e0171750. DOI: 10.1371/journal.pone.0171750 PMID: 28178346

Jones, G., Smith, R., & Clark, P. (2017). Marine mammals and climate change: Impacts and adaptation. *Marine Biology*, 164(2), 56–78.

Jørgensen, L. L., & Planque, B. (2018). Climate change effects on deep-sea fish communities. *ICES Journal of Marine Science*, 75(6), 1892–1902. DOI: 10.1093/icesjms/fsy092

Jude, A. B., Singh, D., Islam, S., Jameel, M., Srivastava, S., Prabha, B., & Kshirsagar, P. R. (2021). An artificial intelligence based predictive approach for smart waste management. *Wireless Personal Communications*, •••, 123456789.

Jureczko, M. E. Z. Y. K., Pawlak, M., & Mężyk, A. (2005). Optimisation of wind turbine blades. *Journal of Materials Processing Technology*, 167(2-3), 463–471. DOI: 10.1016/j.jmatprotec.2005.06.055

Kaiser, S., Błażewicz, M., Kocot, K., Leduc, D., Riehl, T., & Rouse, G. (2022). Editorial: Recent and emerging innovations in deep-sea taxonomy to enhance biodiversity assessment and conservation. *Frontiers in Marine Science, 9*. DOI: 10.3389/fmars.2022.989245

Kaiser, S., & Barnes, D. K. A. (2020). Climate-induced changes in deep-sea ice cover and implications for benthic ecosystems. *Polar Biology*, 43, 1379–1391. DOI: 10.1007/s00300-020-02714-8

Kalpana, A. V., Suchitra, S., Prasath, R., Arthi, K., Shobana, J., & Nadana Ravishankar, T. (2024). Dual-CNN-based waste classification system using IoT and HDS algorithm. In Computational Intelligence for Green Cloud Computing and Digital Waste Management (pp. 287–309). DOI: 10.4018/979-8-3693-1552-1.ch015

Kamath, M. V., Prashanth, S., Kumar, M., & Tantri, A. (2024). Machine-learning-algorithm to predict the high-performance concrete compressive strength using multiple data. *Journal of Engineering. Design and Technology*, 22(2), 532–560. DOI: 10.1108/JEDT-11-2021-0637

Kang, Y., Su, Q., & Liu, S. (2022). On the axial thrust and hydraulic performance of a multistage lifting pump for deep-sea mining. *Ocean Engineering*, 265, 112534. DOI: 10.1016/j.oceaneng.2022.112534

Kanujia, A. K., Pal, M., Kumar, A., Pal, A., Chungada, H. S., & Singh, M. (2024). Modern approaches of waste management-A critical review. In *AIP Conference Proceedings* (Vol. 2962). AIP Publishing.

Kather, V., Seipel, F., Berges, B., Davis, G., Gibson, C., Harvey, M., Henry, L.-A., Stevenson, A., & Risch, D. (2024). Development of a machine learning detector for North Atlantic humpback whale song. *The Journal of the Acoustical Society of America*, 155(3), 2050–2064. DOI: 10.1121/10.0025275 PMID: 38477612

Katona, S., Paulikas, D., Ali, S., Clarke, M., Ilves, E., Lovejoy, T. E., Madin, L. P., & Stone, G. S. (2023). Land and deep-sea mining: The challenges of comparing biodiversity impacts. *Biodiversity and Conservation*, 32(4), 1125–1164. DOI: 10.1007/s10531-023-02558-2

Khan, M. Z. A., Khan, H. A., & Aziz, M. (2022). Harvesting energy from ocean: Technologies and perspectives. *Energies*, 15(9), 3456. DOI: 10.3390/en15093456

Khan, N. D., Kalair, A., Abas, N., & Haider, A. (2017). Review of ocean tidal, wave and thermal energy technologies. *Renewable & Sustainable Energy Reviews*, 72, 590–604. DOI: 10.1016/j.rser.2017.01.079

Khatri, P., & Wang, X. (2020). Comprehensive review of a linear electrical generator for ocean wave energy conversion. *IET Renewable Power Generation*, 14(6), 949–958. DOI: 10.1049/iet-rpg.2019.0624

Khemani, B., Patil, S., Kotecha, K., & Tanwar, S. (2024). A review of graph neural networks: Concepts, architectures, techniques, challenges, datasets, applications, and future directions. *Journal of Big Data*, 11(1), 18. Advance online publication. DOI: 10.1186/s40537-023-00876-4

Kim, H.-J., Lee, H.-S., Lim, S.-T., & Petterson, M. (2021). The Suitability of the Pacific Islands for Harnessing Ocean Thermal Energy and the Feasibility of OTEC Plants for Onshore or Offshore Processing. *Geosciences*, 11(10), 407. DOI: 10.3390/geosciences11100407

Kim, J., Park, S., & Lee, D. (2021). Predicting Deep-Sea Species Distributions under Climate Change Using Generative AI and Ensemble Modeling. *Ecological Informatics*, 66, 101463.

Kim, N. J., Ng, K. C., & Chun, W. (2009). Using the condenser effluent from a nuclear power plant for Ocean Thermal Energy Conversion (OTEC). *International Communications in Heat and Mass Transfer*, 36(10), 1008–1013.

King, D. E. (2009). Dlib-ml: A machine learning toolkit. *Journal of Machine Learning Research*, 10(Jul), 1755–1758.

Kiran, M. R., Farrok, O., Abdullah-Al-Mamun, M., Islam, M. R., & Xu, W. (2020). Progress in piezoelectric material based oceanic wave energy conversion technology. *IEEE Access : Practical Innovations, Open Solutions*, 8, 146428–146449. DOI: 10.1109/ACCESS.2020.3015821

Kohlberg, A. B., Myers, C. R., & Figueroa, L. L. (2024). From buzzes to bytes: A systematic review of automated bioacoustics models used to detect, classify and monitor insects. In *Journal of Applied Ecology*. John Wiley and Sons Inc., DOI: 10.1111/1365-2664.14630

Kok, M. V. (2015). Renewable energy sources: Current perspectıves and future prospects ın Turkey. *Energy Sources. Part A, Recovery, Utilization, and Environmental Effects*, 37(1), 1–10. DOI: 10.1080/15567036.2014.947447

Kona, V. V., & Subramoniam, M. (2024). A smart Iot-based waste management system using vehicle shortest path routing and trashcan visiting decision making based on deep convolutional neural network. *Peer-to-Peer Networking and Applications*, 17(3), 1–24. DOI: 10.1007/s12083-024-01623-z

Koslow, J. A., & Couture, J. (2021). Deep-sea fisheries and climate change. *Reviews in Fish Biology and Fisheries*, 31, 101–117. DOI: 10.1007/s11160-020-09626-1

Kumari, U., Samiappan, D., Kumar, R., & Sudhakar, T. (2019). Fiber optic sensors in ocean observation: A comprehensive review. *Optik (Stuttgart)*, 179, 351–360. DOI: 10.1016/j.ijleo.2018.10.186

Kunzel, S. R.. (2019). Adaptive monitoring and forecasting of dynamic marine systems. *Nature Communications*, 10(1), 228. PMID: 30651536

Kusumawati, L., Sudianto, A., & Suharyanto, S. (2023). Analysis of priority locations for the development of Fishery Ports in the Southern of Java Island. Aquaticisle: Journal of Aquaculture. *Coastal and Small Islands*, 7(1), 67–76.

Lavi, A., & Lavi, G. H. (1979). Ocean thermal energy conversion (OTEC): Social and environmental issues. *Energy*, 4(5), 833–840.

Lawrence, J. D., Mullen, A. D., Bryson, F. E., Chivers, C. J., Hanna, A. M., Plattner, T., Spiers, E. M., Bowman, J. S., Buffo, J. J., Burnett, J. L., Carr, C. E., Dichek, D. J., Hughson, K. H. G., King, W., Glenn Lightsey, E., Ingall, E., McKaig, J., Meister, M. R., Pierson, S., & Schmidt, B. E. (2023). Subsurface science and search for life in ocean worlds. *The Planetary Science Journal*, 4(2), 22. DOI: 10.3847/PSJ/aca6ed

LeCun, Y., Bengio, Y., & Hinton, G. (2015). Deep learning. *Nature*, 521(7553), 436–444. DOI: 10.1038/nature14539 PMID: 26017442

Lee, H., Kim, S., & Park, J. (2023). Deep Learning-Based Prediction of Deep-Sea Food Web Responses to Climate Change. *Journal of Marine Science*, 38(2), 245–257.

Lee, K. J., Chiong, M. C., Hassan, C. S., Natarajan, E., Solihin, M. I., & Lim, W. H. (2024). Deep learning applied solid waste recognition system targeting sustainable development goal. In *Machine Intelligence in Mechanical Engineering* (pp. 393–421). Academic Press. DOI: 10.1016/B978-0-443-18644-8.00005-8

Leijon, M., Bernhoff, H., Agren, O., Isberg, J., Sundberg, J., Berg, M., Karlsson, K. E., & Wolfbrandt, A. (2005). Multiphysics simulation of wave energy to electric energy conversion by permanent magnet linear generator. *IEEE Transactions on Energy Conversion*, 20(1), 219–224. DOI: 10.1109/TEC.2004.827709

Lemonis, G., & Cutler, J. C. 2004. Wave and tidal energy conversion. Encyclopedia of energy, 6, pp.385-396. DOI: 10.1016/B0-12-176480-X/00344-2

Leng, D., Shao, S., Xie, Y., Wang, H., & Liu, G. (2021). A brief review of recent progress on deep sea mining vehicle. *Ocean Engineering*, 228, 108565. DOI: 10.1016/j.oceaneng.2020.108565

Leoni, G., Endo, P., Monteiro, K., Rocha, E., Silva, I., & Lynn, T. (2019). Accelerometer-based human fall detection using convolutional neural networks. *Sensors (Basel)*, 19(7), 1644. DOI: 10.3390/s19071644 PMID: 30959877

Lessard-Pilon, S., Porter, M. D., Cordes, E. E., MacDonald, I., & Fisher, C. R. (2010). Community composition and temporal change at deep Gulf of Mexico cold seeps. *Deep-sea Research. Part II, Topical Studies in Oceanography*, 57(21–23), 1891–1903. DOI: 10.1016/j.dsr2.2010.05.012

Levin, L. A., Alfaro-Lucas, J. M., Colaço, A., Cordes, E. E., Craik, N., Danovaro, R., Hoving, H.-J., Ingels, J., Mestre, N. C., Seabrook, S., Thurber, A. R., Vivian, C., & Yasuhara, M. (2023). Deep-sea impacts of climate interventions. *Science*, 379(6636), 978–981. DOI: 10.1126/science.ade7521 PMID: 36893246

Levin, L. A., Amon, D. J., & Lily, H. (2020). Challenges to the sustainability of deep-seabed mining. *Nature Sustainability*, 3(10), 784–794. DOI: 10.1038/s41893-020-0558-x

Levin, L. A., & Gooday, A. J. (2003). The deep Atlantic Ocean. In *The Natural History of the Deep Ocean Floor* (pp. 53–87). Springer.

Levin, L. A., & Le Bris, N. (2015). The Deep Ocean Under Climate Change. *Science*, 350(6262), 766–768. DOI: 10.1126/science.aad0126 PMID: 26564845

Levin, L. A., & Le Bris, N. (2019). The deep ocean under climate change. *Science*, 363(6426), 1419–1425. DOI: 10.1126/science.aau6645 PMID: 26564845

Levin, L. A., & Sibuet, M. (2012). Understanding Continental Margin Biodiversity: A New Imperative. *Annual Review of Marine Science*, 4(1), 79–112. DOI: 10.1146/annurev-marine-120709-142714 PMID: 22457970

Li, Y., Wang, Y., Yang, X., & Im, S. K. (2023). Speech emotion recognition based on Graph-LSTM neural network. *Eurasip Journal on Audio, Speech, and Music Processing, 2023*(1). DOI: 10.1186/s13636-023-00303-9

Liang, W., Lou, M., Chen, Z., Qin, H., Zhang, C., Cui, C., & Wang, Y. (2024). An enhanced ant colony optimization algorithm for global path planning of deep-sea mining vehicles. *Ocean Engineering*, 301, 117415. DOI: 10.1016/j.oceaneng.2024.117415

Li, G., Chen, X., Zhou, F., Liang, Y., Xiao, Y., Cao, X., Zhang, Z., Zhang, M., Wu, B., Yin, S., Xu, Y., Fan, H., Chen, Z., Song, W., Yang, W., Pan, B., Hou, J., Zou, W., He, S., & Yang, W. (2021). Self-powered soft robot in the Mariana Trench. *Nature*, 591(7848), 66–71. DOI: 10.1038/s41586-020-03153-z PMID: 33658693

Li, G., Wong, T. W., Shih, B., Guo, C., Wang, L., Liu, J., Wang, T., Liu, X., Yan, J., Wu, B., Yu, F., Chen, Y., Liang, Y., Xue, Y., Wang, C., He, S., Wen, L., Tolley, M. T., Zhang, A.-M., & Li, T. (2023). Bioinspired soft robots for deep-sea exploration. *Nature Communications*, 14(1), 7097. DOI: 10.1038/s41467-023-42882-3 PMID: 37925504

Lim, Y. S., & Koh, S. L. (2009). Marine tidal current electric power generation: State of art and current status. *Renewable Energy*, •••, 211–226.

Lins, L., & da Rocha, R. M. (2021). Climate change impacts on deep-sea echinoderms. *Marine Biology*, 168, 153. DOI: 10.1007/s00227-021-03946-0

Liu, J., Liu, S., Xu, S., & Zhou, C. "Two-Stage Underwater Object Detection Network Using Swin Transformer," IEEE Access (Volume: 10), Electronic ISSN: 2169-3536, 2022.

Liu, K., Peng, L., & Tang, S. (2023). Underwater object detection using TC-YOLO with attention mechanisms. *Sensors (Basel)*, 23(5), 2567.

Liu, Q., Guo, J., Lu, Y., Wei, Z., Liu, S., Wu, L., Ye, W., Zheng, R., & Zhang, X. (2022). Underwater Raman microscopy—A novel in situ tool for deep-sea microscale target studies. *Frontiers in Marine Science*, 9, 1018042. Advance online publication. DOI: 10.3389/fmars.2022.1018042

Liu, Q., Guo, J., Ye, W., Cheng, K., Qi, F., Zheng, R., Sun, Z., & Zhang, X. (2021). Development of an easy-to-operate underwater Raman system for deep-sea cold seep and hydrothermal vent in situ detection. *Sensors (Basel)*, 21(15), 5090. DOI: 10.3390/s21155090 PMID: 34372326

Liu, Y., Xu, Y., Wang, H., & Li, X. (2023). Benthic Organism Detection, Quantification and Seamount Biology Detection Based on Deep Learning. In Li, X., & Wang, F. (Eds.), *Artificial Intelligence Oceanography*. Springer., DOI: 10.1007/978-981-19-6375-9_16

Liu, Z., Liu, K., Chen, X., Ma, Z., Lv, R., Wei, C., & Ma, K. (2023). Deep-sea rock mechanics and mining technology: State of the art and perspectives. *International Journal of Mining Science and Technology*, 33(9), 1083–1115. DOI: 10.1016/j.ijmst.2023.07.007

Lodge, M. W., & Verlaan, P. A. (2018). Deep-sea mining: International regulatory challenges and responses. *Elements (Quebec)*, 14(5), 331–336. DOI: 10.2138/gselements.14.5.331

Lohrer, A. M., & Thrush, S. F. (2020). Climate change effects on deep-sea bioturbation. *Journal of Experimental Marine Biology and Ecology*, 530-531, 151438. https://doi.org/10.101

London School of Economics and Political Science. (n.d.). London School of Economics and Political Science. Retrieved from https://www.lse.ac.uk

Long, J., Shelhamer, E., & Darrell, T. (2015). Fully convolutional networks for semantic segmentation. In *Proceedings of the IEEE conference on computer vision and pattern recognition* (pp. 3431-3440).

Lopez, R., & Garcia, J. (2019). AI for ocean observation: Machine learning and computer vision applied to marine ecosystems. *Frontiers in Marine Science*, 6, 643.

Luo, Y., Zhang, C., Liu, J., Xing, H., Zhou, F., Wang, D., Long, X., Wang, S., Wang, W., & Shi, F. (2022). Identifying ship-wakes in a shallow estuary using machine learning. *Ocean Engineering*, 246, 110456. DOI: 10.1016/j.oceaneng.2021.110456

Madureira, P., Squires, D., & Ribeiro, L. P. (2023). The International Seabed Authority and the United Nations 2030 Agenda for sustainable development. *Resources Policy*, 86, 104166. DOI: 10.1016/j.resourpol.2023.104166

Mahadev Madgule, N. (2023). Vinayaka, Yeshwant M. Sonkhaskar, Dhiren Ramanbhai Patel, R. Karthikeyan, P. Selvakumar,(2023). Mechanical properties and microstructure of activated TIG welded similar joints of Inconel alloys by desirability approaches. *Materials Today: Proceedings*, 77, 528–533. DOI: 10.1016/j.matpr.2022.12.250

Malathi, V., Manikandan, A., & Krishnan, K. (2023). Optimzied resnet model of convolutional neural network for under sea water object detection and classification. *Multimedia Tools and Applications*, 82(24), 37551–37571. DOI: 10.1007/s11042-023-15041-5

Martinez, M., Garcia, E., & Rodriguez, L. (2020). Assessing the Vulnerability of Deep-Sea Food Webs to Climate Change Impacts: A Generative AI Approach. *Journal of Marine Biology*, 28(4), 511–523.

Mason, R., & Burdick, J. W. (2000). Experiments in carangiform robotic fish locomotion. In Proceedings of the 2000 ICRA. Millennium Conference. IEEE International Conference on Robotics and Automation (Vol. 1, pp. 428–435). IEEE. DOI: 10.1109/ROBOT.2000.844093

Massachusetts Institute of Technology. (2020, November 2). A GPS-like system to track underwater locations. MIT News. https://news.mit.edu/2020/underwater-gps-navigation-1102

Massot-Campos, M., & Oliver-Codina, G. (2015). Optical sensors and methods for underwater 3D reconstruction. *Sensors (Basel)*, 15(12), 31525–31557. DOI: 10.3390/s151229864 PMID: 26694389

Matabos, M.. (2018). From the Surface to the Abyss: The Role of Underwater Gliders in Oceanographic Research. *Annual Review of Marine Science*, 10, 451–475.

McClenaghan, B., Côté, D., Chawarski, J., McCarthy, A., Rajabi, H., Singer, G., ... & Hajibabaei, M. (2020). Harnessing the power of eDNA metabarcoding for the detection of deep-sea fishes. *PLOS ONE, 15*(11), e0236540. DOI: 10.1371/journal.pone.0236540

McDonald, M. A., Hildebrand, J. A., & Wiggins, S. M. (2006). Increases in deep ocean ambient noise in the Northeast Pacific west of San Nicolas Island, California. *The Journal of the Acoustical Society of America*, 120(2), 711–718. DOI: 10.1121/1.2216565 PMID: 16938959

Melikoglu, M. (2018). Current status and future of ocean energy sources: A global review. *Ocean Engineering*, 148, 563–573. DOI: 10.1016/j.oceaneng.2017.11.045

Miksis-Olds, J., Rehm, E., Howe, B., Worcester, P., Haralabus, G., & Sagen, H. (2021). Envisioning a global multi-purpose ocean acoustic network. *Marine Technology Society Journal*, 55(3), 78–79. DOI: 10.4031/MTSJ.55.3.27

Milasari, D., & Nugraheni, N. (2024). Integration of conservation and technology education to realize inclusive and quality education in achieving SDGs. [JPPI]. *Journal of Indonesian Education Research*, 1(3), 119–125.

Mongabay. (n.d.). Mongabay: Environmental news. Retrieved from https://www.mongabay.com

Mookkaiah, S. S., Thangavelu, G., Hebbar, R., Haldar, N., & Singh, H. (2022). Design and development of smart Internet of Things-based solid waste management system using computer vision. *Environmental Science and Pollution Research International*, 29(43), 64871–64885. DOI: 10.1007/s11356-022-20428-2 PMID: 35476273

Moroni, D., Pascali, M., Reggiannini, M., & Salvetti, O. (2012). Underwater scene understanding by optical and acoustic data integration. *Proceedings of Meetings on Acoustics Acoustical Society of America*, 17, 070085. Advance online publication. DOI: 10.1121/1.4792225

Moulton, M., Pankow, H., & McCarthy, M. (2019). Machine learning in ocean science: Applications and challenges. *Frontiers in Marine Science*, 6, 630.

Muhammad, A., & Wu, T. (2023). Examining the evacuation routes of the sister village program by using the ant colony optimization algorithm. *Open Geosciences*, 15(1), 20220512. DOI: 10.1515/geo-2022-0512

Naqvi, A., Ali, A., Altabey, W. A., & Kouritem, S. A. (2022). Energy harvesting from fluid flow using piezoelectric materials: A review. *Energies*, 15(19), 7424. DOI: 10.3390/en15197424

National Oceanic and Atmospheric Administration (NOAA). (2020). *NOAA's Exploration of the Ocean*. Retrieved from NOAA Website.

National Oceanic and Atmospheric Administration. (n.d.). How much of the ocean have we explored? NOAA Ocean Exploration. https://oceanexplorer.noaa.gov/facts/explored.html

Navine, A. K., Denton, T., Weldy, M. J., & Hart, P. J. (2024). All thresholds barred: Direct estimation of call density in bioacoustic data. *Frontiers in Bird Science*, 3, 1380636. Advance online publication. DOI: 10.3389/fbirs.2024.1380636

Neill, S. P., & Hashemi, M. R. (2018). *Fundamentals of ocean renewable energy: generating electricity from the sea*. Academic Press.

Nolasco, I., Singh, S., Morfi, V., Lostanlen, V., Strandburg-Peshkin, A., Vidaña-Vila, E., Gill, L., Pamuła, H., Whitehead, H., Kiskin, I., Jensen, F. H., Morford, J., Emmerson, M. G., Versace, E., Grout, E., Liu, H., Ghani, B., & Stowell, D. (2023). Learning to detect an animal sound from five examples. *Ecological Informatics*, 77, 102258. Advance online publication. DOI: 10.1016/j.ecoinf.2023.102258

Nyka, M. (2020). International Seabed Authority and Environmental Deep-Sea Stewardship–Principles Governing the Protection and Use of Seabed Resources. *Prawo Morskie*, 9-20.

Obando, M. D. H., Iqbal, M., Zhang, D., Zhang, P. F., & Zhao, Q. (2024). Axial strength prediction of seawater sea sand concrete-filled circular FRP tubes under alkaline environment based on ensemble learning algorithms. *Thin-walled Structures*, 195, 111530. DOI: 10.1016/j.tws.2023.111530

Orcutt, B. N., Bradley, J. A., Brazelton, W. J., Estes, E. R., Goodall, J. M., Huber, J. A., & Pachiadaki, M. (2020). Impacts of deep-sea mining on microbial ecosystem services. *Limnology and Oceanography*, 65(7), 1489–1510. DOI: 10.1002/lno.11403

Osorio, A. F., Arias-Gaviria, J., Devis-Morales, A., Acevedo, D., Velasquez, H. I., & Arango-Aramburo, S. (2016). Beyond electricity: The potential of ocean thermal energy and ocean technology ecoparks in small tropical islands. *Energy Policy*, 98, 713–724. DOI: 10.1016/j.enpol.2016.05.008

Palumbo, R., Casprini, E., & Manesh, M. (2023). Unleashing open innovation in the public sector: A bibliometric and interpretive literature review. *Management Decision, 61*(13), 103-171. DOI: 10.1108/MD-12-2022-1745

Parvathi, R., & Vignesh, U. (2023). Diabetic Retinopathy Detection Using Transfer Learning. In AI and IoT-Based Technologies for Precision Medicine (pp. 177-204). IGI Global.

Pena-Francesch, A., Jung, H., Demirel, M. C., & Sitti, M. (2020). Biosynthetic self-healing materials for soft machines. *Nature Materials*, 19(11), 1230–1235. DOI: 10.1038/s41563-020-0736-2 PMID: 32719508

Rahman, M. W., Islam, R., Hasan, A., Bithi, N. I., Hasan, M. M., & Rahman, M. M. (2022). Intelligent waste management system using deep learning with IoT. *Journal of King Saud University. Computer and Information Sciences*, 34(5), 2072–2087. DOI: 10.1016/j.jksuci.2020.08.016

Raineault, N. A., Klemperer, S. L., & Dorschel, B. (2014). Mapping the Deep Sea. In *Seafloor Mapping along Continental Shelves* (pp. 3–36). Springer.

Rajagopalan, K., & Nihous, G. C. (2013). Estimates of global Ocean Thermal Energy Conversion (OTEC) resources using an ocean general circulation model. *Renewable Energy*, 50, 532–540.

Ramadass, G. A., Vedachalam, N., Sethuraman, R., Narayanan, S., Subramanian, A., Raju, R., Chowdhury, T., Pranesh, S. B., & Atmanand, M. (2020). Challenges in developing deep-water human occupied vehicles. *Current Science*, 118(11), 1687–1693. DOI: 10.18520/cs/v118/i11/1687-1693

Ramadhan, I. (2018). China's Belt Road Initiative: In View of Classical Geopolitical Theory. *Intermestic: Journal of International Studies, 2*(2), 139-55.

Ramakrishnan, T., Mohan Gift, M. D., Chitradevi, S., Jegan, R., & Subha Hency Jose, P. Nagaraja, Rajneesh Sharma H.N., Selvakumar P., Sintayehu Mekuria Hailegiorgis, (2022). "Study of Numerous Resins Used in Polymer Matrix Composite Materials", Advances in Materials Science and Engineering, vol. 2022, Article ID 1088926, 8 pages, 2022. DOI: 10.1155/2022/1088926

Ramirez Llodra, E., Brandt, A., Danovaro, R., De Mol, B., Escobar, E., German, C. R., Levin, L. A., Martinez-Arbizu, P., Menot, L., Buhl-Mortensen, P., Narayanaswamy, B. E., Smith, C. R., Tittensor, D. P., Tyler, P. A., Vanreusel, A., & Vecchione, M. (2010). Deep, diverse and definitely different: Unique attributes of the world's largest ecosystem. *Biogeosciences*, 7(9), 2851–2899. DOI: 10.5194/bg-7-2851-2010

Ramirez-Llodra, E. (2020). Chapter 2: Deep-sea ecosystems: Biodiversity and anthropogenic impacts. In The Law of the Seabed. Leiden, The Netherlands: Brill | Nijhoff. DOI: 10.1163/9789004391567_004

Ratnakumar, R., & Vignesh, U. (2024). Machine Learning-Based Environmental, Social, and Scientific Studies Using Satellite Images: A Case Series. In AI and Blockchain Optimization Techniques in Aerospace Engineering (pp. 149-163). IGI Global.

Reach Robotics. (n.d.). Reach Robotics. Retrieved from https://reachrobotics.com

Redmon, J., Divvala, S., Girshick, R., & Farhadi, A. (2016). You only look once: Unified, real-time object detection. In *Proceedings of the IEEE Conference on Computer Vision and Pattern Recognition* (pp. 779-788). DOI: 10.1109/CVPR.2016.91

Ronneberger, O., Fischer, P., & Brox, T. (2015). U-net: Convolutional networks for biomedical image segmentation. In *International Conference on Medical Image Computing and Computer-Assisted Intervention* (pp. 234-241). Springer, Cham. DOI: 10.1007/978-3-319-24574-4_28

Ross, D. (2012). *Energy from the Waves*. Elsevier.

Rossel, S., Uhlenkott, K., Peters, J., Vink, A., & Arbizu, P. (2022). Evaluating species richness using proteomic fingerprinting and DNA barcoding-A case study on meiobenthic copepods from the Clarion Clipperton fracture zone. *Marine Biodiversity, 52*(6). DOI: 10.1007/s12526-022-01307-y

Sagawa, T., Yamashita, Y., Okumura, T., & Yamanokuchi, T. (2019). Satellite-derived bathymetry using machine learning and multi-temporal satellite images. *Remote Sensing (Basel)*, 11(10), 1155. DOI: 10.3390/rs11101155

Sakellariadou, F., Gonzalez, F. J., Hein, J. R., Rincón-Tomás, B., Arvanitidis, N., & Kuhn, T. (2022). Seabed mining and blue growth: Exploring the potential of marine mineral deposits as a sustainable source of rare earth elements (MaREEs) (IUPAC Technical Report). *Pure and Applied Chemistry*, 94(3), 329–351. DOI: 10.1515/pac-2021-0325

Salas, J., Patterson, G., & de Barros Vidal, F. (2022). A Systematic Mapping of Artificial Intelligence Solutions for Sustainability Challenges in Latin America and the Caribbean. *Revista IEEE América Latina*, 20(11), 2312–2329. DOI: 10.1109/TLA.2022.9904756

Sallang, N. C. A., Islam, M. T., Islam, M. S., & Arshad, H. (2021). A CNN-based smart waste management system using TensorFlow lite and LoRa-GPS shield in Internet of Things environment. *IEEE Access : Practical Innovations, Open Solutions*, 9, 153560–153574. DOI: 10.1109/ACCESS.2021.3128314

Samaei, S. R., & Ghahfarrokhi, M. A. (2023). Using robotics and artificial intelligence to increase efficiency and safety in marine industries. In 2th International Conference on Creative achievements of architecture, urban planning, civil engineering and environment in the sustainable development of the Middle East.

Saputra, R. A. V. W., Pramono, S. A., & Almunawwaroh, M. (2023). Preserving Our Environment: Promoting Sustainability, Conserving Resources, Protecting Wildlife, and Supporting Green Initiatives in the Foothill Region. *West Science Journal of Service*, 2(05), 316–324.

Saranya, P. V. P, R. Maheswari, P. Palanisamy and J. Mani, "Intelligent System for Improving Scalability by Sharding for Agriculture Supply Chain," 2024 *11th International Conference on Reliability, Infocom Technologies and Optimization* (Trends and Future Directions) (ICRITO), Noida, India, 2024, pp. 1-6.

Saranya, P., Shirly, S., & Logeshwari, V. "Development of a Blockchain Vehicle Manufacturing Network Using Hyperledger Composer," *2023 International Conference on Innovative Computing, Intelligent Communication and Smart Electrical Systems (ICSES)*, Chennai, India, 2023, pp. 1-9, DOI: 10.1109/ICSES60034.2023.10465503

Saridewi, T. R. (2021). Farmers' Perceptions of the Application of Land Conservation Technology in Rice Cultivation in Nagrak District, Sukabumi Regency. *Journal of Agricultural Extension*, 16(1), 37–50.

Sato, T., ... (2019). Exploring hydrothermal deposits with multiple autonomous underwater vehicles. In 2019 IEEE Underwater Technology (UT) (pp. 1–5). IEEE. DOI: 10.1109/UT.2019.8734460

Schall, E., Kaya, I. I., Debusschere, E., Devos, P., & Parcerisas, C. (2024). Deep learning in marine bioacoustics: A benchmark for baleen whale detection. *Remote Sensing in Ecology and Conservation*, 10(5), 642–654. Advance online publication. DOI: 10.1002/rse2.392

Schmidhuber, J. (2015). Deep learning in neural networks: An overview. *Neural Networks*, 61, 85–117. DOI: 10.1016/j.neunet.2014.09.003 PMID: 25462637

Selvakumar, P. (2024). Emerging Engineering Technologies: Design, Development, and Industrial Applications of IoT-Based Aquaculture Monitoring System. In El Kacimi, Y., & Alaoui, K. (Eds.), *Emerging Engineering Technologies and Industrial Applications* (pp. 332–341). IGI Global., DOI: 10.4018/979-8-3693-1335-0.ch016

Selvakumar, P., Mariappan, P., Shunmugam, S., Paul, R., & Jabez, S. (2024). Removal of chemicals from effluent using photobioreactor technology to improve environmental and health impacts. *Zeitschrift für Physikalische Chemie*, 2024(7), 1295–1302. Advance online publication. DOI: 10.1515/zpch-2023-0450

Selvakumar, P., Muthusamy, S., Satishkumar, D., Vigneshkumar, P., Selvamurugan, C., & Satheesh Kumar, P. (2024). AI-Powered Tools. In Satishkumar, D., & Sivaraja, M. (Eds.), *Using Real-Time Data and AI for Thrust Manufacturing* (pp. 20–42). IGI Global., DOI: 10.4018/979-8-3693-2615-2.ch002

Selvakumar, P., Palanisamy, S. K., Cinthaikinian, S., Palanisamy, V., Mariappan, R., & Selvakumar, P. K. (2024). Biosensors and its diverse applications in healthcare systems. *Zeitschrift für Physikalische Chemie*, 2024. Advance online publication. DOI: 10.1515/zpch-2023-0406

Selvi, R. T., Elakya, R., & Vignesh, U. (2024). Securing Sensitive Patient Data in Healthcare Settings Using Blockchain Technology. In Blockchain and IoT Approaches for Secure Electronic Health Records (EHR) (pp. 73-88). IGI Global.

Selvi, S., Aggarwal, K., Pandurangan, R., Vijayan, V. P., Ali, A., & Anuradha, K. (2024). Enhancing the accuracy of target detection in remote video surveillance analytics through federated learning. *Optical and Quantum Electronics*, 56(2), 185. DOI: 10.1007/s11082-023-05664-1

Sharma, R. (2011). Deep-sea mining: Economic, technical, technological, and environmental considerations for sustainable development. *Marine Technology Society Journal*, 45(5), 28–41. DOI: 10.4031/MTSJ.45.5.2

Sharma, R. (Ed.). (2019). *Environmental issues of deep-sea mining: impacts, consequences and policy perspectives*. Springer. DOI: 10.1007/978-3-030-12696-4

Shaw, S. (2021, March). Ocean–A vast reserve of renewable energy. [). IOP Publishing.]. *Journal of Physics: Conference Series*, 1834(1), 012017. DOI: 10.1088/1742-6596/1834/1/012017

Sheets, H. E. (1975). Power generation from ocean currents. *Naval Engineers Journal*, 87(2), 47–56. DOI: 10.1111/j.1559-3584.1975.tb03714.x

Shetty, C., & Priyam, A. (2022). A review on tidal energy technologies. *Materials Today: Proceedings*, 56, 2774–2779. DOI: 10.1016/j.matpr.2021.10.020

Shreyas Dixit K. G, Chadaga M. G, Savalgimath S.S, Ragavendra Rakshith G, and Naveen Kumar M. R, "Evaluation and evolution of object detection techniques YOLO and R-CNN," Int. J. Recent Technol. Eng., vol. 8, no. 2 Special issue 3, pp. 824–829, 2019.

Silva-Send, N. (2024). Deep Sea Mining in the Area Beyond National Jurisdiction–A Lost Opportunity or Yet Another Reason for the United States to Join UNCLOS?. *San Diego Legal Studies Paper*, (24-011).

Singh, H., & King, J. W. (2015). Observations of Vertical Pesticide Fluxes in the Deep Sargasso Sea. *Deep-sea Research. Part II, Topical Studies in Oceanography*, 113, 131–138.

Singh, P. A. (2022). Deep seabed mining and sustainable development goal 14. In *Life Below Water* (pp. 271–283). Springer International Publishing. DOI: 10.1007/978-3-319-98536-7_135

Sivakumar, M., Renuka, P., Chitra, P., & Karthikeyan, S. (2022). IoT incorporated deep learning model combined with SmartBin technology for real-time solid waste management. *Computational Intelligence*, 38(2), 323–344. DOI: 10.1111/coin.12495

Smith, C. R., & Glover, A. G. (2016). The Deep Ocean Carbon Cycle: An Overview. [). CRC Press.]. *Oceanography and Marine Biology - an Annual Review*, 54, 1–34.

Smith, J. K., & Jones, L. M. (2020). Incorporating uncertainty into generative artificial intelligence models of deep-sea ecosystems. Ecological Modeling, 410, 108831. Garcia, M., Martinez, E., & Rodriguez, A. (2023). Machine learning approaches for understanding deep-sea food web dynamics in a changing climate. *Deep-sea Research. Part I, Oceanographic Research Papers*, 168, 103487.

Soemarmi, A., Indarti, E., Pujiyono, P., Azhar, M., & Wijayanto, D. (2020). Vessel monitoring system (VMS) technology as a strategy for the protection and development of the fishing industry in Indonesia. *Legal Issues*, 49(3), 303–313.

Song, H., Ahmad, A., Farooq, F., Ostrowski, K. A., Maślak, M., Czarnecki, S., & Aslam, F. (2021b). Predicting the compressive strength of concrete with fly ash admixture using machine learning algorithms. *Construction & Building Materials*, 308, 125021. DOI: 10.1016/j.conbuildmat.2021.125021

Song, Y. P., Song, L. Y., & Zhao, G. F. (2004). Factors affecting corrosion and approaches for improving durability of ocean reinforced concrete structures. *Ocean Engineering*, 31(5-6), 779–789. DOI: 10.1016/j.oceaneng.2003.07.006

Song, Y., Nakath, D., She, M., & Köser, K. (2022). Optical imaging and image restoration techniques for deep ocean mapping: A comprehensive survey. PFG –. *Journal of Photogrammetry, Remote Sensing and Geoinformation Science*, 90(3), 243–267. DOI: 10.1007/s41064-022-00206-y

Song, Y., Zhao, J., Ostrowski, K. A., Javed, M. F., Ahmad, A., Khan, M. I., Aslam, F., & Kinasz, R. (2021a). Prediction of compressive strength of fly-ash-based concrete using ensemble and non-ensemble supervised machine-learning approaches. *Applied Sciences (Basel, Switzerland)*, 12(1), 361. DOI: 10.3390/app12010361

Sosa-Trejo, D., Bandera, A., González, M., & Hernández-León, S. (2023). Vision-based techniques for automatic marine plankton classification. *Artificial Intelligence Review*, 56(11), 12853–12884. DOI: 10.1007/s10462-023-10456-w

Soundarya, B., Parkavi, K., Sharmila, A., Kokiladevi, R., Dharani, M., & Krishnaraj, R. (2022, January). CNN based smart bin for waste management. In 2022 4th International Conference on Smart Systems and Inventive Technology (ICSSIT) (pp. 1405-1409). IEEE.

Spearman, J., Taylor, J., Crossouard, N., Cooper, A., Turnbull, M., Manning, A., Lee, M., & Murton, B. (2020). Measurement and modeling of deep-sea sediment plumes and implications for deep sea mining. *Scientific Reports*, 10(1), 5075. DOI: 10.1038/s41598-020-61837-y PMID: 32193479

Sundaramurthy, A., Manogaran, G., & Thota, C. (2018). Application of artificial intelligence techniques in the maritime industry. *Maritime Informatics*, 45(2), 123–145.

Sungheetha, A. (2023). Revolutionizing Underwater Exploration of Autonomous Underwater Vehicles (AUVs) and Seabed Image Processing Techniques. arXiv preprint arXiv:2402.00004.

Sun, K., Cui, W., & Chen, C. (2021). Review of underwater sensing technologies and applications. *Sensors (Basel)*, 21(15), 7849. DOI: 10.3390/s21237849 PMID: 34883851

Sunny, M. S. H., Dipta, D. R., Hossain, S., Faruque, H. M. R., & Hossain, E. (2019, May). Design of a convolutional neural network based smart waste disposal system. In 2019 1st International Conference on Advances in Science, Engineering and Robotics Technology (ICASERT) (pp. 1-5). IEEE.

Suryantoro, A., & Siswanto, B. (2010). Correlation Analysis of Surface Air Temperature and Rainfall in Jakarta and Pontianak with Tropical Indian and Pacific Ocean Sea Surface Temperature Anomalies in the Framework of Tropospheric Biennial Oscillation (TBO). *Aerospace Science Journal, 6*(1).

Tang, Y., Liu, C., & Yuan, X. (2024). Recognition of bird species with birdsong records using machine learning methods. *PLoS ONE, 19*(2 February). DOI: 10.1371/journal.pone.0297988

Thomas, E. 2020. Green Technology Products for Sustainable Development. In Chemistry and Chemical Engineering for Sustainable Development (pp. 29-82). Apple Academic Press. Charlier, R.H. and Finkl, C.W., 2009. Ocean energy: tide and tidal power. Springer Science & Business Media. DOI: 10.1201/9780367815967-3

Thomsen, L., & Booth, J. A. (2020). Deep-Sea Mining: Technology, Environmental, and Socio-Economic Considerations. *Ocean and Coastal Management*, 184, 105019.

Thresher, R., & Musial, W. (2010). Ocean renewable energy's potential role in supplying future electrical energy needs. *Oceanography (Washington, D.C.)*, 23(2), 16–21. DOI: 10.5670/oceanog.2010.39

Thurber, A. R., Levin, L. A., & Rowden, A. A. (2014). Benthic Oxygen Consumption in Oxygen Minimum Zone Regions: A Global Comparison of Fluxes and Ecosystem Impacts. *Deep-sea Research. Part I, Oceanographic Research Papers*, 94, 173–183.

Thurber, A. R., Sweetman, A. K., Narayanaswamy, B. E., Jones, D. O., Ingels, J., & Hansman, R. L. (2014). Ecosystem function and services provided by the deep sea. *Biogeosciences*, 11(14), 3941–3963. DOI: 10.5194/bg-11-3941-2014

Tom, N., & Yeung, R. W. (2015). Experimental confirmation of nonlinear-model-predictive control applied offline to a permanent magnet linear generator for ocean-wave energy conversion. *IEEE Journal of Oceanic Engineering*, 41(2), 281–295. DOI: 10.1109/JOE.2015.2439871

Toro, N., Robles, P., & Jeldres, R. I. (2020). Seabed mineral resources, an alternative for the future of renewable energy: A critical review. *Ore Geology Reviews*, 126, 103699. DOI: 10.1016/j.oregeorev.2020.103699

Udupi, P. K., Jose, M., & Ullah, A. (2024). AI-Enabled Smart City Waste Management System. In Handbook of Artificial Intelligence for Smart City Development (pp. 76–99). CRC Press.

Ukaogo, P. O., Ewuzie, U., & Onwuka, C. V. (2020). Environmental pollution: causes, effects, and the remedies. In *Microorganisms for sustainable environment and health* (pp. 419–429). Elsevier. DOI: 10.1016/B978-0-12-819001-2.00021-8

Ullah, H. S., Khushnood, R. A., Farooq, F., Ahmad, J., Vatin, N. I., & Ewais, D. Y. Z. (2022). Prediction of compressive strength of sustainable foam concrete using individual and ensemble machine learning approaches. *Materials (Basel)*, 15(9), 3166. DOI: 10.3390/ma15093166 PMID: 35591498

Van Der Grient, J. M. A., & Drazen, J. C. (2021). Potential spatial intersection between high-seas fisheries and deep-sea mining in international waters. *Marine Policy*, 129, 104564. DOI: 10.1016/j.marpol.2021.104564

VanZwieten, J. H., Rauchenstein, L. T., & Lee, L. (2017). An assessment of Florida's ocean thermal energy conversion (OTEC) resource. *Renewable & Sustainable Energy Reviews*, 75, 683–691.

Veerabhadrappa, K., Suhas, B. G., Mangrulkar, C. K., Kumar, R. S., Mudakappanavar, V. S., & Seetharamu, K. N. (2022). Power generation using ocean waves: A review. *Global Transitions Proceedings*, 3(2), 359–370. DOI: 10.1016/j.gltp.2022.05.001

Verma, B. G. K. Yudheksha, P. Sanjana Reddy, and U. Vignesh. "An Intelligent Flood Automation System Using IoT and Machine Learning." In *Recent Developments in Electronics and Communication Systems*, pp. 444-449. IOS Press, 2023.

Verma, B., Yudheksha, G. K., Sanjana Reddy, P., & Vignesh, U. (2023). An Intelligent Flood Automation System Using IoT and Machine Learning. In *Recent Developments in Electronics and Communication Systems* (pp. 444–449). IOS Press. DOI: 10.3233/ATDE221295

Viet, N. V., Wu, N., & Wang, Q. (2017). A review on energy harvesting from ocean waves by piezoelectric technology. *Journal of Modeling in Mechanics and Materials*, 1(2), 20160161. DOI: 10.1515/jmmm-2016-0161

Viet, N. V., Xie, X. D., Liew, K. M., Banthia, N., & Wang, Q. (2016). Energy harvesting from ocean waves by a floating energy harvester. *Energy*, 112, 1219–1226. DOI: 10.1016/j.energy.2016.07.019

Vignesh, P., Maheswari, R., Vijaya, P., & Vignesh, U. (2024). Machine Learning for Aerospace Object Categorization. In AI and Blockchain Optimization Techniques in Aerospace Engineering (pp. 164-180). IGI Global.

Vignesh, U., & Ratnakumar, R. (2024). An Empirical Review on Clustering Algorithms for Image Segmentation of Satellite Images. AI and Blockchain Optimization Techniques in Aerospace Engineering, 33-52.

Vignesh, U., Parvathi, R., & Goncalves, R. (2023). *"Structural and Functional Data Processing in Bio-Computing and Deep Learning." Structural and Functional Aspects of Biocomputing Systems for Data Processing 2023*. IGI Global.

Vignesh, U., Ratnakumar, R., & Al-Obaidi, A. S. M. (2024). *AI and blockchain optimization techniques in aerospace engineering*. AI and Blockchain Optimization Techniques in Aerospace Engineering. DOI: 10.4018/979-8-3693-1491-3

Vijayakumar, G., Rajkumar, M., Rajiv Chandar, N., Selvakumar, P., & Duraisamy, R. (2024). Muniyandi Rajkumar, Rajiv Chandar N, Selvakumar P and Ramesh Duraisamy, (2024). Environmentally friendly TDS removal from waste water by electrochemical ion exchange batch-type recirculation (EIR) technique. *Environmental Science. Water Research & Technology*, 10(4), 826–835. DOI: 10.1039/D3EW00793F

Wahab, M. H. A., Kadir, A. A., Tomari, M. R., & Jabbar, M. H. (2014, October). Smart recycle bin: A conceptual approach of smart waste management with integrated web based system. 2014 International Conference on IT Convergence and Security (ICITCS). Presented at the 2014 International Conference on IT Convergence and Security (ICITCS), Beijing, China. DOI: 10.1109/ICITCS.2014.7021812

Wang, C., Qin, J., Qu, C., Ran, X., Liu, C., & Chen, B. (2021). A smart municipal waste management system based on deep-learning and Internet of Things. *Waste Management (New York, N.Y.)*, 135, 20–29. DOI: 10.1016/j.wasman.2021.08.028 PMID: 34461487

Wang, L., Zhang, Y., & Liu, Q. (2023). Integrating Generative AI and Climate Models to Predict Deep-Sea Ecosystem Changes. *Environmental Modeling and Assessment*, 45(1), 113–126.

Wang, R., Chen, G., Zhang, N., & Zhang, H. (2023). Study on storm evacuation operations in deep-sea mining. *Ocean Engineering*, 281, 114690. DOI: 10.1016/j.oceaneng.2023.114690

Wang, Z., Liu, Y., & Zhang, X. (2020). Deep Learning-Based Prediction of Climate Change Effects on Deep-Sea Food Webs: A Case Study in the Pacific Ocean. *Deep-sea Research. Part II, Topical Studies in Oceanography*, 172, 104694.

Wang, Z., & Zou, L. (2022). Natural frequency analysis of deep-sea mining riser considering varying tension and buffer station. *Ocean Engineering*, 264, 112372. DOI: 10.1016/j.oceaneng.2022.112372

Ward, D., Melbourne-Thomas, J., Pecl, G. T., Evans, K., Green, M., McCormack, P. C., Novaglio, C., Trebilco, R., Bax, N., Brasier, M. J., Cavan, E. L., Edgar, G., Hunt, H. L., Jansen, J., Jones, R., Lea, M. A., Makomere, R., Mull, C., Semmens, J. M., & Layton, C. (2022). Safeguarding marine life: Conservation of biodiversity and ecosystems. *Reviews in Fish Biology and Fisheries*, 32(1), 65–100. DOI: 10.1007/s11160-022-09700-3 PMID: 35280238

Wear, S. L., Acuña, V., McDonald, R., & Font, C. (2021). Sewage pollution, declining ecosystem health, and cross-sector collaboration. *Biological Conservation*, 255, 109010. DOI: 10.1016/j.biocon.2021.109010

Wernberg, T., Bennett, S., Babcock, R. C., de Bettignies, T., Cure, K., Depczynski, M., Dufois, F., Fromont, J., Fulton, C. J., Hovey, R. K., Harvey, E. S., Holmes, T. H., Kendrick, G. A., Radford, B., Santana-Garcon, J., Saunders, B. J., Smale, D. A., Thomsen, M. S., Tuckett, C. A., & Wilson, S. (2016). Climate-driven regime shift of a temperate marine ecosystem. *Science*, 353(6295), 169–172. DOI: 10.1126/science.aad8745 PMID: 27387951

White, M., & Rycroft, S. (2016). Remote Sensing of the Deep Ocean and Sea Ice. In Remote Sensing of the Cryosphere (pp. 189-212). Wiley.

Wilberforce, T., El Hassan, Z., Durrant, A., Thompson, J., Soudan, B., & Olabi, A. G. (2019). Overview of ocean power technology. *Energy*, 175, 165–181. DOI: 10.1016/j.energy.2019.03.068

Williams, S., & Mahon, I. (2004). Simultaneous localisation and mapping on the Great Barrier Reef. *Proceedings of the 2004 IEEE International Conference on Robotics and Automation*, 1771–1776. DOI: 10.1109/ROBOT.2004.1308080

Wiraguna, S. A., & Purwanto, L. M. F. (2024). Implementation of Digital Technology in the Construction Stage of Oceanarium in Indonesia. *GEWANG: Gate of Architectural Discourse and Design*, 6(1), 1–5.

Woods Hole Oceanographic Institution. (n.d.a). Three types of vehicles. Dive and Discover. Retrieved from https://divediscover.whoi.edu/underwater-vehicles/three-types-of-vehicles/

Woods Hole Oceanographic Institution. (n.d.b). Know Your Ocean: Ocean Tech - Acoustics & Sonar. Retrieved from https://www.whoi.edu/

Woods Hole Oceanographic Institution. (n.d.c). Sonar: Single-beam and multibeam. WHOI Know Your Ocean. https://www.whoi.edu/know-your-ocean/ocean-topics/ocean-tech/acoustics/sonar-single-beam/

World Resources Institute. (n.d.). World Resources Institute. Retrieved from https://www.wri.org

Wu, N., Gao, J., Ennaanay, D., Mendoza, G. F., Luo, Z., Li, D., & Tian, M. (2011). Evaluation of ecosystem service for water supply and power generation and its economic value: A case study of Ertan hydropower on Yalong river. [Journal of Hydroelectric Engineering]. *Shuili Fadian Xeubao*, 30(4), 56–64.

Wu, N., Wang, Q., & Xie, X. (2015). Ocean wave energy harvesting with a piezoelectric coupled buoy structure. *Applied Ocean Research*, 50, 110–118. DOI: 10.1016/j.apor.2015.01.004

Wu, Q., Yang, J., Guo, X., Liu, L., Lu, W., & Lu, H. (2022). Experimental study on dynamic responses of a deep-sea mining system. *Ocean Engineering*, 248, 110675. DOI: 10.1016/j.oceaneng.2022.110675

Wursig, B., & Jefferson, T. A. (1990). Methods of photo-identification for small cetaceans. *Report of the International Whaling Commission. Special Issue*, 12, 43–52.

Wynn, R. B., Huvenne, V. A. I., Le Bas, T. P., Murton, B. J., Connelly, D. P., Bett, B. J., Ruhl, H. A., Morris, K. J., Peakall, J., Parsons, D. R., Sumner, E. J., Darby, S. E., Dorrell, R. M., & Hunt, J. E. (2014). Autonomous underwater vehicles (AUVs): Their past, present and future contributions to the advancement of marine geoscience. *Marine Geology*, 352, 451–468. DOI: 10.1016/j.margeo.2014.03.012

Xie, X. D., Wang, Q., & Wu, N. (2014a). Energy harvesting from transverse ocean waves by a piezoelectric plate. *International Journal of Engineering Science*, 81, 41–48. DOI: 10.1016/j.ijengsci.2014.04.003

Xie, X. D., Wang, Q., & Wu, N. (2014b). Potential of a piezoelectric energy harvester from sea waves. *Journal of Sound and Vibration*, 333(5), 1421–1429. DOI: 10.1016/j.jsv.2013.11.008

Xu, K.. (2015). Show, attend and tell: Neural image caption generation with visual attention. In *Proceedings of the 32nd International Conference on Machine Learning* (Vol. 37, pp. 2048-2057).

Xu, Y., Ahmad, W., Ahmad, A., Ostrowski, K. A., Dudek, M., Aslam, F., & Joyklad, P. (2021). Computation of high-performance concrete compressive strength using standalone and ensembled machine learning techniques. *Materials (Basel)*, 14(22), 7034. DOI: 10.3390/ma14227034 PMID: 34832432

Yang, H., Liu, P., Hu, Y., & Fu, J. (2020). Research on underwater object recognition based on YOLOv3. *Microsystem Technologies*, 27(4), 1837–1844. DOI: 10.1007/s00542-019-04694-8

Yang, P., Wu, H., He, C., & Luo, S. (2023). Underwater image restoration for seafloor targets with hybrid attention mechanisms and conditional generative adversarial network. *Digital Signal Processing*, 134, 103900. DOI: 10.1016/j.dsp.2022.103900

Yan, J., Zhou, Z., Zhou, D., Su, B., Xuanyuan, Z., Tang, J., Lai, Y., Chen, J., & Liang, W. (2022). Underwater object detection algorithm based on attention mechanism and cross-stage partial fast spatial pyramidal pooling.". *Frontiers in Marine Science*, 9, 1056300. DOI: 10.3389/fmars.2022.1056300

Yeh, C. H., Lin, C. H., Kang, L. W., Huang, C. H., Lin, M. H., Chang, C. Y., & Wang, C. C. (2021). Lightweight deep neural network for joint learning of underwater object detection and color conversion. *IEEE Transactions on Neural Networks and Learning Systems*, 33(11), 6129–6143.

Yeh, R. H., Su, T. Z., & Yang, M. S. (2005). Maximum output of an OTEC power plant. *Ocean Engineering*, 32(5-6), 685–700.

Yoshida, T., Kawato, M., Fujiwara, Y., Nagano, Y., Tsuchida, S., & Yabuki, A. (2023). Optimization of environmental DNA analysis using pumped deep-sea water for the monitoring of fish biodiversity. *Frontiers in Marine Science*, 9, 965800. Advance online publication. DOI: 10.3389/fmars.2022.965800

Zahid, S. M., Najesh, T. N., Ameen, S. R., & Ali, A. (2023). A Multi Stage Approach for Object and Face Detection using CNN. In *2023 8th International Conference on Communication and Electronics Systems (ICCES)* (pp. 798-803). IEEE. DOI: 10.1109/ICCES57224.2023.10192823

Zhang, C., Lin, Y., & Pan, W. (2018, July). Research on real-time processing system of deep sea biological image for manned submersible vehicle. In *2018 37th Chinese Control Conference (CCC)* (pp. 4006-4010). IEEE. DOI: 10.23919/ChiCC.2018.8483593

Zhang, J., Ning, P., Zhang, X., Wang, X., & Zhang, A. (2023). Deriving sea subsurface temperature fields from satellite remote sensing data using a generative adversarial network model. Earth and Space Science, 10(4), e2022EA002804.

Zhang, D., Zhao, Y. X., Cui, Y. F., & Wang, P. C. (2021). A visualization analysis and development trend of intelligent ship studies. *J. Transp. Inf. Saf.*, 39(01), 7–16.

Zhang, P. F., Iqbal, M., Zhang, D., Zhao, X. L., & Zhao, Q. (2024). Bond strength prediction of FRP bars to seawater sea sand concrete based on ensemble learning models. *Engineering Structures*, 302, 117382. DOI: 10.1016/j.engstruct.2023.117382

Zhang, T., Liu, B., & Liu, Y. (2018). Positioning systems for Jiaolong deep-sea manned submersible: Sea trial and application. *IEEE Access : Practical Innovations, Open Solutions*, 1, 71644–71650. Advance online publication. DOI: 10.1109/ACCESS.2018.2881390

Zhang, W., Gao, X.-Z., Yang, C.-F., Jiang, F., & Chen, Z.-Y. (2020). A object detection and tracking method for security in intelligence of unmanned surface vehicles. *Journal of Ambient Intelligence and Humanized Computing*.

Zhou, T., Si, J., Wang, L., Xu, C., & Yu, X. " Automatic Detection of Underwater Small Targets Using Forward-Looking Sonar Images," IEEE Transactions on Geoscience and Remote Sensing (Volume: 60), Print ISSN: 0196-2892, 2022.

Zhou, G., Li, C., Zhang, D., Liu, D., Zhou, X., & Zhan, J. (2021). Overview of underwater transmission characteristics of oceanic LiDAR. *IEEE Journal of Selected Topics in Applied Earth Observations and Remote Sensing*, 14, 8144–8159. DOI: 10.1109/JSTARS.2021.3100395

About the Contributors

U. Vignesh is currently an Assistant Professor Senior Grade 2 in School of Computer Science and Engineering, Vellore Institute of Technology (VIT) - Chennai. Prior to his recent appointment at the VIT, he was a Post-Doctoral Fellow in National Institute of Technology (NIT), Trichy – India. Dr. Vignesh received his undergraduate degree in B.Tech (IT) as well as his M.Tech (IT) degree from Anna University - Chennai, and his PhD in Computer Science and Engineering from VIT University - Chennai. Dr. Vignesh published several papers in preferred Journals, patents and chapters in books, and participated in a range of forums on computer science, social science, etc. He also presented various academic as well as research-based papers at several national and international conferences. His research activities are currently twofold: while the first research activity is set to explore the developmental role that society needs with technology such as, Artificial Intelligence; the second major research theme that he is pursuing is focused on the bioinformatics and data mining.

Parvathi R. is a Professor in School of Computer Science and Engineering, Vellore Institute of Technology - Chennai Campus, Chennai-600127, India and she Research and Professional Experience of 25 Years in the same field.

***

Chitradevi D. was born in Trichirappalli, India, in 1980. She received the B.Sc. Chemistry and M.C.A degrees in Bharathidasan University, M.E. (CSE) in Anna University and Ph.D in Hindustan University Chennai. Her research interests include image Processing, Biomedical Instrumentation, Biometric Security and Cloud Computing. She has published more than 15 research articles in International Journals and 15 Conferences. She is a reviewer in Applied Soft Computing, SN Computer Science, Neural computing and many reputed International Journals.

She has around 17 years of experience in teaching. Currently she is working as an Associate Professor in SRM Institute of Science and Technology.

S. Geetha received the B.E., from the Madurai Kamaraj University, M.E., and Ph.D. degrees in Computer Science and Engineering from Anna University, Chennai, in 2000, 2004 and 2011 respectively. She has 14+ years of teaching experience. Currently, she is a professor at School of Computing Science and Engineering at VIT-University, Chennai Campus. She has published more than 50 papers in reputed IEEE International Conferences and refereed Journals. She joins the review committee for IEEE Transactions on Information Forensics and Security and IEEE Transactions on Image Processing, Springer Multimedia Tools and Security, Elsevier – Information Sciences. She was an editor for the Indian Conference proceedings of ICCIIS 2007 and RISES-2013. Her research interests include multimedia security, intrusion detection systems, machine learning paradigms and information forensics. She is a recipient of University Rank and Academic Topper Award in B.E. and M.E. in 2000 and 2004 respectively. She is also a pride recipient of the "Best Academic Researcher Award 2013" of ASDF Global Awards.

Sagar Kashinath Khokale is Assistant professor in the Pharmaceutics Department of Sandip Institute of Pharmaceutical Sciences, Nashik, India. He received his Masters in Pharmaceutics. Currently pursuing Doctor of Philosophy-Pharmacy from Sunrise University, Alwar, Rajasthan. His research interest is in Pharmaceutical Sciences. His research has been disseminated in several internationally reputed journals. Orcid id: 0000-0002-7351-0613.

Swati Kumari is a final-year student at Vellore Institute of Technology, Chennai, pursuing a B.Tech degree in Computer Science and Engineering. Her academic journey has been distinguished by her continuous outstanding performance and guided by dedicated mentorship that has fostered her personal and professional growth. Her research interests include Machine Learning, Artificial Intelligence, etc.

Destriman Laoli, S.Pi., M.Si, Lecturer in the Aquatic Resources Study Program, Faculty of Science and Technology, University of Nias. The author was born in Onozikho, Gunungsitoli City, precisely on Nias Island, North Sumatra Province on December 10, 1987. The author is a permanent lecturer at the Aquatic Resources Study Program, Faculty of Science and Technology, University of Nias. The author completed his undergraduate education in the Aquaculture Study Program at the University of Riau. Riau University and continued his Masters Education in the Marine Science Study Program Marine Science Study Program Concentration in Aquaculture at Riau University. The author lectures on Fundamentals of aquatic microbiology, Disease of Aquatic Organisms, Aquatic Productivity. Aquatic Organisms, Aquatic Productivity.

Ayler Beniah Ndraha is a government scholar and practitioner who has made valuable contributions to the field of Human Resource Management. He is known as someone who combines deep academic knowledge with rich practical experience, thus providing deep insights and practical solutions in human resource planning. human resource planning. With a strong educational background, Dr. Ayler Beniah Ndraha earned a Bachelor of Applied Science in Government (S.STP) and a Master of Science (M.Si) degree as well as a Doctor of Government Science (DIP), all of which were taken at the Institute of Domestic Government (IPDN) Jatinangor-Bandung. His experience working in government agencies such as Bappeda, BKPSDM, Kominfo, and Kesbangpol makes his expertise in government science provide a solid foundation for understanding organizational dynamics and governance in collaboration with empirical practices. His contribution to the Human Resource Management book can be seen through the chapter titled "Human Resource Planning." In this section, the author brings an in-depth understanding of how to manage important aspects of HR planning, including workforce needs analysis, forecasting, and human resource development strategies. His emphasis on integration between theory and practice provides valuable guidance for readers in facing the complex challenges of human resource management. The author is currently active as a civil servant lecturer with a functional position of Lektor at Nias University. Although still relatively young, the author has also contributed to various seminars and scientific discussions both on a regional, national and even international scale sharing his valuable views on the dynamics of government and human resource management. His expertise and dedication in developing government science as well as his deep knowledge in human resource management make him an important figure in inspiring and guiding future generations in this field.

Selvakumar P. is currently an Associate Professor in the Department of Science and Humanities at Nehru Institute of Technology, Coimbatore, Tamil Nadu, India. Since 2023, he has served as the Department IQAC Coordinator at the institute. Prior to joining Nehru Institute of Technology, Dr. Selvakumar held positions as an Assistant Professor in the Department of Science and Humanities at Info Institute of Engineering and Dr. N.G.P Institute of Technology, both located in Coimbatore, Tamil Nadu, India, spanning from 2011 to 2023. Dr. Selvakumar's academic and research interests lie primarily in the field of phytochemistry. He has made significant contributions to his field, with an impressive record that includes 30 journal publications, 12 patents, and 20 conference papers. Additionally, he has organized 5 seminars and conferences, further demonstrating his commitment to academic and scientific exchange. Throughout his career, Dr. Selvakumar has been recognized for his dedication and excellence in teaching, having been honored with several awards, including the Best Teacher Awards. He holds a prominent position in the academic community, contributing actively to research.

Lakshmi Harika Palivela completed her B.Tech in Computer Science and Engineering from Jawaharlal Nehru Technological University Kakinada , M.Tech in Computer Science and Engineering from Jawaharlal Nehru Technological University Kakinada, and Ph.D. in the faculty of Information and Communication Engineering MIT Campus, Anna University, Chennai. She is currently working as an Assistant Professor (Sr.) at Vellore Institute of Technology, Chennai. Her research interests include Machine Learning, Distributed systems, Image processing, Data Analytics and Video Analytics.

S. Prabha completed her Ph.D. degree at College of Engineering Guindy campus, Anna University in the field of "Analysis of Breast Thermograms using Adaptive Level Set and Riesz Transform". She has previously worked as an Associate Professor at Hindustan Institute of Technology and Science, Chennai, India, and is currently a Researcher at Department of Radiology, Mayo Clinic, Florida, USA. Her research interests include image and signal processing, biomedical instrumentation and biometric security. She has well published with Covid-19 book by CRC press, nine chapters in edited Books, 15 research articles in International Journals and 24 articles in conference proceedings. She has secured two best paper awards and is a member of IEEE, IET, and ISOI.

Shrinwantu Raha is working as a faculty member in Department of Geography, Bhairab Ganguly College, Belgharia, India since 2018. He has published more than 20 research articles and book chapters from several national and international publishing houses. He has also presented over 15 research papers in national and international seminars. He has a consistent good academic record with zeal to learn new concepts quickly and apply innovative ideas for achieving best solutions. Motivated, self-starter with a passion to succeed and desire to excel in the areas of drought monitoring, water resource management, remote sensing and GIS, oceanography and coastal management, environmental conservation, safety in disaster emergency, human Geography, and Sustainability. He is looking forward to work on a challenging environment that encourages learning and creativity provides exposure to new ideas and stimulates professional growth. ORCID id: 0000-0003-4883-8657.

Thenmozhi S. is working as Professor, Department of Civil Engineering at St.Joseph's College of Engineering, OMR, Chennai. She has 26 years of teaching experience. She has obtained bachelor's degree in Civil Engineering from Thiagarajar College of Engineering, Madurai and Post graduation from Annamalai University. She has been awarded with Ph.D in Civil Engineering from Anna University, Chennai. Her research interests are Green Concrete and Soil Stabilization. She has published many research papers in National and International Journals and attended National and International Conferences.

Betzy Victor Telaumbanua, S.Pi., M.Si., is a permanent lecturer at the Aquatic Resources Study Program, Faculty of Science and Technology, Nias University. He completed his undergraduate education at the Aquatic Resources Management Study Program, Faculty of Agriculture, University of North Sumatra in 2013 and completed his Masters education at the Marine Science Study Program (Concentration in Aquatic Resources Management), Riau University in 2017. The author has published several research articles in accredited national journals and reference books.

Gowri V. is currently working as Associate Professor, Department of Civil Engineering at SRM Madurai college for Engineering and Technology, Madurai. She has 18 years of teaching experience. She has obtained bachelor's degree in civil engineering from PSG College of Technology, Coimbatore and post graduation from College of Engineering, Guindy. She has been awarded with Ph.D, in Civil Engineering from Anna University, Chennai. Her research interests are Data fusion, Image Processing and Remote sensing application in water resources. She has published many research papers in national and international journals and attended national and international conferences. Orcid id:0000-0002-7847-9774.

Ratna Dewi Zebua was born in Tumori Balohili, Gunungsitoli City, precisely on Nias Island, North Sumatra Province on January 16, 1991. The author is a permanent lecturer at the Aquatic Resources Study Program, Faculty of Science and Technology, University of Nias. The author completed his undergraduate education in the Aquaculture Study Program at the Sibolga Fisheries College and continued his Masters Education in the Marine Science Study Program with a concentration in Aquaculture at Riau University. The author is a lecturer in Fundamentals of Aquaculture, Fisheries Biology.

Asokhiwa Zega is a scholar and practitioner in the field of Aquatic Resources who has made valuable contributions to the management of deep-sea biodiversity. He is recognized as someone who combines deep academic knowledge with rich practical experience, thus being able to provide profound insights and innovative solutions in the conservation of deep-sea resources. With a strong educational background, Asokhiwa Zega is currently a student of the Aquatic Resources Study Program at Nias University's Faculty of Science and Technology. His experience in writing for Sinta-indexed journals shows his dedication and expertise in quality scientific research. His contribution to the book chapter titled "Designing and Managing Deep Sea Biodiversity" is seen through an in-depth and detailed chapter on how to design and manage deep sea biodiversity. In this chapter, the author brings an in-depth understanding of various scientific techniques and methods, such as the use of underwater cameras and pressure sensors, to monitor and sustainably manage deep-sea ecosystems. His emphasis on integrating theory and practice provides valuable guidance for readers facing the complex challenges of deep-sea conservation. Although still a student, Asokhiwa Zega has contributed to numerous regional and national scientific seminars and discussions, sharing his valuable insights on aquatic resource dynamics and management. His expertise and dedication to developing aquatic resource science and his in-depth knowledge of deep-sea biodiversity management make him an important figure in inspiring and mentoring future generations in this field.

Index

A

B

C

D

P

R

S

T

U

V

W

www.ingramcontent.com/pod-product-compliance
Lightning Source LLC
LaVergne TN
LVHW080339011026
849161LV00009B/2125
9798369366707